SHOCKING FROGS

Shocking Frogs

GALVANI, VOLTA, AND THE ELECTRIC ORIGINS OF

NEUROSCIENCE

Marco Piccolino

Marco Bresadola

Translated by Nicholas Wade

OXFORD
UNIVERSITY PRESS

OXFORD
UNIVERSITY PRESS

Oxford University Press is a department of the University of Oxford.
It furthers the University's objective of excellence in research, scholarship,
and education by publishing worldwide.

Oxford   New York
Auckland   Cape Town   Dar es Salaam   Hong Kong   Karachi
Kuala Lumpur   Madrid   Melbourne   Mexico City   Nairobi
New Delhi   Shanghai   Taipei   Toronto

With offices in
Argentina   Austria   Brazil   Chile   Czech Republic   France   Greece
Guatemala   Hungary   Italy   Japan   Poland   Portugal   Singapore
South Korea   Switzerland   Thailand   Turkey   Ukraine   Vietnam

Oxford is a registered trademark of Oxford University Press in the UK and certain other
countries.

Published in the United States of America by
Oxford University Press
198 Madison Avenue, New York, NY 10016

© Oxford University Press 2013

All rights reserved. No part of this publication may be reproduced, stored in a
retrieval system, or transmitted, in any form or by any means, without the prior
permission in writing of Oxford University Press, or as expressly permitted by law,
by license, or under terms agreed with the appropriate reproduction rights organization.
Inquiries concerning reproduction outside the scope of the above should be sent to the Rights
Department, Oxford University Press, at the address above.

You must not circulate this work in any other form
and you must impose this same condition on any acquirer.

Library of Congress Cataloging-in-Publication Data
Piccolino, Marco. [Rane, torpedini e scintille. English] Shocking frogs : Galvani, Volta,
and the electric origins of neuroscience / Marco Piccolino, Marco Bresadola ; translated
by Nicholas Wade.pages cmSummary: "Frogs, Torpedoes, and Sparks: Galvani, Volta,
and Animal Electricity is an English translation of Rane, torpedini e scintille. Galvani,
Volta e l'elettricità animale (Torino, Italy: Bollati-Boringhieri, 2003)"—Provided by publisher.
ISBN 978–0–19–978216–1 (hardback)
1. Electrophysiology—History.   2. Galvani, Luigi, 1737–1798.
3. Volta, Alessandro, 1745–1827.   I. Bresadola, Marco.   II. Title.
QP341.P68513 2013
612'.01427—dc23
2012046275

9 8 7 6 5 4 3 2 1
Printed in the United States of America
on acid-free paper

The translation of this book has been funded by SEPS—SEGRETARIATO
EUROPEO PER LE PUBBLICAZIONI SCIENTIFICHE

Via Val d'Aposa 7 - 40123 Bologna - Italy
seps@seps.it - www.seps.it

*To Anita, Anna Maria,
Giulia, and Laura*

My Master likes frogs:
every night he sends me to the river,
but he does not give them to the maid for frying.
Instead of treating his patients
he hangs them up to the balcony,
he skins them, he tortures them with a pin,
and spends the day seeing them dancing,
and writes some letters in Latin.
What does he hope to draw from them!
—*from Casa Galvani, by Primo Levi (1984)*

Contents

Foreword ix
Authors' Preface to the Italian Edition xi
Authors' Preface to the English Edition xv
Acknowledgments xix
Bibliographical Note xxi

1. Galvani, Volta, and the Forgotten Electrophysiology 1

2. "Truth and Usefulness": Medicine and Natural Philosophy
   in the Eighteenth Century 26
   2.1. Galvani's Education in Bologna: The University, the Institute of Sciences,
       and the Hospitals 26
   2.2. Galvani's Professional Career 36
   2.3. Galvani's Early Anatomo-Physiological Investigation 40

3. Animal Spirits, Vital Forces, and Electricity: Nervous Conduction and
   Muscular Motion in the Eighteenth Century 42
   3.1. The Debate on Hallerian Irritability 43
   3.2. The Study of Electricity in the Eighteenth Century 48
   3.3. "Artificial" Electricity, "Natural" Electricity, and Their
       Role in the Human Body 56
   3.4. Electric Fish 62

4. Artificial Electricity, the Spark, and the Nervous Fluid:
   Galvani's Early Research on Muscular Motion 69
   4.1. The Beginning of Electrophysiological Experimentation 71
   4.2. A "Problematic" Turn: The Observation of Contractions at a Distance 88
   4.3. Galvani's Saggio sulla forza nervea *of 1782* 99

Contents

5. A "Fortunate" Discovery: Galvani's Theory of Animal Electricity 108
    5.1. The Study of "Airs" in Relation to the Living Organism 109
    5.2. The Effects of Atmospheric Electricity on Muscular Motion and the Discovery of Metal Arcs 116
    5.3. The Model of the Muscle as an Animal Leyden Jar 127
    5.4. The Final Elaboration of the Theory of Animal Electricity 132

6. The Controversy Between Galvani and Volta Over Animal Electricity: The First Stage 141
    6.1. Galvani's Work in the Scientific Culture of the Late Eighteenth Century 142
    6.2. Volta's Early Research on Animal Electricity: Quantification, Muscular Physiology, and the "Special Theory of Contact Electricity" 152
    6.3. Galvani's Trattato dell'arco conduttore: The Criticism Against Volta and the Notion of a Circuit of Animal Electricity 162

7. The Controversy Between Galvani and Volta Over Animal Electricity: The Second Stage 177
    7.1. Volta's "General Theory of Contact Electricity" 179
    7.2. Galvani's Reply to Volta's Criticisms and the 1797 Memorie Sulla Elettricità Animale 186
    7.3. Galvani's Research on Electric Fish and the Various Forms of Electricity 201
    7.4. The Conclusion of the Galvani-Volta Controversy 211

8. The Electrophysiological Work of Alessandro Volta 215
    8.1. Volta and Life Sciences 217
    8.2. Volta's Research on Sensations 229
    8.3. Sensation and Muscular Motion in Volta's "Chain" Experiments 241
    8.4. Volta's Research on Electric Fishes and the Invention of the Electric Battery 249

9. From Galvani to Hodgkin and Beyond: The Central Problem of Electrophysiology in the Last Two Centuries 269
    9.1. Measuring Animal Electricity 270
    9.2. Nervous Conduction: Propagated Electric Signal and the Firing of a Train of Gunpowder 280
    9.3. The Involvement of Animal Electricity in Nerve Conduction Demonstrated 289

10. Neuromuscular Excitability: The Modern Explanation 299
    10.1. Cell Membrane and Ions: A Machine Generating Electric Potentials 300
    10.2. The Electric Mechanism of Nerve Conduction and Muscle Excitation 310

11. Concluding Remarks 319

BIBLIOGRAPHY 335
INDEX 361

Foreword

ELECTRICITY IS AN important element of modern life. We encounter it in appliances, electronic equipment, and power lines. We also have a clear perception of "bioelectricity," the electrical signals that control our body and are at the basis of our thoughts. This was quite different at the end of the eighteenth century. Our predecessors were puzzled by a variety of phenomena: lightning; sparks generated by machines called electrophores; Leyden jars, which had the capacity to store electricity; and, after Galvani's discoveries, animal electricity, which seemed to exist independently of other forms of electricity. Lichtenberg, the sharp-witted physics professor at Göttingen, describes eloquently in one of his famous letters the joy of doing experiments together with Alessandro Volta on the electricity of the fire, the electricity of the air, and on other forms of electrical phenomena. Scientists of that time made great efforts to sort out the observations, and to develop a unifying concept, which, of course, was accompanied by heated debates and extreme positions. The chapters of *Shocking Frogs* vividly illuminate these struggles and the origins of our present concept of electricity. They show that the perception of electricity was radically different from the present one: Not currents and circuits, but sparks, electric shocks, and qualities of the atmosphere dominated the thoughts about electricity. The chapters also show that, nevertheless, Alessandro Volta's discovery of his battery constructed in the effort to better evoke "animal electricity" (as witnessed by the phrase "*organe électrique artificial*" first used by Volta to denote it) provided the first source of electricity capable of producing a sustained current. It opened the door for the study of electrical phenomena, such as electromagnetism, which subsequently dominated the concept and use of electricity. The main topic of *Shocking Frogs*, however, regards the development of our knowledge on "animal electricity" and its maturation into electrophysiology. Together, the chapters illuminate in a fascinating way the thoughts and disputes of scientists two centuries ago and demonstrate the origin of concepts that are essential elements of modern technology, as well as of our understanding of ourselves.

Erwin Neher
Max Planck Institute for Biophysical Chemistry, Göttingen
Göttingen, September 26, 2012

Authors' Preface to the Italian Edition

THIS BOOK IS a result of a series of encounters and discoveries. First is our encounter with the figure of Luigi Galvani, which was the point of departure for a long research work that has ranged from the Bologna physician to various aspects and figures of eighteenth-century science and culture. Though this encounter was very different for each of us, it had for both the character of a discovery (and we may say of a "revelation"). We knew that the story of Galvani and his frogs had been told so many times that what remained to be written were apparently only marginal aspects or intellectual wits. In fact, we discovered with surprise that the story waited to be told in all its fundamental features, such as the scientific problems Galvani was facing, the questions he asked, the experimental and theoretical path of his research, and the meaning of the hypotheses he formulated. This was for us a true revelation, at the same time shocking and fascinating, and marked the beginning of a journey into largely unexplored lands.

We met each other during this journey almost by chance, while following different paths, and with some initial wariness, as each of us believed he had the cultural background and conceptual tools to put the story of the Bologna physician and his frogs under the brightest light. One of us, an experimental scientist engaged in the measurement and study of electric potentials generated by nerve cells, was convinced that what really mattered in the story was the fact that Galvani's hypothesis about the existence of an imbalanced electricity in the excitable tissues of the organism, which he formulated after more than 10 years of experimental investigation, was still completely valid after two centuries and could be regarded as the founding principle of modern electrophysiology. This sort of "epiphanic" message had to be revealed to the public, after removing all the incrustations deposited by generations of scholars of various disciplines that had obscured the revolutionary role of Galvani's work in the development of neuromuscular physiology, a field which at the end of the eighteenth century was still populated by mysterious entities such as "animal spirits." The other of us, an historian of science with a humanistic formation, much more used to archives and libraries than to the laboratories of experimental science, was convinced of the importance of the historical documentation about Galvani's life and work within the social and cultural context of eighteenth-century Bologna, and of the need to discuss the methodological aspects of Galvani's investigation on animal electricity from a different perspective.

After a difficult beginning, our encounter became the point of departure for a long collaboration, which has lasted several years and brought a deep transformation of both of us. It was first a linguistic and conceptual transformation based, on the one hand, on the need of communicating to each other our knowledge, hypotheses, and conclusions, which had to be freed of the technicalities of a language that in both cases had become too specialized and "esoteric," and on the other hand, on the purpose of recovering the deep cultural values of our disciplines. It was also a "methodological" transformation that led the scientist to convince himself of the need to relate scientific concepts and knowledge to the historical and cultural context in which they were formulated, and of the importance to search for original documents, especially in a case such as Galvani's historiography, which had been largely dominated by the repetition of stereotypes and a scarce interest in primary sources (many of which still lay unexplored in various archives and libraries). This sort of transformation is such that the scientist begins to feel the fascination coming out not only from the discoveries made in a laboratory populated by instruments and animal preparations but also from those which can be made rummaging in old documents turned yellow by time and dust, in which he himself is able to find important manuscripts that no one has noted before. On the other hand, the historian becomes persuaded that, in order to understand the scientific work of a past scholar, and especially to correctly interpret experiments and laboratory arrangements, it is very useful to have a deep knowledge of the scientific problems implicated in that research work; that is, that it is not possible to do history of science without knowing, or taking into consideration, science, as sometimes unfortunately happens. For the historian the transformation is such that he himself enters the laboratory of the scientist and participates in the repetition of some of Galvani's experiments, an activity that, beyond allowing him to feel the fascination of doing the same kind of encounter with nature of Galvani and his contemporaries, proved useful to understand the difficulties implied in preparing the frogs "in the usual manner" and to interpret various passages and illustrations of Galvani's writings.

In a specific field such is that of Galvani and his frogs, the encounter between the experimental scientist and the "humanistic" historian thus becomes a meeting of those "two cultures" that in our cultural tradition have long coexisted separately, reciprocally respecting but ignoring each other. This meeting is favored by the period in which Galvani's work (and that of the other scholars to whom we have extended our reconstruction) took place—the eighteenth century—a time in which "men of letters" were interested in science, as well as in poetry and the arts; in which ladies studied not only music and painting but also mathematics and "natural philosophy," opening their salons to the surprising demonstrations of the newborn experimental science; and in which military governors and even emperors assisted with pleasure to the demonstrations of the extraordinary powers of those fishes which were to be named electric. It was certainly a time in which no real gap existed between the two cultures.

We think that what we have said so far can introduce the reader to our book, whose character is somehow difficult to define with the classical categories used in a well-established tradition, as it is the outcome of multiple encounters.

As someone has claimed, in writing a book an author is intimately convinced that his or her work comes out at the right time, when readers feel the need for it and the world is ready to catch its message. Rather than this conviction, we have the hope that those who will read our book can feel at least partially the fascination we have felt in telling, in our way, the stories of Galvani and his frogs, Volta and his battery, and many other stories which have accompanied us during these years of study.

Marco Piccolino
Marco Bresadola
July 2002

Authors' Preface to the English Edition

AT THE END of the "first day" of his masterpiece, the *Dialogue Concerning the Two Chief World Systems*, Galileo Galilei wrote a wonderful praise of the human achievements and inventions that concluded on considerations about the art of writing:

> But surpassing all stupendous inventions, what sublimity of mind was his who dreamed of finding means to communicate his deepest thoughts to any other person, though distant by mighty intervals of place and time! Of talking with those who are in the Indias; of speaking to those who are not yet born and will not he born for a thousand or ten thousand years; and with what facility, by the different arrangements of twenty characters upon a page!

If writing is this wonderful invention, capable of allowing for a distant communication through times and spaces of the "deepest thoughts," then it might be considered a natural wish the authors' desire to use those particular forms of writing potentially more efficacious in the attempt of disseminating more widely the fruits of their intellectual work, condensed as they are in the written pages of a text. Nowadays, English is by far the language most apt to ensure a most widespread diffusion of culture and science among a learned readership throughout the world. This explains our joy in seeing in print, after 10 years from the original edition, the translation of our book concerning the research on animal electricity carried out more than two centuries ago by Luigi Galvani and Alessandro Volta. We are not going to repeat the reasons and the meaning of our historical research, which, we hope, shall appear clear from the text of our book. We wish only to stress here that, though Galvani and Volta were both Italians, the history of their research is certainly not restricted to Italy and has undoubtedly a worldwide dimension. This is so not only because of the boundless significance of their respective scientific achievements (the demonstration of the involvement of a biological form of electricity in nerve and muscle excitability on Galvani's side; and the invention of the electric battery on Volta's side). It is also because of the widespread connections of science and culture underlying the scientific endeavor of the two scholars. Galvani's experimental investigation blossomed within a scientific milieu, that of the Bologna Institute of Sciences, deeply rooted in the science of Boyle and Newton not less than in that of Galileo and Malpighi. Among the starting references for Galvani's research path on animal electricity were, on one side, the doctrine of "irritability" of Albrecht von Haller, a Swiss-born scholar, professor at the German University of Göttingen; and, on the other side, the experimental demonstration of the electrical nature of the shock of the torpedo and the eels of Surinam, obtained between France and England, by an Englishman born in India, John Walsh. Among the tools used by Galvani in his research there were the "Leyden jar" and Franklin's "magic square," two instruments that broaden the views toward, on the one hand, the Low Countries, and particularly toward the old town and University of Leyden, a center of electric studies in the mid-eighteenth century, where the first electric capacitor of the history was invented; and, on the other hand, toward the American colonies, the homeland of Benjamin Franklin, who was not only one of the Founding Fathers of the United Stated and a statesman, inventor, and diplomat but also—among many other things—a scientist and the author of one of the most generally accepted theories of electricity. Franklin, who would direct the research program on electric fishes of John Walsh, had been attracted to electric experimentation by reading on a London periodical, the *Gentleman's Magazine*, the translation of a review article on electric science written by Haller. Franklin's electric theory was endorsed and diffused throughout Europe by Giambattista Beccaria, an Italian scholar (with strong international connections and a Fellow of the Royal Society of London) who became one of the main correspondents of the young Volta, the physicist of Como, bound to become one of the most internationally famous scientists of his epoch (the Newton of electricity, as it was said).

Far from being the chronicle of a scientific discussion of a provincial character between two Italian scholars, the story of Galvani and Volta has therefore a very ample, worldwide dimension, and this warrants the effort we have made, with the linguistic collaboration of our colleague and friend Nicholas (Nick) Wade, to present it to the wide readership acquainted with the English language around the world. To Nick we express our deep gratitude for his hard work in rendering into a fluent and readable English a text certainly complex and multifarious in both its historical and scientific dimensions, requiring a somewhat difficult harmonization of terms and expressions belonging to different epochs and disparate cultural fields.

A final point: The original book was very rich in quotations from the works of eighteenth-century scientists (and, of course, particularly of Galvani and Volta), an option adopted also in order to give to the readers a fresh and direct vision of the style and of the personality of the protagonists of our story. With the help of Nick we have made our best effort to keep many of these quotations in the present edition, even though rendering into modern English the Italian of the eighteenth century is by no means a simple task, with particular difficulties in the case of texts (as, for instance, many nonliterary writings, such as letters and notes, particularly in Volta's case) written in a personal and lively style, which are very difficult to convey into a different language. Notwithstanding these difficulties, we hope that the readers will be pleased to come close, through our book, to some of the original writings of the players of a story which has been mostly narrated as a second- and thirdhand tale, with often a total disregard to both the real facts and the original documents. As a matter of fact more than a story, Galvani and Volta's case has been told as a legend and—as it happens for any legend—its narration has been based largely on more or less "apocryphal" texts. We hope that by offering the readers of our book a rather direct access to some of the texts of the two scholars, we would help eradicate from a rather widespread imagery the story of the frogs' broth having triggered Galvani's interest in the electrical mechanisms of muscle and nerve functions and the subsequent image of the Bologna doctor as a "frogs dancing master," two stereotypes born from the rhetorical fantasy of scholars unacquainted with Galvani's texts. We hope, moreover, also to present a different view of Volta and of a fundamental phase of science history with important consequences for the progress of both life and physical sciences.

Marco Piccolino
Marco Bresadola
August 2012

Acknowledgments

WHILE WORKING ON this book we have received help, encouragement, and assistance from many friends, colleagues, collaborators, and other people to whom we wish now to express our gratitude. Some have read parts of the book or the whole of it, giving us many useful suggestions, which we hope we were able to follow at least partially. Among these people, we wish to give special mention to Marta Cavazza, Gianluigi Goggi, Anna Guagnini, Giacomo Magrini, and Giuliano Pancaldi. Our relatives (Anna Maria, Laura, Rosanna, and Giulia), who have often read our writings, not only suggested important readings but have long tolerated with great patience the mess of books and papers that have long encumbered our houses. Thanks to the people whom we frequented in the institutions where we have worked in the last years, the Departments of Biology and of Humanities of the University of Ferrara, the International Centre for the History of Science and Universities of the University of Bologna, and the Department of Oftalmology and Visual Sciences of the Nebraska Medical Center. Persons who have contributed in several ways to the birth and growth of our book include Ferdinando Abbri, Giulio Barsanti, Walter Bernardi, Paola Bertucci, Paolo Brenni, Michelangelo Ferraro, Robert Fox, Paola Govoni, Larry Holmes, Raffaella Seligardi, Angela Pignatelli, Andrea Moriondo, Bruna Pelucchi, Antonio Passarelli, Augusto Foà, Cristiano Bertolucci, Ottorino Belluzzi, Lucia Cadetti, Wallace Thoreson, Gerald Cristhensen, and Jacques Neyton. A special thank-you to those librarians, colleagues, and friends who helped us approach the sources, allowing us to find or to consult ancient works, often difficult to get, and to reproduce parts or all of them, and in particular Ottavio Barnabei and Massimo Zini of the Academy of Sciences of Bologna, Jean Flouret of the Académie des Sciences et Belles Lettre of La Rochelle, Mara Miniati of the Museo Galileo of Florence, Paolo Tongiorgi of the University of Modena and Reggio Emilia, Paola Manzini of the Centro Studi Lazzaro Spallanzani of Scandiano, Gabriele and Enrico De Angelis and Marco Grondona of the University of Pisa; and also David Rhees, Elisabeth Ihring e Ellen Kuhfeld of the Bakken Library and Museum of Electricity in Minneapolis, Livia Iannucci of the Library "G. Moruzzi" of the University of Pisa, Rupert Baker of the Library of the Royal Society of London, Adele Bianchi Robbiati and Corrado Vailati of the Istituto Lombardo Accademia di Scienze e Lettere in Milan, Vincenzo Reale of the Library "Romiti" of the University of Pisa, Alessandro Leoncini of the Archivio Storico of the University of Siena, Pier Luigi Gasco of the Centro Studi Monregalesi in Mondovì, Sarah Bakewell of the Wellcome Institute for the History of Medicine of London, Luigi Ponziani of the County Library in Teramo, Lorenzo Cangiano of the Nobel Institute of Neurophysiology of Stockholm, Carlo Alberto Segnini of the Domus Galileana in Pisa, Dino Bravieri and Andrea Cecconi of Archivio Storico degli Scolopi della Provincia Toscana, Danielle Etherington of the John Rylands Library in Manchester, John Simmonds of the Crawford Hansford & Kimber in London, Stefano Casati and Elena Montali of the Museo Galileo in Florence, Franco Ruggeri of the Institute of Human Anatomy of the University of Bologna, Marco Bortolotti of the Archivio Storico of the University of Bologna, the master and fellows of the Trinity College of Cambridge, Paola Ferretti Dalpiaz of the Department of Physics of the University of Ferrara, Viktor Govardovskii of the Russian Academy of Science, Andrey Dmitriev of the Department of Neurobiology of the University of Alabama, the librarians of the Archiginnasio and of the Biblioteca Universitaria in Bologna, of the Archivio di Stato in Bologna, of the Biblioteca Ariostea in Ferrara, of the Biblioteca Municipale in Reggio Emilia, of the Biblioteca Estense in Modena, of the Biblioteca Comunale in Mantova, of the Biblioteca di Filosofia, di Economia, di Antichistica e di Storia dell'Arte of the University of Pisa, of the Biblioteca di Fisica of the University of Ferrara, of the Libraries of the Nebraska University Medical Center, and of the Creighton University of Omaha, Nebraska, and many other librarians and archivists who have helped us in our long and difficult search for works and manuscripts around the world.

Finally, a very special thanks to Nick Wade, our longtime colleague and friend, who has assisted us with intelligence and patience in the long and demanding task of rendering into the language of Shakespeare and Newton a book written in the language of Dante and Galileo.

BIBLIOGRAPHICAL NOTE

Following is a list of abbreviations that we have used for the quotations from Galvani's and Volta's works:

GAC L. Galvani, *Dell'uso e dell'attività dell'arco conduttore nelle contrazioni dei muscoli*, in Bologna, a S. Tommaso d'Aquino, 1794.
GDV L. Galvani, *De viribus electricitatis in motu musculari commentarius*, in "De Bononiensi Scientiarum et Artium Instituto atque Academia commentarii," vol. 7, 1791, pp. 363–418.
GEA L. Galvani, *Memorie sulla elettricità animale... al celebre abate Lazzaro Spallanzani*, Bologna, per le stampe del Sassi, 1797.
GME L. Galvani, *Memorie ed esperimenti inediti*, Bologna, Cappelli, 1937.
GMS Luigi Galvani's manuscripts kept in the archives of the Academy of Sciences of Bologna (Fondo Galvani).
GOS L. Galvani, *Opere scelte*, a cura di Gustavo Barbensi, Torino, Utet, 1967.
GSA [L. Galvani], *Supplemento al Trattato dell'uso e dell'attività dell'arco conduttore nelle contrazioni de' muscoli*, Bologna, 1794.
VEN A. Volta, *Le opere di Alessandro Volta. Edizione nazionale sotto gli auspici della Reale Accademia dei Lincei e del Reale Istituto Lombardo di Scienze e Lettere*, Milano, Hoepli, 1918–1929, 7 vols.
VEP A. Volta, *Epistolario di Alessandro Volta. Edizione Nazionale*, Bologna, Zanichelli, 1949–1955, 5 vols.

The following abbreviations have been used to indicate archives and libraries in which the quoted documents are to be found:

AASB Archive of the Accademia delle Scienze—Bologna
ASB   Archivio di Stato—Bologna
BAB   Biblioteca dell'Archiginnasio—Bologna
BUB   Biblioteca Universitaria—Bologna

The Bibliography includes works to which we refer in the book and other studies that have been particularly important for our research. It is not intended to be a complete bibliography of the works on the topics we have dealt with but a first guide to help the reader who wants to deepen his or her knowledge of the themes of the book.

As we have stressed in the Foreword, and shall repeat in many points of the book, this study is the product of an intense and prolonged collaboration between the two authors, who have already published, both together and separately, several studies on Galvani, Volta, and the history of science (see the Bibliography). As it often happens in this kind of collaborative work, there has been a division of tasks: Marco Bresadola wrote chapters 2, 3.1–3.3, 4, 5, and 6, while Marco Piccolino wrote chapters 1, 3.4, 7, 8, 9, and 10.

Ambiguous frogs
drying in the wind
floundering, crying
without an electric soul.

1
Galvani, Volta, and the Forgotten Electrophysiology

LUIGI GALVANI IS generally characterized by negative stereotypes, both by his contemporaries as well as physicists in later generations. To elucidate this, we can quote from the *Cours de Physique* by Adolphe Ganot. This work, in its various forms and many editions, was widely read in the period between the nineteenth and the twentieth centuries, and it was translated into several languages. In the seventh French edition, published in 1859, he wrote:

> It is Galvani, professor of anatomy in Bologna, who in 1790 discovered the new electrical phenomena. By chance he noticed that a recently killed frog, put near an electric machine, produced lively contractions when sparks were drawn from the machine. This was a simple phenomenon. It was nothing else than the effect of a return shock, and a physicist would have given no importance to it. Unfortunately, Galvani, a clever anatomist, was poorly versed in electric knowledge. He varied his experiments in thousands of ways to find out the explanation of the phenomenon. Eventually, looking for what a physicist would have not deemed worthy of investigation, he made one of those discoveries which—as properly said—come more from a shaking ignorance than from a resting knowledge. One day he saw a dead frog suspended from a copper hook on the railings of a balcony, and noticed muscular contractions whenever the wind blew the lower extremities against the iron bars of the balcony. This was a strange, inexplicable fact, because there was no intervention of any external electricity, with no electric machine or return stroke. "The astonishment of the Bologna professor—as Arago writes—was then totally legitimate, all

Europe shared it." Galvani's experiment had a great echo. Everywhere people hurried to repeat it, and it can be said that that was for the frog a fatal date. Fig. 283 [our Fig. 1.1] shows how this experiment can be set up. Having cut off the frog's head, we suspend the hind half of the body by a copper hook passed between the back bone and the nerve filaments which run on each side of it; then holding a small plate of zinc in the hand, we bring one end of it in contact with the copper stem that holds the hook, and then touch the legs of the frog with the other end, At every contact we see the legs bending and shaking, thus reproducing all the motion of life.

To account for this phenomenon, Galvani assumed the existence of an electricity proper to the animal tissues, which, passing from the nerves to the muscle, through the metals through the metallic arc, produced the contractions. Under the phrase *animal electricity* or *vital fluid* this theory was adopted especially by physicians and physiologists, who supposed that the secret of life had been discovered. However, this brilliant idea faded away soon with the rigorous experiments of Volta. (Ganot, 1859, pp. 478–479)

In the lines that follow, Ganot alludes to the experiments of Alessandro Volta, the celebrated scientist of Como and professor of physics at the University of Pavia, who pursued Galvani's trail from its outset. In the period of 1791–1792, Galvani published a work entitled *De viribus electricitatis in motu musculari*; it appeared in the "Commentaries" of the Academy of Sciences of his hometown, Bologna. At the time this work was published Europe was in social and political turmoil as a consequence of the French revolution; *De viribus* added an element of intellectual turmoil. By resuming the results of more than 10 years of

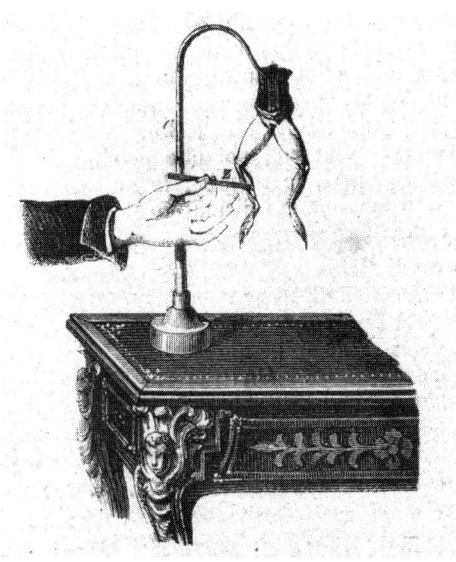

FIGURE 1.1. The experiment of Galvani with the metallic arc according to Ganot (from Ganot, 1859).

physiological experiments, Galvani reached the conclusion that nervous conduction and muscular contraction depend on the flux of an electricity intrinsic to the animal organism. This electricity would be accumulated in a condition of "disequilibrium" between the interior and the exterior of muscle fibers. Galvani referred to this as "animal electricity" by borrowing an expression already present in the scientific literature of the period.

Ganot described how Volta extended the experiments initiated by Galvani. Circuits composed of two different metals proved particularly efficacious in producing frog contractions. Accordingly, Volta assumed, in contrast to Galvani, that the contraction effects originated from the contact between two different metals (the "special theory of contact electricity"). In his view, this electricity had nothing to do with the animal organism. To prove his hypothesis, he made recourse to an especially sensitive instrument he had invented (the "*condensatore*" or "microelectroscope"). In this way Volta succeeded in measuring the electricity consequent upon contact between two metals (zinc and copper).

Afterward Ganot wrote:

> By developing his contact theory Volta went on (in 1800) to invent that wonderful instrument which won him an immortal fame, known also nowadays as the *pile of Volta*. Wishing to multiply the contacts and sum up together the electricity produced by every one of them, the celebrated physicist contrived to put together one above the other, a zinc disk, a copper disk, and afterwards a small piece of cloth moistened with acidulated water; and again a zinc disk, a copper disk, and a small piece of cloth, and again in the same way, with the caution of never inverting the order, as shown in Fig. 345 [our Fig. 1.2]. What could be expected from such an

FIGURE 1.2. Volta's battery according to Ganot (from Ganot, 1859).

arrangement? "Indeed—wrote Arago—I do not hesitate in saying it, this apparently inert stuff, this bizarre assemblage, this *pile* of so many different metallic couples separated by a tiny liquid, is, as far as the effects are concerned, the most wonderful instrument ever invented with no exception, not even that of the steam engine and the telescope." (p. 481)

From the aforementioned passages, it is well evident that the importance of Galvani's research is seen by Ganot exclusively with reference to the fact that it opened, fortuitously, the path to Volta's studies, which culminated in the invention of the battery. The battery was described as "the most wonderful instrument ever invented" by François-Jean Arago (the great French physicist who is the main reference for Ganot's account of the Galvani-Volta story). Galvani, the pioneer, whose ignorance of physics is given more prominence than his scientific merits, is thus made an emblem of "shaking ignorance"— ignorance that is still ignorance, even though it propitiates the progress of scientific knowledge.

It is worth recognizing here that Ganot's caricature of Galvani had its antecedents. The charge of being ignorant of the laws of physics, taken from the *éloge* of Volta delivered by Arago at the Paris Académie des Sciences (see Arago, 1854), was already implicit in Volta's *Memoria seconda sull'elettricità animale* (*Second Memoir on Animal Electricity*). Volta wrote this in May 1792, a few months after commencing his research on frog contractions. With respect to the contractions appearing in the frog's leg when a spark was drawn from an electric machine (which had greatly surprised and stimulated Galvani), Volta wrote, with a clear reference to the "return stroke" theory elaborated in 1779 by Charles Stanhope (Lord Mahon):

> How could that happen, and from what causes […] it is easy to answer if one know the theory of the Electric Atmospheres. The phenomenon is due to the fluid, pushed and chased away from the conductors that are influenced by the electrified one, i.e. that are immerged in its sphere of activity, which extends itself to a large range. This fluid flows back and comes to its proper place through the same way, i.e. through the series of conductors laid down on the table etc. This happens when one produces in any way and from whatever places the big spark from the electric Conductor [of the machine], and, by consequence, destroys that Atmosphere, causing the fluid, I say, to flow back instantaneously. (VEN, I, p. 46)

It is clear that, for Volta, Galvani's surprise was the consequence of his ignorance of the correct "theory of the Electric Atmospheres" and of the "return stroke" phenomenon "put in the better light by Milord Mahon," as he specifies afterward. In the previous part of his *Memoir*, Volta had written that it was not worthwhile for him to deal extensively with the laws that are at the basis of the electric influence. This was because—he said— either these laws are already known and "it is therefore not necessary to say more," or they are not known. Thus, "in order to make them comprehensible to those who are not aquainted with the matter, it would be necessary to begin from the first principles." This implied that Galvani, who had been surprised by the occurrence of contractions, belonged to the category of the people who "are not acquainted with the matter."

As for the role of chance in Galvani's research (alluded to by Ganot), it was the doctor from Bologna himself who remarked on its importance in the opening of *De viribus*. There he wrote that an attendant warned him of the occurrence of the phenomenon while "I was completely engrossed and contemplating other things" (GDV, p. 364; translated in Galvani, 1953a, p. 47). We will see that the account where Galvani emphasised the role of chance in his investigation is only a partial representation of the true path leading to his interest in the involvement of electricity in muscular motion. As we know, chance is an important element in scientific discovery. This process is rarely linear and rational, although it is often presented this way in textbooks, in some historical reconstructions, and occasionally even in the accounts by scientists themselves.

In this context it is certainly worth asking why Galvani opens *De viribus* with the "chance" observation; that is, contraction of the frog's leg "situated at a rather long distance" from the conductor of the electric machine at the instant of the spark. It would, however, be legitimate to enquire how "by chance" a "prepared frog" and a charged electric machine were both together in the laboratory of a scientist? And how could it be, also by chance, that a collaborator was manipulating the frogs while somebody else was drawing a spark from the machine?

We will consider in some detail the peculiar meaning of "chance" in Galvani's research. It was nothing other than what would be expected of a scientist setting up a special animal preparation for studying the electric phenomena involved in muscular motion. In his laboratory, together with several collaborators, he was investigating the effects of artificial electricity on muscle contraction, in order to verify some physiological hypotheses circulating at that time. Galvani had been concerned with investigating this subject for a long while. It is certainly possible that he had not consciously programmed that particular experiment, and therefore the occurrence of the contraction at the moment of the spark could be properly defined as a "chance observation." It was, nevertheless, an observation presented to a prepared mind, to somebody ready to pick up the element of novelty and interest present in an event that could easily be neglected by others.

This interpretation is contrary to the traditional one of how chance was involved in Galvani's research, and it is certainly a more trivial one. By relying on an anecdote narrated (and probably invented) by Jean-Louis Alibert in his *Eloge historique de Galvani*, a legend has been created: The jumping frogs were not there to investigate the role of electricity in muscle contraction but for Galvani to prepare a tonic *bouillon*, a broth, for his beloved wife, Lucia, who was of a frail constitution (Alibert, 1801).

It is also significant how Ganot refers to another of Galvani's fundamental observations, which many in Italy know well because, until some years ago, it was almost unfailingly included in textbooks of physics, often with an illustration depicting an ancien régime scientist manipulating frog preparations hanging from the iron railings of a balcony.

Speaking of Galvani, Ganot writes:

> One day he saw a dead frog suspended from a copper hook at the balcony of a window, and noticed violent muscular contractions whenever the wind blew the lower extremities against the iron bars of the window. (Ganot, 1859, p. 479)

Also in this account it seems that the observation had occurred just by chance, as if the frog was simply hanging from the balcony railings, without any clear purpose, or at least not with a scientific one. We might even suppose that it had not been Galvani himself who hung up the frog, but a servant or a waitress, perhaps in order to prepare a delicacy to stimulate the appetite of the unhealthy spouse of the Professor. This dead frog, hanging on a copper hook, and suddenly shaking whenever the wind blew its legs against the iron railings, reminds us of the succinct description of the experiment reported in the *Dictionary of Scientists* published in 1996 by Cambridge University Press: "dead frogs being dried by fixing by brass skewers to an iron fence showed convulsions" (Millar et al., 1996). The new colorful detail in the Cambridge dictionary, that the frogs were "being dried," should not be ascribed to the creative imagination of the authors because we find it, in almost the same wording, in another *Dictionary of Scientists*, published by Larousse in English in 1994: "dead frogs undergoing drying by being fixed to an iron fence by skewers suffered convulsions" (Muir, 1994).

The descriptive concordance is indicative of how such "dictionaries of scientists" (and similar reference works, which have been of fundamental importance for the cultural imagery of laymen—before the Internet-Wikipedia era) are compiled. On the other hand, it is difficult to believe that the authors of such texts consulted the original sources or specialized works on the scientists. The entries could indeed include such disparate people as Galen and Einstein, Lorente de No and the Becquerels, Vesalius and Doty, Millikan and Schaudinn, the Herschels, and Volta or Galvani (just to mention the names that came to our attention when more or less casually browsing the two dictionaries). It is therefore understandable that one would refer to what was found in previously published works. The outcome of this kind of transmission (which has a long tradition dating back at least to the Greco-Roman era) is that stereotypes are perpetuated, passing from one work to the other with only minor variations.[1] As acutely remarked by the English biochemist David Keilin, the problem is that these kinds of inaccuracies tend to remain in the cultural imagery for a long time, even when the original erroneous statement is shown to be wrong by accurate historical studies (Keilin, 1966).

---

[1] For a recent discussion about transmission of stereotypes in the case of Galvani and his frogs, see Piccolino and Wade (2012).

The bicentenary of Galvani's death was marked in 1998. It might be hoped that critical re-examination of the work of the scientist stimulated by the occasion would have an echo in the dictionaries of scientists published thereafter, but this is not apparent. In the third edition of the *Hutchinson Dictionary of Scientific Biography* published in 2000, under the direction of a famous historian, Roy Porter, we find many of the classical stereotypes of Galvani. As well-consolidated concretions, they seem to resist the assaults of the time:

> In 1786 Galvani noticed that touching a frog with a metal instrument during a thunderstorm made the frog's muscles twitch. He later hung some dissected frogs from brass hooks on an iron railing to dry them, and noticed that their muscles contracted when the legs came into contact with the iron—even if there was no electrical storm. Galvani concluded that electricity was causing the contraction and postulated (incorrectly) that it came from the animal's muscle and nerve tissues. He summarized his findings in 1791 in a paper, "De viribus electricitatis in motu musculari commentarius," that gained general acceptance except by Alessandro Volta, who by 1800 had constructed electric batteries consisting of plates (electrodes) of dissimilar metals in a salty electrolyte, thus proving that Galvani had been wrong and that the source of the electricity in his experiment had been the two different metals and the animal's body fluids. But for many years current electricity was called Galvanic electricity and Galvani's name is preserved in the word galvanometer, a current-measuring device developed in 1820 by André Ampère, and in galvanization, the process of covering steel with a layer of zinc (originally by electroplating). Galvani's original paper was translated into English and published in 1953 as "Commentary on the effect of electricity on muscular motion." (Porter & Ogilvie, 2000, vol. 1, p. 406)[2]

Again we read of frogs drying in the wind and that Galvani was wrong in attributing the origin of the electricity involved in the phenomena of muscle and nerve excitation to animal tissues. Moreover, Volta is considered to be the only one dissenting from Galvani on that point, as well as proving (with the invention of the battery) that the electricity responsible for the contractions in the experiments with metals originated from the contact between the metals and the animal liquids.

The mechanisms responsible for the contractions in Galvani's and Volta's experiments in which bimetallic arcs were used to connect the excitable tissues of the animal can be understood only in the light of research carried out in England by Alan Hodgkin and Andrew Huxley in the 1950s. This research still represents one of the most fascinating aspects of modern experimental research, and we will deal with it in Chapters 9 and 10.

---

[2] Needless to say, the text is absolutely the same in the following editions of the work, including the ebook first issued in 2009 and unmodified in 2010, when the English edition of our book was being prepared.

For the moment, it is appropriate to provide an idea of the degree of error contained in statements like that of the *Hutchinson Dictionary*. Namely, that Galvani was wrong in attributing the origin of the electricity responsible for the contractions to the animal, and also that Volta was right in attributing it to the contact between metals and between metals and animal liquids. To that end, let us refer to an example from everyday life—a car approaching a gate that can be opened by remote control. The driver presses a button and the gate opens. To say, as Volta did, that the electricity generated by the metallic contact excites the nerves and muscles of the frog preparation is equivalent to saying that it is the tiny electric power of the battery contained in the hand-operated remote control system that provides the energy for the opening of the gate. As we know well, in the case of the gate, the remote control system only provides the tiny energy that enlists the substantial electric energy of the mains to activate a motor. This is the real effective cause of the gate's opening.

In a similar way, in the case of Galvani's and Volta's experiments with frogs, the electricity generated by the metallic contacts only operates by activating a very sophisticated molecular switch, situated within the membrane of nerve or muscle cells. This, in turn, releases the electrochemical energy stored at the sides of the membrane and produces nerve conduction and, eventually, muscle contraction. In both the natural, physiological system and in the artificial technological apparatus, the initial electricity responsible for the beginning of the action provides only a control command, not the power involved in the final effect (i.e., nerve conduction and muscle in one case, gate movement in the other). In a way somewhat analogous to the complex system involved in opening the gate, in muscle and nerve fibers the electricity is stored in a condition of disequilibrium at the two sides of the fiber membrane and is ready to flow, once a command control (also electrical in nature) acts on the membrane.

The idea of electricity accumulated in a condition of disequilibrium between the interior and the exterior of the excitable fibers was just the idea expounded in 1791 by Galvani in his *De viribus*. As we shall discuss at length, from the historical point of view it is comprehensible that Volta attributed the electric power of dissimilar metals to the muscle contractions produced by the metallic arcs. In a similar way, it could be understood why he rejected the existence of electricity in a condition of disequilibrium inside animal tissues (Piccolino, 1997, 1998).

However, at present, to maintain a similar view would be—to use Ganot's words—more the expression of a "shaking ignorance" than of the scientific knowledge necessary to understand the intellectual intricacies that underlie an extraordinary phase of scientific development more than two centuries ago.

The hypothesis of animal electricity is the strong idea that emerged from Galvani's experiments, and it is the historical evolution of this idea that has led to the development of electrophysiology. Electrophysiology is a fundamental discipline of contemporary science and one of the main foundations of modern neurosciences. It is the science that enables us to study the functioning of the circuits of our brain. These circuits underlie not only our movements but also our sensations, perceptions, and even our most subtle mental activities.

As pointed out with poetical prose by the great English physiologist Charles Scott Sherrington: Every sensation, every perception, even that which arises in us from the observation of a beautiful landscape, a beautiful painting, or from seeing a friend's visage, is due to the flow of tiny electric signals. Like an "electric storm," these signals incessantly flow along the fibers of our peripheral nerves or the central circuits of our brain (Sherrington, 1949, p. 113). If these signals did not exist, we could experience no sensation whatsoever; we could not speak, write, think, or have feelings. These signals arise because the electricity existing in the organism in a condition of disequilibrium (as proposed by Galvani) moves suddenly, when excited by internal or external stimuli. Through a series of physiological events, now known precisely down to the molecular level, this electric motion brings about the production of short-duration electric signals known as nervous impulses or action potentials. These signals propagated along the membranes of nerve fibers are the basis of our brain function. Comparable electric impulses play a fundamental role in the phenomena leading to the excitation of muscle fibers, and, moreover, also in the rhythmic activity of the heart (and thus they are basic to life itself).

As mentioned earlier, electrophysiology is one of most important fields of modern biology, and its relevance goes even beyond the boundaries of life science. For some decades, the results obtained in the electrophysiological study of nerve circuits have also become an essential reference in technological and applied areas of modern science. These include so-called machine vision, neural networks, artificial intelligence, several areas of robotics, development of high-performance sensors, and also various branches of nanotechnologies. All these fields are having a significant, and rapidly growing, impact on the lifestyle of modern people.

The figure of Galvani has certainly attracted the interest of various scholars, and many pages have been written on his frog experiments and on his controversy with Alessandro Volta. Surprisingly, however, scant attention has been given to the problem of the validity of his hypothesis of animal electricity. As already noted, this hypothesis is one of the main foundations of modern electrophysiology.

As an example, let us consider *The Ambiguous Frog*, written by the philosopher of science Marcello Pera, the first Italian edition of which was published in 1986 (Pera, 1986/1992). The reader of this book would try in vain to answer the question of whether Galvani's animal electricity has some scientific reality or is an obsolete notion passed over by the progress of human understanding. Similar examples of the latter would be the celestial spheres or the epicycles of classical cosmology; the humors, virtues, or sympathies of ancient medicine; Descartes's vortices; Stahl's phlogiston; or in more modern times, cold fusion or tachyons.

As a matter of fact, the declared aim of Pera's book is not to investigate the scientific questions underlying the Galvani-Volta controversy but to demonstrate that this controversy was basically due to an opposition of two contrary mental attitudes (or, to use Pera's language, of two opposite *Gestalten*): the electrobiological (or electrophysiological) attitude of Galvani, and the electrophysical one of Volta. In the attempt to substantiate his view, Pera tends to undermine the importance of the scientific problems that were at the heart of the argument between Galvani and Volta. Pera's limited interest in exploring in depth the work of the two scientists is therefore understandable. This also applies to his disregard for the subsequent scientific developments, which, as we shall see in detail, have allowed for the exposure of the intellectual enigmas surfacing in Galvani's and Volta's research.

From another point of view, the distinction between electrophysical and electrobiological (or electrophysiological) perspectives, a leitmotif in Pera's book, appears to be poorly justified. This is mainly because it is referred to as a historical period (the Enlightenment) characterized by the absence of rigid barriers between different fields of science (and culture). As we shall discuss later, the distinction between different perspectives did not really concern Galvani and Volta, but it has been a central theme for some historians and philosophers who have dealt with the two scientists. They, like Pera, have represented the debate between the two scientists as based mainly on the "persuasive argumentations" typical of rhetorical language. In this way the scientific discussion is rather like a case study of some modern philosophical theories that pay less heed to the weight of experimental evidence and scientific objectivity. It also concerns those who, like Giovanni Polvani (still considered an authority on Volta's studies), are in no way interested in the electrophysiological aspects of both Galvani's and Volta's research. This occurs to such a degree in Polvani's biography of Volta that some important historical elements for the comprehension of the two are totally ignored, according to a process that could be qualified as a Freudian lapsus.

Galvani's research flourished within a milieu characterized by the great interdisciplinarity of the scientific approach, the Institute of Sciences of Bologna, which was one of the first modern experimental institutions in Europe. This Institute housed laboratories of physics, chemistry, and anatomy, as well as an astronomical observatory and even an art academy; these coexisted in close association, both geographically and culturally. Within the Bologna Institute, scholars specializing in different fields could easily exchange the fruits of their research and knowledge.

This felicitous interchange (rather than a rigid opposition) between the study of the animate world and inanimate nature, represented by the Galvani-Volta case, was certainly not a unique occurrence in the eighteenth century. As a matter of fact, a crucial event in the chemical revolution of the period was Joseph Black's research on *magnesia alba* (Black, 1777). This research was the outcome of the need to find the most suitable substance to dissolve kidney stones (the subject of Black's medical dissertation). This research allowed the Scottish doctor to discover the chemical mechanism underlying the transformation of carbonate in hydroxides. With the studies of Joseph Priestley, Henry Cavendish, and many others, Black's investigation opened the path to the discoveries of Antoine-Laurent Lavoisier on combustion phenomena and thus led to modern chemistry.

From the onset of research into electricity, those involved in it had the characteristic of interdisciplinarity. The book that heralded the modern era, *De magnete* (1600) was written by William Gilbert. The great English scholar clearly recognized that the attracting powers of some bodies after rubbing does not belong exclusively to amber (ἤλεκτρον in Greek). It was also a characteristic of other substances, like glass and resin, which he defined as "electrical." Gilbert was the personal physician of Queen Elizabeth I.

If a rigid distinction is drawn between physics and biology (or medicine), how could we classify a scientist like Stephen Hales, who in the eighteenth century gave fundamental contributions to the physics of gas, to physiology, and to medicine? What about Jan Ingenhousz, the personal physician to the Austrian Empress Maria Theresa, who made important studies in the field of electricity, heat conduction, and chemistry, and is particularly famous for his discovery of photosynthesis? Where would we place Thomas Young, who first demonstrated the undulatory nature of light, gave important contributions to ophthalmology, and even started deciphering hieroglyphs?

A clear distinction between the various fields of science started developing only in the nineteenth century. This notwithstanding, we would still have difficulties in categorizing a scientist like Hermann von Helmholtz. He graduated in medicine and eventually became professor of physics at the University of Berlin. In his very prolific and varied scientific life he made fundamental contributions to physics (with the formulation of the first principle of thermodynamics), to physiology (where he was the first to measure the conduction speed of nerves, elaborated the trichromatic theory of vision, and enunciated a modern theory of hearing), to medicine (where he invented the ophthalmoscope), and he even contributed to the development of musical theory.

A common principle of historiography is that one should not interpret an event or an idea of the past with reference to subsequent categories of historical development. There are consequences to strictly adhering to this principle, partially derived from the work of Thomas Kuhn (see Kuhn, 1962). It often leads historians of science to assume that they should ignore modern science when trying to interpret phases of scientific progress, paths to discoveries, or controversies among scientists from the recent or remote past. It is as though the scientific laws themselves, not only the theories or interpretative categories, would change with the time. Personally we are suspicious of the possible drawbacks of such an attitude. We believe that a deep knowledge of the scientific problems underlying the research or the debates of the past should be cognizant of recent scientific advances, as this is of fundamental importance for a perceptive historical comprehension of them. This is at least what will emerge in our book for the electrophysiological research of Galvani and Volta. Only with reference to modern electrophysiological investigation, and particularly to the achievements of Hodgkin, Huxley, and their followers, will it be possible to understand why some of Galvani's and Volta's experimental results changed drastically with apparently minor, and sometimes unnoticed, variations of the experimental arrangements and procedures.

To speak about Galvani (and also Volta) without considering the problem of the validity of the hypothesis on animal electricity and the subtleties of the involvement of electricity in the function of excitable fibers is like speaking of Copernicus, Galileo, Kepler, and their adversaries without considering whether the Earth rotates around the Sun or vice versa. Only by knowing the precise orbits of the planets as calculated by modern astronomers can we understand the possible relevance of some relatively minor deviations in the astronomical measurements made by sixteenth- and seventeenth-century astronomers, as, for instance, Kepler with the orbit of Mars, with regard to the predictions of Ptolemaic cosmology.

To say, as has been said, that before Galileo there were only oscillating bodies but not pendulums as we see them now should not mean that the laws of physics have changed in the meantime and, even worse, that it is totally irrelevant to know them.

As we shall see in our book, the lack of an electrophysiological perspective in the analysis of Galvani's studies makes it impossible to situate his endeavor within the most relevant scientific context. It also frustrates any attempts to understand properly the development of his investigation, the problems that he had to face in his experiments, and also to clarify some of important points of his controversy with Volta. It makes it, moreover, difficult to understand Volta's scientific endeavor and particularly the path leading him to his landmark achievement, the invention of the battery. As we shall see, the debate over the role of electricity in "animal economy" was of great importance in Volta's path to the battery. Even more important, Galvani's prepared frogs enabled Volta to detect the small electric currents generated by contact between two metals (less than 1 volt as we now know). By using Galvani's frogs, Volta could achieve this in a period when the physical electroscopes could only detect potentials of hundredths of a volt.

The reason for the high sensitivity of the animal "electroscope" of Galvani's frogs is because, at some point during animal evolution, nature needed to contrive a very sensitive electric amplifier. This was necessary in order to endow excitable fibers with the capability to generate and conduct effectively electric signals for relatively long distances. The amplifier of molecular size present in the excitable fibers of animals, which is still unsurpassed in its efficiency and structural designs by humanmade devices, was (as we will discuss in detail in Chapter 10) the device somewhat inadvertently used by Volta to demonstrate the minute electricity of metals. It is a biological amplifier that Volta happened to use eventually because of his wide scientific interests, spanning into the life sciences. All this certainly stands in opposition to a rigid separation between physiology and physics in Volta's scientific work and for an adequate regard of his interest in the phenomena of life.

There is even more. It is particularly remarkable how Volta presented his newly invented device, in communicating his discovery to the Royal Society of London on March 20, 1800. In addition to reporting and discussing the physiological effects of electric currents, he proposed to call his invention an *organe électrique artificiel* in order to underline the similarity to the *organe électrique naturel* of the electric fishes. As will be discussed, an important reason for Volta's pride in presenting the invention was his supposition that he had discovered the way in which animals could produce and use electricity for their physiological needs. It is perhaps not far from the truth to say that, in Volta's opinion, the battery was more the culminating phase of a chapter of history of physiology than the starting point of a new era of physical and technological sciences. This was because the possibility, illustrated by the new device of an "electricity excited by the mere contact of conductive substances" (as indicated by the title of Volta's letter to the Royal Society), undermined the main difficulty that electricity could be stored within animal organisms. Since animal tissues were known to be conductive, it was argued they would rapidly dissipate any electrical disequilibrium possibly produced inside them. The difficulty was even more stringent for the case of electric fishes, which, in addition to having electric conductive tissues—as other animals do—also lived in a conductive habitat.

Electric fish research was a fundamental aspect of the eighteenth-century science, particularly with the studies performed by John Walsh, an English gentleman and amateur scientist, who succeeded in the period 1772–1776 in discovering the electric nature of the shock produced by the torpedo ray and the electric eel of the Guyana (Finger & Piccolino, 2011; Piccolino, 2003a; Piccolino & Bresadola, 2002).

It has been said that electrophysiology was born with the experiment in which Walsh obtained decisive evidence of the electrical nature of the fish's shock (Wu, 1984). This is in part an overstatement, but it points to the importance of studies of electric fish in the emergence of a discipline that, in the last few decades, has undergone a development comparable to that of quantum mechanics in Max Planck's day. In 1991 the Nobel Prize for Medicine has been jointly awarded to two German scientists, Erwin Neher and Bert Sakmann, for the invention of a new electrophysiological technique (the *patch clamp*). This extraordinary technique allows for the recording of the tiny current passing through a single ionic channel; that is, across the molecular structure of the cell membrane which underlies nervous conduction and muscle excitability.

The patch-clamp technique enables electrophysiologists to study the "animal machine" at the molecular level. This is what Galvani hypothesised with great perspicacity more than two centuries ago. According to the Bologna doctor, this machine is responsible for the electrical disequilibrium underlying the phenomena of animal electricity. With its momentous advancements in the fields of electrophysiology and related fields (biophysics, molecular biology), modern science can thus give an answer, well beyond the expectations of a "prudent philosopher" of the Enlightenment, to the questions aroused by Galvani when he wrote with reference to the electric disequilibrium of excitable fibers:

> Such a disequilibrium in the animal either must be there naturally or should result from artifice. If it is there naturally, we should admit that in the animal there is a particular machine capable of generating such a disequilibrium, and it will be convenient to refer to this form of electricity as to an animal electricity in order to

denote, not an electricity whatsoever, but a particular one referred to a particular machine. [...] But what will it be this animal machine? We cannot establish it with certitude; it remains totally occult to the most acute sight; we can do nothing else than figure out its properties, and, from these, somewhat envision its nature. (GAC, pp. 70–71 and 76)

As we shall see in detail in Chapter 10, the animal machine hypothesised by Galvani is a complex and integrated structure based on the presence in cell membranes of molecular structures (ionic pumps) capable of separating ions and thus creating concentration gradients between the interior and exterior. It is also based on the presence of other molecular structures on the membrane, the ionic channels, capable of transforming the chemical energy of these gradients in a difference of electrical potential. In other words, is that electric disequilibrium that Galvani had assumed to be present in animal tissues. It is the current that passes across a single one of these channels (i.e., a single protein) that can be recorded, with the time resolution of a microsecond, by using Neher and Sakmann's patch-clamp technique.

We have seen how Volta opposed Galvani's hypothesis of animal electricity and proposed the theory of metallic contact. In Volta's view there was no electrical disequilibrium within the animal tissues, and the disequilibrium arose from the contact between the two different metals used to connect the muscle and nerve of the frog preparation (or two points of a nerve). This metal electricity was the effective cause of nerve excitation. If I can show—Volta said—that metals are "electromotors," why should we assume, in order to account for nervous conduction, an electric equilibrium intrinsic to animal tissues? Galvani replied: If, as my research clearly indicates, electricity is accumulated in a condition of disequilibrium and is ready to move along the nerve in consequence of this disequilibrium, why should we look for another reason for a phenomenon that is already accounted for a known cause? A fundamental principle of science is that one should not multiply the cause. The first *regula philosophandi* that, in Ockham's footsteps, Newton declared in his *Principia*, is as follows: "*Causas rerum naturalium non plures admitti debere, quam quæ & veræ sint & earum phænomenis explicandis sufficient* [...] *Natura enim simplex est & rerum causis superfluis non luxuriat*" ("We are not admit no more causes of natural things than such as are both true and sufficient to explain their appearance [...] for Nature is pleased with simplicity and does not pomp of superfluous causes" (Newton, 1687, p. 402).

In this apparently insurmountable dilemma between animal and metallic electricity, as the only possible and sufficient cause of nervous conduction, resides the heart of the problem in the Galvani-Volta controversy. Surprisingly, the progress of electrophysiology, and particularly Hodgkin-Huxley studies in the mid-twentieth century, showed that the dilemma was only apparent, or—to use a philosophical expression—was a false dilemma. This is because a third possibility appeared to exist (*tertium datur*). Electricity is indeed accumulated within animal tissues in a condition of disequilibrium (as Galvani assumed). In normal conditions, however, it cannot flow because its passage across the cell membrane is hindered by the nonconductive condition of the membrane in the resting state. In order for the membrane properties to change (thus allowing for ionic fluxes), it is necessary that a triggering stimulus, itself of an electrical nature, act on it, such as to change the electrical potential between the interior and the exterior of the cell. This is because the passage from the nonconductive to the conductive state of the membrane is controlled by the electric potential across the membrane. In the conditions of Galvani's and Volta's experiments with metals, this triggering stimulus was provided by the electricity excited by the metallic contacts (i.e., Volta's type of electricity). As a consequence, ions could flow under the action of the preexisting disequilibrium (Galvani's type of electricity), thus bringing about nerve conduction.

There are, as we shall see in Chapters 9 and 10, fascinating reasons why evolution has contrived such a complex mechanism for generating and propagating nerve signals. This complexity accounts for the very unexpected chance by which animal electricity (i.e., that underlying vision, motion, and life itself) encountered, on the same experimental table (that of Galvani and Volta), the physical electricity in motion. This is the form of electricity, which, besides being at the inception of one of most important developments of modern science, has marked, with its technological applications the lifestyle of humankind.

Returning to the story of Galvani and Volta, it is important to note that there was no real possibility of solving the apparent dilemma underlying their controversy within the boundaries of eighteenth-century science. This was because neither of the two scholars could anticipate the complexity and subtlety of electrical processes involved in nerve conduction. While that is certainly the case, knowledge of these mechanisms is commonplace in the recent decades for anybody having some familiarity with biological science. For instance, in Curtis and Barnes's *Biology*, a high-school textbook first published in America in 1972 (and also available in Italian and Spanish), we can find a statement of this type: that for at least 200 years it is known that nervous conduction is associated with electric phenomena. In the following pages of this same book, we find a succinct, and yet accurate, description of the process whereby an electric difference is generated between the interior and the exterior of a cell; and, moreover, of the molecular mechanisms by which this biological electricity makes possible the generation and propagation of nerve impulses.

We are therefore somewhat surprised at the way Galvani's story is represented in the volume *Personaggi e scoperte nella fisica contemporanea*, first published by the Nobel laureate Emilio Segrè in 1976 (and afterward re-edited several times with somewhat different titles and translated into various languages, including English). After having ascribed to Galvani the merit of opening two great chapters of science, Segrè writes that while Volta had a clear vision of only one part of the complete problem, "the remaining part has not yet completely clarified" (Segrè, 1996, vol. 1, p. 157).

A great disregard, not to say ignorance, concerning the scientific aspect of Galvani's work is not only a characteristic of the dictionaries of scientists and of other works of scientific dissemination. In an important volume published in 1967, and completely devoted to Galvani's life and research, the Italian scholar Gustavo Barbensi comments on Galvani's and Volta's different interpretations of the frog contractions in their experiments with metallic conductors. He concludes by saying that, because of the difficulties in the interpretation of the results, "many experiments were necessary for Volta to find out the true cause of the muscle contractions that Galvani attributed to an electricity intrinsic to the animal." This evidently means that, since Volta had found the correct cause of the event, Galvani's explanation was certainly wrong, a view in line with the general representation of Galvani's research portrayed by Barbensi in his work (see Barbensi 1967, p. 26).

The evaluation of Galvani's work implicit in Barbensi's text is based on a knowledge of the post-Galvanian electrophysiology as matured at the beginning of the nineteenth century. This is almost generally the case for most of the historians and philosophers of science in even more recent times who have written about Galvani and Volta. To fully appreciate the anachronism of Barbensi (and also of many of his followers), we should consider that he wrote more than one century after Leopoldo Nobili and Carlo Matteucci had measured the electricity intrinsic to animal tissues with a physical instrument (the galvanometer), and that Emil du Bois-Reymond has detected the electric changes in nerve conduction (the *negative Schwankung*, i.e., negative "oscillation" or "variation"). Barbensi wrote almost exactly one century after Julius Bernstein had provided the first recording of the "action current" in nerve with a high temporal resolution graphical method. Barbensi's volume appeared about 30 years after the first recording obtained with an intracellular microelectrode of the electric potential across the cell membrane and of its modification during the nervous impulse. As we shall see in Chapter 9, these last experiments carried out on the giant nervous fiber (or axon) of the squid would inaugurate the modern era of electrophysiology, when the squid assumed the position that Galvani's frogs had occupied in electrophysiological experiments for more than 150 years.

In present times, the evaluation of Galvani's electrophysiological work, particularly with reference to his controversy with Volta, still oscillates between two main attitudes. The first, similar to that held by Barbensi, concludes that Galvani was wrong in his interpretation of the origin of the electricity involved in frog contraction and that Volta was right. This interpretation emerged, as already mentioned, at the beginning of the nineteenth century on the wave of Volta's triumph with the invention of the battery.

The second attitude is that which considers that both Galvani and Volta were right in the "assertive" part of their respective statements (i.e., Galvani by asserting animal electricity and Volta by invocating metal electricity), while they were both wrong in the "negative" part of their propositions (i.e., Galvani negating metal electricity and Volta refuting animal electricity). This interpretation, which recognizes the validity of the animal electricity hypothesis, has been formulated by historians with a physiological background (like John Fulton, Giulio Cesare Pupilli, and Ettore Fadiga; see Fulton, 1926,

1930, and Pupilli & Fadiga, 1963). Although generally correct, this attitude attributing to Galvani and Volta a comparable share of truth and error remains on the surface of the discussion between the two scholars and does not make apparent some of the most salient aspects of the electrophysiological research that emerged from their endeavors. With this view in mind, on the one hand, we could not understand the difficulty that they had in accepting the opinion of the assertive part of the antagonist's opinion. On the other hand, we could not figure out why animal and metallic electricity were discovered at once, during research aimed at ascertaining the role of electricity in muscle contraction and nerve conduction.

Among the various reasons that make it difficult to analyze critically Galvani's research is the opinion that in his experiments he was not investigating a clearly defined physiological problem but was searching for the mysterious principle of life, for immaterial forces that could not be investigated with scientific methods. This opinion has been favored by several factors. One is purely linguistic and due to the relation in Latin (the language of *De viribus*) and correspondingly in Italian (the language of Galvani's subsequent works) between the words *animalis* (in the sense of pertaining to living beings) and *anima* meaning soul, and therefore evoking the metaphysical principle of life. In Galvani, the adjective *animalis* is used without any vitalistic or transcendent connotation, and less than so, animistic. The complete title of the first published work on the subject reads *De viribus electricitatis in motu musculari, Commentarius* and refers to "forces" (*viribus*) and "motion" (*motu*), two well-defined categories of post-Newtonian science. In the 50 or so pages of this memoir, the word *anima* appears very rarely and only in a speculative context.

Another important element that has contributed to a vitalistic reading of Galvani's work is the interpretation proposed by a certain tradition of German culture flourishing at the turn of the eighteenth and nineteenth centuries, the *Naturphilosophie* with its strong Romantic connotations. This was the case particularly for Johann Wilhelm Ritter (and also, to a minor extent, for Alexander von Humboldt; see Humboldt, 1797, and Ritter, 1798). As a matter of fact, the idea suggested by Galvani's experiments that it was possible to obtain, through an electrical stimulation, a life reaction from a preparation of a dead animal, made a deep mark on collective imagery, suggesting a possible identification of electricity with vital "spirit" or "fluid." The macabre experiments performed in London at the beginning of nineteenth century by Galvani's nephew Giovanni Aldini also contributed to this. He applied electricity to the head or to the body of decapitated individuals and obtained movements, sometimes particularly horrifying, which seemed to indicate a temporary revitalization of the dead person (Aldini, 1804).

This tradition inspired a vast literature that flourished outside Germany (in France, Italy, England, and America—including authors like Nathaniel Hawthorne and Edgar Allan Poe). The myth of Frankenstein, given cinematic substance several times, is linked to the idea that electricity, particularly the violent and primeval energy associated with thunderstorms, could infuse life in the monsters fabricated by humans.

It is well known that the novel *Frankenstein, or the Modern Prometheus*, was written by Mary Wollstonecraft (later Shelley) in the period of 1816–1818, following a terrifying dream in which a student tried to infuse life into a monstrous creature (Wollstonecraft Shelley, 1818). This nightmare also entertained the possibility of restoring life in a corpse by applying galvanic current. This possibility had been evoked the night before in a conversation between her; her future husband, Percy Bysshe Shelley; Lord Byron; and John Polidori. One of the themes of this conversation had been Aldini's macabre experiments in London.

Modern historiographic research has made clear that the opposed terms of "vitalism" and "mechanism" are more complex and more difficult to distinguish than traditionally assumed. In any case the two concepts are of limited usefulness when trying to understand the work of Galvani and Volta, and they may even be misleading. As we shall see, similar to other great scholars of their time, both Galvani and Volta passed with ease from the study of inanimate, physical objects to that of the animate, living organisms, in an attitude of fruitful exchange between "physical" and "physiological" investigation. In particular, Galvani approached the study of the electrical phenomena involved in nerve conduction and muscle excitability as a true experimental problem, asking himself precise scientific questions, to which he attempted to give answers with scientific experiments and logic. In his research he made constant reference to well-defined physical models. One of Galvani's greatest merits is just that, of having contributed in bringing to the domains of modern science the theme of the relation of electricity to life, and away from the domain of speculations tainted with vitalistic connotations.

Galvani's most celebrated work, *De viribus*, which was first published in 1791, at the beginning of a revolutionary period of research involving various European countries and also America, stands out in the scientific panorama of the time as an avant-garde opus in its content and also in its style. This becomes evident particularly when comparing it with previous writings on the subject, like those of Francesco Giuseppe Gardini in Italy and Pierre Bertholon in France both first published in 1780, at the beginning of Galvani's electrophysiological research. In contrast to Galvani's *De viribus*, these texts still give the impression of the philosophico-scientific treatises with strong conjectural character written before the great scientific revolution of Galileo and Newton.

Galvani opens *De viribus* with an unexpected incipit, almost coup de théâtre for his contemporary readers (to whom it probably appeared too abrupt, justifying the addition of a vast foreword to the second edition, authored by his nephew Aldini). Without any preamble, and with a rapid and concise style, Galvani describes the experimental observation which would stimulate his interest in the role of electricity in muscular motion: that concerning the contractions excited by the spark of an electrical machine located at some distance from the animal preparation. He portrays his laboratory with his instruments, collaborators, and the intellectual effervescence dominating it:

> The course of the work has progressed in the following way. I dissected a frog and prepared it as in Fig. 2, Tab. 1 [see Fig. 1.3]. Having in mind other things, I placed

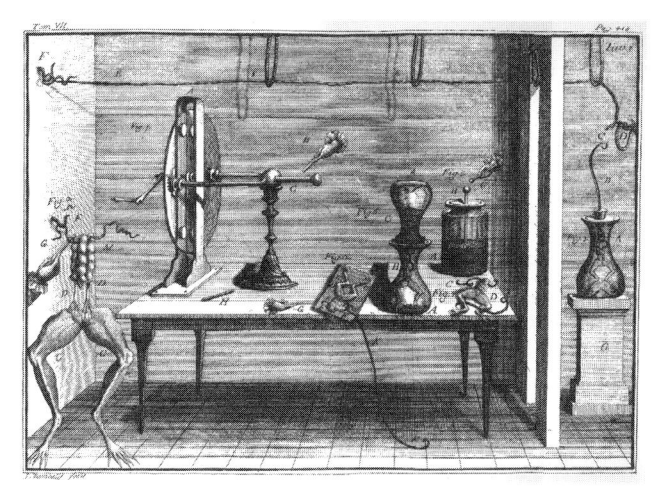

FIGURE 1.3. The first plate of *De viribus* (from Galvani, 1791).

the frog on the same table as an electrical machine, Fig. 1, Tab. 1, so that the animal was completely separated from and removed at a considerable distance from the machine's conductor. When one of my assistants by chance lightly applied the point of a scalpel to the inner crural nerves, DD, of the frog, suddenly all the muscles of the limbs were seen so to contract that they appeared to have fallen into violent tonic convulsions. Another of those who were helping us when we were performing electrical experiments thought he observed that this phenomenon occurred when a spark was discharged from the conductor of the electrical machine, Fig. 1, B. Marvelling at the novelty of the phenomenon, he immediately brought it to my attention while I was completely engrossed and contemplating other things. Hereupon I became inflamed by an incredible ardour and desire to repeat the experiment so as to bring to light what was obscure in the matter. I myself, therefore, applied the point of the scalpel first to one then to the other crural nerve, while at the same time one of the assistants produced a spark; the phenomenon repeated itself in precisely the same manner as before. Violent contractions were induced in the individual muscles of the limbs and the prepared animal reacted just as though it were seized with tetanus at the very moment when the sparks were discharged. (GDV, p. 364; translated with revisions from, Galvani, 1953a, p. 47)

Afterward Galvani goes on to describe in detail a very potent series of experiments on the effects of various forms of electricity on muscular contractions. In these experiments he varied the experimental conditions and the animal preparations in subtle ways.

It is experimental science that is portrayed in the scene from *De viribus*—not the science of "sophisms" and endless "conjectures," but that of Galileo and the great protagonists of the scientific revolution. This is a science that does not emerge from collating books but from reading Galileo's *Book of the Universe*, that unique volume alluded to by Marcello Malpighi (a constant reference for Galvani) in his *De renibus*, the work in which the structures of kidney filtration (later referred to as Malpighi's glomeruli) are described for the first time. In his introduction (*proemium*) to this work, the Italian anatomist writes, "Be sure of only one thing: I have not found this structure of kidneys by collating books, but instead through a patient, long and varied use of the microscope" (Malpighi, 1666/1669, p. 66).

In our book, we will discuss the detail of Galvani's experimental methods, of his hypotheses, of his attitude in raising questions that could be answered by experiments, of his capability of developing coherent scientific models. With this aim we will refer both to the published works and to Galvani's manuscripts (to a large extent still unedited and residing at the Bologna Institute of Sciences). The need to make recourse to original documents, in all historical research and particularly in the history of science, cannot be overstated.

Galvani's case is in some respects an epitome of the scientific enterprise. The nineteenth-century Italian physicist Silvestro Gherardi is one of the few scholars who surely had a direct knowledge of Galvani's works, having published in 1841 an ample edition of his writings under the auspices of the Institute. In commenting on this edition, he recognized that initially he had shared the negative interpretation of Galvani's work with "the heavy assertions [...] of physicists on him." However, on the basis of a detailed examination of Galvani's works, he changed his mind, acknowledging the importance of his research: "I am now confident and fully convinced of that, having previously been of the common persuasion concerning Galvani before I had the valuable opportunity of being involved with the studies and doctrines of this great Man" (Gherardi, 1841, p. 275, footnote).

An in-depth study of Galvani's works, and of the development of electrophysiology in the last two centuries, largely confirms Gherardi's assessment, which was expressed when the study of animal electricity was being rekindled, after the debacle of "galvanism" due to the success of Volta's ideas (Gherardi was writing when Carlo Matteucci was obtaining the first instrumental recording of animal electricity in frog muscles).

The reference to Gherardi allows us to introduce one of the weighty elements favoring the negative judgements of Galvani's works. We are obliged here to invoke Emil du Bois-Reymond, the German scientist who between 1848 and 1884 published the monumental *Untersuchungen über thierische Elektricität*, a work that has registered the virtues (as well as the vices) of much of the historiography concerning Galvani.

It is to du Bois-Reymond that we owe the first measurement of animal electricity in nerve and the first demonstration of the change of nerve electric condition associated to signal conduction (*negative Schwankung*). Besides presenting a detailed account of the author's experiments, the *Untersuchungen* also aims at offering to the reader an accurate and vast historical outline of the previous studies on the role of electricity in nervous conduction and muscle contraction. Du Bois-Reymond was undoubtedly an admirer of the Bologna scholar. He ascribes to Galvani (with some ambiguity) the discovery of animal electricity and he recognizes the crucial importance of the experiments, published in the period of 1794–1797, in which Galvani was able to obtain frog contraction without using any metal.

The problem is, however, that du Bois-Reymond's history of electricity was written to advance his claim, and not Galvani's, for first demonstrating that electricity is the force used by the organism for nerve conduction and muscle excitation.[3] As he was well aware, this is a fundamental achievement that somehow defines the end of an era in classical physiology and medicine, the era in which the mysterious and elusive agents "the animal spirits" were considered the carriers of nervous messages. With this view du Bois-Reymond is interested in underplaying the value of Galvani's statements, at the end of *De viribus*, in which he proudly vindicated his rights of discoverer of the role of electricity in nervous function:

> If it will be so, then the electrical nature of animal spirits, until now unknown and for a long time uselessly investigated, perhaps will appear in a clearer way. Thus, after our experiments certainly nobody would, in my opinion, cast doubt on the electrical nature of such spirits. [...] we could never suppose that fortune were to be so friendly to us, such as to allow us to be perhaps the first in handling, as it were, the electricity concealed in nerves, in extracting it from nerves, and, in some way, in putting it under everyone's eyes. (GDV, p. 402; translated with revision from Galvani, 1953a, p. 79)

To attribute the merit of this discovery to himself rather than to Galvani, du Bois-Reymond tries to amplify the experimental uncertainties and the theoretical weaknesses of the Bologna doctor. As the German scientist says, Galvani is a pioneer, and—like all pioneers who move toward unknown territories—he progresses with difficulty, often groping blindly. Undoubtedly we need—du Bois-Reymond says—to forgive his errors and difficulties, but we should not overestimate his work. In particular, he writes:

> No one, who has read Galvani's writings, can, without reverence, turn away from the simple picture of that man, whose restless, yet blind labours and naïve desire for knowledge were destined to bear such fruits. Every one will easily excuse his having wandered in that way which we shall soon see him take. The problem presented to him was an equation with two unknown quantities, one of which was the galvanism which Volta discovered, the other animal electricity, which latter, after half a century, now again appears claiming its proper place. (Du Bois-Reymond, 1852, p. 3)

---

[3] Even though a historical account of previous research written by a scientist in presenting his own work is justified and useful, it might become a surreptitious way to affirm the importance of the personal work, when the roles of the historian and that of the scientist are not kept sufficiently distinct. In the nineteenth century, a particularly relevant case of the use of historical texts in order to boost the value of personal achievements was that of the French physiologist Marie Jean Pierre Flourens (see Levinson, 2012).

In the aim of playing down Galvani's achievements, du Bois-Reymond compares his work with that of Volta, arriving to an exaltation of the physicist of Pavia, which is surprising for the overtly exaggerated tones characterizing his words:

> No one who wishes to judge impartially of the scientific history of those times and of its leaders, will consider Galvani and Volta as equals, or deny the vast superiority of the latter over all his opponents or fellow-workers, more especially over those of the Bologna school. We shall scarcely again find in one man gifts so rich and so calculated for research as were combined in Volta. He possessed that "incomprehensible talent," as Dove has called it, for separating the essential from the immaterial in complicated phenomena; that boldness of invention which must precede experiment, controlled by the most strict and cautious mode of manipulation; that unremitting attention which allows no circumstance to pass unnoticed: lastly, with so much acuteness, so much simplicity, so much of conception, combined with such depth of thought, he had a hand which was the hand of a workman. It is strange, indeed, that a German voice should call an Italian philosopher to order for depreciating the great Volta. (pp. 4–5)

The other scientist who was the target of du Bois-Reymond's criticism was Carlo Matteucci, another redoubtable competitor for the German scientist in his attempt for primacy as the true discoverer of animal electricity. Matteucci was the first to demonstrate with a physical instrument (the *astatic* galvanometer contrived by Leopoldo Nobili) the existence of electricity intrinsic to the animal organism. Of particular significance was that Matteucci was able to exclude experimentally any possibility of artifactual origin of the recorded electricity.

A few years before the first volume of du Bois-Reymond's *Untersuchungen* appeared in print, Gherardi had published Galvani's works and studied his extant manuscripts at the Bologna *Istituto*. In doing so, he had realized how untenable the common judgements on Galvani's' works were, particularly those of physicists (like Volta and Arago) who had accused Galvani of ignorance of the laws of physics. Gherardi argued for a substantial revaluation of Galvani versus Volta, an attitude that created some disappointment in du Bois-Reymond, who—as we have seen—had attempted to downplay the achievements of the Bologna doctor. It is with reference to Gherardi that du Bois-Reymond concluded his eulogy of Volta by saying: "It is strange, indeed, that a German voice should call an Italian philosopher to order for his depreciation of the great Volta."

Later in this book we will evaluate du Bois-Reymond's scientific work and we will see how he must be ascribed some important discoveries in the field of electrophysiology. This notwithstanding, we will also see that with his scientific and cultural influence, and with his stubborn attitude in support of his scientific ideas, even when contradicted by the available experimental evidence, du Bois-Reymond slowed the scientific progress in this field. Du Bois-Reymond's theory of "electric molecules" elaborated in order to account for the electric phenomena of living organisms is, in many regards, outdated with respect to Galvani's hypothesis of electricity accumulated between the interior and the exterior of the muscle fiber (as in "a minute Leyden jar," see Chapter 10).

Our digression on du Bois-Reymond is in no way aimed at highlighting his theoretical (and experimental) drawbacks. What we wish to state is that his judgement on Galvani has heavily influenced the historiography concerning the Bologna scholar. The reason is that because of its vastness and erudition, du Bois-Reymond's *Untersuchungen* has been a reference work not only for physiologists but also for physicists and science historians who have dealt with Galvani. It will be rather easy to show that du Bois-Reymond's texts have been the most cited work dealing with the progress of electrophysiology read by most historians of science who, even in recent years, have dealt with Galvani. This has happened mainly because the separation between different scientific and cultural disciplines that started in the nineteenth century has made mutual communication progressively more difficult. This is also true for those who are involved in pursuing a scientific discipline and those who write its history.

Among the reasons that have contributed to the difficulty of appreciating Galvani's character and work is the stereotype transmitted by certain personal hagiographies. Galvani was a pious, religious man, full of solicitude and affection for his wife, the beloved Lucia, who would die young, leaving her poor husband in inconsolable sorrow. Galvani, because of his religious conviction, refused the oath to the Napoleonic authorities of the Cisalpine Republic and was thus obliged to leave his teaching post and to lose his salary. Many elements seem to support this view and certainly nobody can say that he was not a deeply religious man, endowed with high moral standards. It is, however, important to consider that Galvani was a man of ambition. He ascended almost the entire cursus honorum of a Bologna scientist, becoming professor both at the University (the oldest University of the Western World) and at the Institute of Sciences. Of this prestigious establishment, a model for subsequent European institutions, like the Institute of France and the Royal Institution of London, Galvani was a full ("Benedictine") member, with the right to a pension; for many years he belonged to the College of Physicians as an effective fellow (i.e., having the right of vote), and of this institution he also held the prestigious charge of *Protomedicus* and of *Prior*. For someone like him, of a relatively modest birth (he was the son of a goldsmith), his career was certainly favored by his marriage with Lucia Galeazzi, the daughter of one of the most influential scholars of the city, Domenico Gusmano. Besides having held the most prestigious positions within Bologna medical and scientific milieu, Lucia's father had organized the most prestigious anatomical school of the town in his palace (Bresadola, 1997, 2011a). We do not make these points with the aim of depriving Galvani of his scientific and human merits. We want simply to remark that as highly religious and virtuous as Galvani might have been, he was, on the other hand, fully free and modern as a scientist and a man. Nor was he submissive and placid, showing, particularly regarding his arguments with Volta, a notable amount of sagacity and combativeness, as we shall see later (particularly in Chapters 2, 6, and 7).

The aim of this book is to contribute to the knowledge of Luigi Galvani, a highly celebrated scientist, who has extensively struck the collective imagery and has been the subject of many written works but has very rarely been studied in depth and without preconceptions. His discovery of animal electricity marked a crucial phase of modern science. This is why at the end of the eighteenth century it was compared to Harvey's discovery of blood circulation.

In dealing with Galvani's work, we must necessarily consider his controversy with Volta. This will not be done to grant him a sort of revenge against his scientific adversary. It will be, instead, a way to understand better the problems underlying Galvani's research and, moreover and importantly, to address some significant, and yet neglected, aspects of Volta's work.

Our book will proceed along a rather defined logical course, which we hope the reader will have no difficulty in following. We would like sometimes to take the opportunity of making a variety of digressions, which will allow us to look at interesting characteristics of the conduct of science in Galvani's and Volta day, as well as to other aspects of the culture and society of their times. In some parts of our exposition, we will be obliged to abandon the chronological sequence of events and to pass to a more logical treatment in order to discuss the historical and scientific problems of Galvani's and Volta's research.

Another warning has to do with the possible usage of some terms without any strict correspondence to the historical period to which they are referred. This is done in order to make our writing a little less heavy and tortuous. We will certainly discuss in some detail the meaning of words like "physicist" and "physiologist" in the eighteenth century. On the other hand, we could not avoid using words that at the time still did not exist or had different meaning, as, for instance, "scientist" and "biology."

This book is addressed to all the readers who wish to know more about Galvani and his frogs, Volta and his battery, and a phase of profound transformation of eighteenth-century "natural philosophy" that was laying down the basis of modern science of electrical phenomena. In reconstructing these important episodes of scientific progress, we have made recourse to the writings, published and unpublished, of these two great Italian scholars, of their antecedents, contemporaries, and successors. We have taken into account the results of modern historiography, which has tried to provide a more accurate vision of Galvani, Volta, and their research but has not succeeded in fully undermining the stereotypes and concretions accumulated over the last two centuries. To this end, it was necessary, in our opinion, to put together the scientific and the humanistic dimensions and to make them work side by side. To understand the work of two scientists and to appreciate the scientific problems before them, we have attempted to unite, around their working benches, the "two cultures" so distinct now but so deeply interconnected in Galvani's and Volta's time. It is worth recalling the scientific interests of poets like Goethe and Leopardi, and the literary connotations present in the texts of scientists like Albrecht von Haller, Erasmus Darwin, and Alexander von Humboldt. In this respect our book does not want to be simply a history of science in the classical sense.

We hope that, in addition to assisting the understanding of Galvani's and Volta's story, the interaction between the two cultures we have attempted will also help to recover the fascination of the scientific endeavor. This fascination could be lost with the excessive specialization and sectionalism that characterizes science and technology at the beginning of the third millennium.

2

"Truth and Usefulness"

MEDICINE AND NATURAL PHILOSOPHY IN THE EIGHTEENTH CENTURY

ON AUGUST 2, 1794, in the midst of the controversy triggered by the publication of his research on animal electricity, Luigi Galvani wrote a letter to Lazzaro Spallanzani, one of the most important naturalists of the time. In the letter Galvani thanked his correspondent for his support of animal electricity, which made Galvani confident about the "truth" and "usefulness" of his discovery of a new "force" in the animal body on which muscular motion depended (Spallanzani, 1985, p. 44).

Truth and usefulness were two important values of the culture of Enlightenment. In the opinion of Galvani and many other naturalists, they distinguished scientific discourse, on the one hand, from magic and myth and, on the other hand, from speculation and metaphysics. It was a view that Galvani derived from the scientific and medical milieu of Bologna, the city where he was born on September 9, 1737, and where he lived all his life. It is to this milieu, in which Galvani was educated and built his scientific and professional career, that we shall devote our attention in the present chapter.

2.1. GALVANI'S EDUCATION IN BOLOGNA: THE UNIVERSITY, THE INSTITUTE OF SCIENCES, AND THE HOSPITALS

In the eighteenth century, Bologna was the most important city of the Papal States after Rome. Charles de Brosses—president of the Dijon Parliament who visited Bologna in September 1737—described it as an "excellent town, the most beautiful one after Genoa from a material point of view." De Brosses was particularly impressed by the number of "fine churches and handsome buildings," its magnificent porticoes, the hospitality of the people and—last but not least—its women "intelligent, good-looking, and full of coquetry; they are well read, and quote Racine and Molière" (De Brosses, 1799, pp. 341–351). Though Bologna appeared to De Brosses as a "rich, commercial, populous" city, being referred to as *"la grassa"* ("the opulent," literally "the fat,") because of its location in one of the most fertile regions on the Italian peninsula, it was then going through difficult times. The crisis of the silk industry, which formed the main pillar of its economy, together with a number of serious famines, epidemics, and inundations which took place at the end of the seventeenth century, had much reduced its productive capacity and wealth. Nor was the situation of its cultural institutions, and especially the University, much better, due to continuing tensions between modernizers and conservatives. In 1736 an important reform of the University was approved, aimed at re-establishing its prestige and capacity to attract foreign students (who were a relevant source of income) and at improving its teaching methods. New disciplines such as chemistry and "surgical operations" were introduced, and greater importance was attributed to mathematics. Thanks to these reforms, there was an increase in the number of students who took their degree and an improvement in teaching methods, favored also by the presence of the Institute of Sciences and of teachers who worked in both institutions (Baldelli, 1984; Prosperi, 2008).

One of the principal actors of this reformist movement was Prospero Lambertini, archbishop of the city and then—from 1740—pope with the name of Benedict XIV (see Fig. 2.1). Lambertini was a supporter of progressive ideals mainly derived from the work of Ludovico Antonio Muratori, one of the most important Italian scholars in the first half of the eighteenth century. Though these ideals did not imply any overthrow of absolutist regimes or of the political and cultural role of the Church, they produced a number of projects that aimed at improving the wealth of the lower social strata, reforming economic institutions such as the land register, the bank, and the taxation system, and promoting public works such as drainage. These projects, inspired by Muratori's theory of "public happiness," had to be accompanied by analogous efforts on the religious ground, favoring the practice of a "regulate devotion" deprived of the excesses of Baroque superstition and guided by the value of charity and charitable action instead, as well as in the educational system. A leading figure of this movement, defined by historians as "Catholic Enlightenment," was Lambertini, who focused his action on the reform of religious practices such as canonization processes and the cult of relics, and of cultural institutions such as the University "La Sapienza" in Rome. But it was in Bologna that his reforming activity proved especially important with the reform of the University, the support of the Institute of Sciences, and, more generally, the patronage of scientific culture (see Cecchelli, 1982; Mazzotti, 2007; Rosa, 1981).

In his early years Galvani absorbed the values and ideals of Catholic Enlightenment, thanks to his family and the education received in religious places such as the "Oratorio" of the Philippine Fathers, a congregation that shared Lambertini's views (see Bresadola,

FIGURE 2.1. Prospero Lambertini (1675–1758), archbishop of Bologna since 1731 to 1740 (and afterward pope, with the name of Benedict XIV, until his death), promoted science and culture in his native town and contributed particularly to the development of the Institute of Sciences by generously funding it and endowing it with various teaching and research facilities. He donated to the Institute his rich library and instituted a class of academicians receiving a regular stipend (named "Benedictines" after him).

2011a). It was probably the same cultural milieu that favored the family decision to send Galvani to the University, as had already happened with an elder brother. Moreover, university studies carried a promise of social prestige and economic prosperity for people belonging to the merchant classes, like Galvani. Indeed, he was the son of Domenico Galvani, a goldsmith, and Barbara Caterina Foschi, who came from a local merchant family herself. His father could afford the direct and indirect expenses of sending a son to the University. In the middle of the eighteenth century the final examination cost 500–750 Bolognese lire, which did not include the expenses for celebrations and presents to various teachers and other people. If we consider that the annual stipend of a university professor was about 400–500 lire, the annual rent for a three-room house was 20–30 lire, and that for a shoemaker's shop a little more than 10 lire, we can say that taking a university degree was not a cheap affair. In fact, tax exemptions were granted for special cases, and Galvani was among the students who could benefit from it.

Galvani matriculated into the Faculty of the Arts around 1755, when he was 17 or 18 years old. The "artists" were one of the two main groups of students of the Bologna

University—the other group being the "jurists"—and could obtain a degree either in philosophy and/or in medicine. Galvani began his studies with philosophy by attending the course in metaphysics offered by Ercole Antonio Cussini (or Cossini), canon of St. Petronius. Then he went on to medicine and natural sciences, as he explained in his curriculum vitae: "He [Galvani] studied medicine with Mr. Dr. Beccari, while attending also the courses in chemistry, experimental physics, and natural history which were offered at the Institute of Sciences" (Malagola, 1879, pp. 10–11).

These were the main subjects that made up Galvani's university education. But what did these labels exactly mean in eighteenth-century Bologna? In Bologna, as in other European universities, medicine was mainly taught through the reading and comment of classical sources, in particular the writings of Hippocrates, Galen, and Avicenna. These authors formed the basis of the three areas in which the medical curriculum was divided, that is, medical theory, practice, and surgery. Theory was about "the nature and natural constitution of the human body, preternatural [i.e., pathological] effects, and signs"; thus, it basically included physiology, pathology and semiotics. Practical medicine was "that part [of medicine] which aims at preserving health and expelling diseases," while surgery related to the understanding and treatment of wounds, fractures, dislocations, and a few other problems such as ulcers and stones (Beccari, 1955, pp. 32–33).

These definitions belonged to Jacopo Bartolomeo Beccari, anatomist *emeritus* of Bologna University and Galvani's teacher of medicine. He was one of the prominent figures in the local cultural and scientific milieu, and a member of the Royal Society of London. In the 1750s he was over 70 years old, but he was still teaching a course of "medical institutions" at home, in which he integrated the traditional sources with recent knowledge and theories. For instance, when considering the body's "teguments" (cutis and fat matter), he supplemented the classical notions with "modern" discoveries on glands. Beccari, who was clearly influenced by the mechanical philosophy of the seventeenth century, conceived the human body in terms of "a sort of machine composed of solid and fluid parts, which are so interconnected and ordered to produce many marvellous motions." His view of medicine was thus that of an "art" which investigated the solid parts through anatomy and mechanics, while taking from chemistry the methods to understand the bodily fluids.

The conception of medicine Beccari taught Galvani and his other students derived directly from the work of Marcello Malpighi, the great Bolognese anatomist and physician who lived in the second half of the seventeenth century and ended his career as a papal archiater in Rome (see Fig. 2.2). Beyond his extraordinary contributions to microscopic, comparative, and plant anatomy—his discoveries include the alveolar structure of the lungs and the capillaries—Malpighi was a great supporter of a "rational" view of medicine. In his view, the physician "must study and learn natural things—that is philosophy, physiology, anatomy and animal economy in health and disease states—and should try to acquire all possible notions of mechanics, chemistry, optics and the like" (Malpighi, 1697, p. 147.). In other words, medical practice, that is, the diagnosis and treatment of diseases, should be based on the knowledge of the structure and functioning of the living body,

FIGURE 2.2. Portrait of Marcello Malpighi (1628–1694) from the frontispiece image of his *Opera posthuma*, published by the Royal Society of London in 1797, 3 years after the author's death. Malpighi, one of the greatest anatomists of his age, was a follower of the principles of the Galilean science that he endeavored to apply to biology and medicine, using the microscope as a discovery tool in a somewhat analogous way to Galileo's use of the telescope in astronomical observations.

both in health and disease. This knowledge depended on anatomical and physiological investigation, which rested on operative instruments—such as the microscope, anatomical dissection, and animal experimentation—and cognitive methods taken from other fields of natural knowledge, like mechanics and chemistry (see Adelman, 1966; Belloni, 1975; Bertoloni Meli, 1997, 2011; Bresadola, 2011b; Piccolino, 1999).

Malpighi's view of medicine, which appears quite obvious to our eyes, was in fact controversial in his time or even in Galvani's. His approach was criticized by both the defenders of traditional learning and the supporters of a more empirical medicine, based on bedside observation and aimed at a more precise classification of diseases and the search for specific remedies. These "empirics" argued that all the discoveries made by recent anatomists, from Harvey's circulation to Malpighi's capillaries, had not affected the knowledge of diseases and especially the quality of treatments (Cavazza, 1997b). Indeed, early modern medical paraphernalia were not very different from those in the Middle Ages, diagnosis was still mainly based on Hippocratic anamnesis, and a great role in prevention and cure was still played by the Galenic "Six Non-Naturals" (air, food and drink, exercise and rest, sleep and wakefulness, retention and evacuation of wastes, and perturbation of the mind and emotions). The universal therapeutic method remained bloodletting, which was practiced to restore the humoral balance, to contrast fevers (a very common but ill-defined condition), and to prevent illness. Beyond bloodletting, therapeutics used a number of vegetal substances, mainly known since ancient times, which were classified on the basis of their effects: palliatives, emetics, "vesicants" (to expel malignant fluids), purges, anodynes, and so on.

The inefficacy and immobility of practical medicine contributed to generate satirical representation of physicians, such as Moliere's *Le malade imaginaire* (*The Imaginary Invalid*) but also many criticisms and polemics within the medical profession. A prominent critic in the latter part of the seventeenth century was the English physician Thomas Sydenham, who considered anatomical and botanical investigations as a "non sense" for medicine, more suitable—he wrote—for butchers and florists at Covent Garden than for physicians (see Sydenham, 1700). The latter should instead work at the bedside, the only place where knowledge of diseases could be gained. Sydenham's insistence on the importance of bedside observation and disease classification started to become common practice at the beginning of the nineteenth century, when the movement of clinical medicine established itself in Paris hospitals and, then, all over Europe.

Equally relevant for the development of modern medicine, however, were investigations in anatomy and physiology carried out by people like Malpighi, as well as the application of physical and chemical knowledge and methods to the study of the human body. This approach characterized the great nineteenth-century epoch of "laboratory medicine" and the birth of experimental physiology. Claude Bernard, the French physiologist who made a fundamental contribution to the birth of experimental medicine, underlined the role of physico-chemical sciences and of the knowledge of organic functions in the foundation of diagnosis and therapy (Bernard, 1865). Bernard's approach shared its fundamental traits with the view that, about two centuries before, physicians like Malpighi and his follower Giovanni Battista Morgagni—the founder of pathological anatomy—had developed in their work. For Malpighi, "by examining the bodily parts through anatomy, philosophy, and mechanics, [we] have learnt their structure and use, and by proceeding in *a priori* way, we have formed models with which we can visualize the causality of the effects [...]; so that by understanding the way in which nature operates, we can fund physiology and pathology, and then the art of medicine" (Malpighi, 1697, p. 110).

Malpighi's view of "rational medicine" took root in Bologna and in naturalists such as Beccari or Domenico Gusmano Galeazzi, who from 1739 gave a university course of anatomy carried out in his house through human dissections. Moreover, in 1714 a new scientific institution opened in Bologna thanks to the activity of Count Luigi Ferdinando Marsili, a former pupil of Malpighi's (Cavazza, 1990; Stoye, 1994; see Fig. 2.3). Marsili was a soldier who took part in a number of military campaigns in Europe in the emperor's army, but his interests were mainly scientific and directed toward the study of geography and natural history (he wrote a very important work on the phenomena of the sea). When

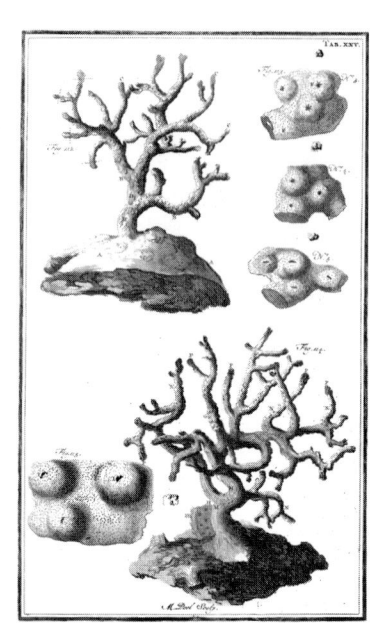

FIGURE 2.3. The founder of the Bologna Institute of Sciences Count Luigi Ferdinando Marsili (1658–1730) in military uniform with, on the right, an image with corals from his *Histoire physique de la mer*, published in 1725, which is considered one of the first works of marine biology. The great Dutch physician Herman Boerhaave, who wrote the introduction to this book, praised the author as "the philosopher not in the museum but in the sea," thus remarking on the experimental approach adopted by Marsili in his marine research. It is possible, however, that Boerhaave was also alluding to Marsili as the founder of the Institute.

he went back to Bologna after his military campaigns, he became a critic of the cultural situation of the city, and especially of the conservative and rearguard forces that directed the University. He then proposed a number of reforms, aimed at renovating the teaching methods, improving the quality of professors, and eliminating privileges and deficiencies, but his ideas did not find their ways into the rulers' politics. He thus decided to establish a new institution and called it the Institute of Sciences of Bologna (see Fig. 2.4).

In Bologna, Marsili and other reformers—Beccari and Galeazzi were among them—wanted to realize the Baconian ideal of a public and collective scientific organization, aimed at the knowledge of nature and social usefulness (see Angelini, 1993; Cavazza, 1990; Cremante & Tega, 1984; Heilbron, 1991; Tega, 1986–1987). The Institute of Sciences was to look like "Salomon's House" described by Bacon in his *New Atlantis*, a utopian place dedicated to "the study of the works and creatures of God," provided with "caves" for chemical investigations, "high towers" for astronomical and meteorological observations, "perspective houses" for experiments on light and colors, "engine-houses" for the building of machines and instruments, and "parks and enclosures of all sorts of beasts" for the study of anatomy, physiology, and for the improvement of the medical art. The fellows of the house had several roles, from collecting discoveries and inventions made elsewhere to performing experiments, from elaborating research outcomes, "to draw things of use and practice for man's life and knowledge" to promoting "the inventions and experiences which [we] have discovered" (Bacon 1626/1974, pp. 255, 265–275).

FIGURE 2.4. Palazzo Poggi in Bologna, the building where in 1714 the Institute of Sciences was established (frontispiece image of *De Bononiensi Scientiarum et Artium Instituto atque Academia Commentarii*).

Like Salomon's House, Marsili's institution needed books, instruments, and laboratories; a stable group of paid researchers; and independent financial resources. Thanks to the donations of Marsili himself and the intervention of other patrons—the most important being Lambertini—the Bologna Institute of Sciences became one of the main scientific centers in the eighteenth century. Throughout its active life of almost 100 years (which substantially ended with the advent of Napoleonic regime), it represented a point of reference for Italian, and in some cases, European science, and it can be considered a model for the emergence of nineteenth-century scientific laboratories and institutions in the fact that it united teaching and research. This was an original characteristic compared to the most important scientific centers of the time, the Royal Society of London and the Académie des Sciences in Paris.

The Institute's activities were organized around several disciplines, which in the mid-eighteenth century included antiquities, military art, geography and navigation, astronomy, optics, chemistry, physics, anatomy, natural history, and obstetrics. Each discipline had its own spaces, instrumentation, and a paid professor, who was responsible for the research and teaching done in that subject. Many of the professors were also lectors at the University; for instance, Beccari taught chemistry at the Institute and medicine at the University, Galeazzi physics at the Institute and anatomy at the University, and Giovanni Antonio Galli (another teacher of Galvani) obstetrics at the Institute and surgery at the University.

The connection between the Institute and the University was not limited to the sharing of professors. The former was conceived as complementary and not alternative to the Studium, offering its courses in the form of "exercises" instead of lectures and carrying out its teaching activity when the University was closed, usually on Thursdays. What differentiated the two institutions was, in fact, the teaching method: While at the University the teachers used to read and commentate on the authoritative texts, the Institute's professors taught their disciplines in an "ostensive" and practical manner, using the instruments and material that they had at their disposal. The "rooms" of physics, for instance, hosted a magnificent collection of machines and apparatuses manufactured by the famous Dutch instrument-maker Jan van Musschenbroek and acquired thanks to the intervention of Lambertini, who had become Pope Benedict XIV. The physics professor Galeazzi used the instruments to repeat, in front of the students, the Newtonian experiments on the properties of light and colors, or those of Robert Boyle on the vacuum pump.

The Newtonian experimental tradition had a deep influence on the scientific approach adopted by the Institute of Sciences, the motto of which was "an Institute which would teach through the eyes, not the ears." Its members shared the idea that natural philosophy was aimed at the discovery of natural laws and regularities, not at the identification of ultimate causes; Galileo, Boyle, Newton, and Malpighi were their "authorities," and mechanics, electricity, and the study of life were the areas that most attracted their attention. At least two of these areas were frequented by Galvani, whose work can be considered a sort of crowning achievement of the experimental tradition developed by the Institute of Sciences of Bologna.

Beside the Institute of Sciences and the University, other places played an important role in Galvani's education. These were the hospitals, which Galvani frequented both as a student and as a young graduate in medicine. During the eighteenth century in Bologna, as well as elsewhere in Europe, the hospital gained an increasing role in medical education, which was part of a process of "medicalization" of the social management of illness. Although hospitals continued to be run mainly by religious confraternities, they transformed their organization on the basis of a classification of the typologies of patients admitted, the creation of a paid and stable body of healers, and the establishment of direct links with the local medical authorities. In the middle of the eighteenth century, Bologna could count on at least nine different hospitals, which were devoted to the treatment of different sorts of disease: For instance, those of Santa Maria della Morte and Santa Maria della Vita treated wounds and fevers, San Giobbe syphilis and San Lazzaro leprosy, while Sant'Orsola specialized in surgical diseases. Some of these hospitals offered teaching courses to students of medicine and surgery, and organized anatomical lessons based on the dissection of corpses (Comitato, 1960).

Galvani attended Santa Maria della Morte's hospital during his university years, following the treatment of patients and anatomical dissections. In 1757, when he was 20 years old, he applied for a post as assistant in the same institution, a role which was assigned to a student of medicine who had completed his studies but had not been given his degree yet. His application was rejected in favor of another student, but he was able to join another hospital—Sant'Orsola—as an assistant and, when needed, he acted as a substitute for the primary surgeon-physician Giovanni Antonio Galli (Bresadola, 2011a). Surgery had been traditionally considered an ancillary and inferior art to medicine, practiced by uninstructed "barbers" and devoted to the treatment of the exterior, material part of the human body. Surgeons offered their care to wounds, sores, burns, gangrenes, dislocations, and "cancers" (i.e., swelling or outgrowth on the skin); their instrumentation consisted of knives, pincers, bandages, and their main remedies were liniments and creams. They were also deputed to practising bloodletting in the many cases in which this remedy was prescribed. In fact, the post of surgeon was a wide and important professional occupation, especially at a time when wounds and infections were much more frequent and widespread than today. However, these healers could not operate without the permission of a physician, nor could they give pills or other remedies aimed at curing the interior and more essential parts of the body, which was an exclusive domain of medical professionals. Moreover, at least in Bologna, they lacked a specific professional body, and their activity depended on the issuing of a specific license, given by the medical authority run by physicians (Pomata, 1994).

During the eighteenth century the rigid distinction between medicine and surgery increasingly faded—thanks, on the one hand, to a separation between the professions of barber and surgeon and, on the other hand, to a better appreciation of surgery by physicians. In 1742 a reform of surgery in Bologna was approved that obliged would-be surgeons to study anatomy, medical theory, and practice for a 1- to 3-year period. Moreover, on direction from Pope Benedict XIV's will, the teaching of "surgical operations" was introduced in the university curriculum, which consisted of a series of practical exercises made on cadavers in order to instruct physicians and surgeons to operate on the living body.

One of the main supporters of a tighter connection between medicine and surgery was Giovanni Antonio Galli, lecturer of surgery at the University, professor of obstetrics at the Institute of Sciences, and surgeon-physician of Sant'Orsola. Galli taught Galvani the theory and practice of surgery and took him as his assistant and substitute at Sant'Orsola's. Here Galvani could widen his knowledge of illness and treatment but also develop his skills in managing and operating on the body. Although this aspect of Galvani's education is rarely considered by historians, it was in fact fundamental for the development of his investigative approach and for the research that led him to the discovery of animal electricity. Indeed, surgical practice implied the use of interventionist techniques in order to modify the bodily conditions, and thus it required a special attention for the reactions of the organism to its manipulation. These characteristics made surgery particularly relevant for the emergence of experimental physiology in the nineteenth century (Lesch, 1984). But the case of Galvani shows that this process had already started in the eighteenth century, and that the habits and skills implied by surgery could play an important role in the development of an experimental and dynamic approach to the study of the living being.

2.2. GALVANI'S PROFESSIONAL CAREER

On July 15, 1759, at the age of 21, Galvani took his degree (*Laurea*) in philosophy and medicine at the University of Bologna. The example, and probably the advice, of his teachers induced him to seek both an academic career and a medical practice. Beccari, Galeazzi, and Galli—his main teachers and mentors—were indeed all physicians *and* natural philosophers who, inspired by the Malpighian view of rational medicine, considered the study of natural phenomena fundamental to the understanding of the human body in health and disease. However, as these activities required much time and effort, a decision about preeminence had to be made, and Galvani—again in the steps of his masters—chose to concentrate his energies primarily on the academic path.

Galvani was an ambitious man, and he could not be satisfied with the common title of lecturer, a title which was granted to anybody graduating from Bologna University and being a native of the city. He thus directed his efforts to joining the Academy of Sciences, which was part of the Institute of Sciences together with the Academy of Arts, and which was the most important place of élite culture in Bologna. The Academy was also well connected with the main European centers of scientific culture and had among its foreign members celebrities such as d'Alembert, Bonnet, Boscovich, Buffon, Haller, Lavoisier, Nollet, and Voltaire. Its secretary kept an extensive correspondence with institutions and people throughout Europe and received the official publications of the other academies as well as the most important scientific journals and books, which enriched the library of the Institute (Cremante & Tega, 1984; Tega, 1986–1987).

At the meetings of the Academy, which were usually on a weekly basis and open to members only, the research of the academics was discussed, the relevant scientific literature was presented, and the most interesting scientific topics were debated. These research works and discussions could be made public, even if only in part and often with delay, through the official publication of the Institute of Sciences, the *Commentarii*. The Academy was thus one of the nodes of that network of exchanges which formed the eighteenth-century "Republic of Letters," allowing its members, even those reluctant to travel and maintain correspondence, like Galvani, to be in touch with what was going on in European culture and to make their own research internationally known (Daston, 1991).

Galvani became an "alumnus" of the Academy of Sciences in 1761, while, in the following year, he publicly defended a series of theses in the palace of the Studium in order to gain a post as university lecturer. He chose to defend 20 "physico-medico-surgical theses" on bones, in which he took advantage of the interdisciplinary education he had received from the university, the Institute of Sciences, and the hospitals. In particular, he not only treated bones from an anatomical perspective but also from the point of view of their chemical composition and physiological function. Then he took into consideration some diseases of bones, such as fractures and dislocations, together with their remedies, both medical and surgical, like amputation and reduction (Galvani, 1967, pp. 50–55). As we shall see, the clarification of the structure and function of the human body, the adoption of methods taken from chemistry and experimental physics, and the application of this knowledge to medical and surgical practice, were all characteristic traits of Galvani's most important research.

Having successfully defended his public theses, Galvani obtained his first "honorary lectureship" (i.e., unpaid) in surgery and, more important, enrolment among the anatomists of the university. The latter was a very prestigious position, which cast Galvani among the élite of the academic corpus together with his teachers Beccari and Galeazzi, but involved also some demanding duties. The most important one was to deliver, at regular intervals, the annual "Public Anatomy" lectures in the anatomical theatre of Archiginnasio, that is, the main university building. This Anatomy course consisted of a series of public lectures on the cadaver, which the anatomist presented in front of colleagues, students, local authorities, and a general public. It was a major event in the cultural life of Bologna and took place in the Carnival period for several reasons. During Carnival the university was closed; thus, all students and teachers could assist in the public lectures. Moreover, this period of the year was particularly apt for the conservation of the cadaver, which was one of the protagonists of the Anatomy. Finally, Carnival was a very popular secular holiday, which fell just before the long penitential period of Lent and permitted the expression of attitudes and events that were normally not accepted. In fact, Public Anatomy was not only an academic event but an extraordinary show in which foreign visitors and Bolognese citizens could admire—perhaps with a bit of repulsion—the interior of a human body exposed in all its most intimate details (Ferrari, 1987).

In the eighteenth century the lectures on Public Anatomy had almost lost their initial purpose of instructing students. The teaching of anatomy using the cadaver was carried out in hospitals or in the house of professors such as Galeazzi. What went on in the anatomical theatre was instead a very ritualized ceremony, which served different aims. In the eyes of the university authority, the Anatomy was a very important event of self-promotion mainly directed at foreign visitors and potential students, while for the anatomist it was a fundamental opportunity to establish his prestige and to make his progress in the academic career. In Galvani's time, Public Anatomy lasted about 2 weeks and each day the anatomist examined a part of the human body—an organ or a system, such as bones, muscles, heart, brain, and so on—by discussing the theoretical views on it and by showing it on the cadaver (or by guiding the demonstration of a sector). At the end of the lesson, the other lectors asked questions and made comments about what had been said by the anatomist in the form of the traditional academic dispute. The dispute was the crucial and most demanding part of the event, an occasion for confrontation and sometimes for verbal conflict, in which dialectical skills and rhetorical strategies played a major role in deciding the success or defeat of the anatomist.

Holding the Public Anatomy lectures was thus a very demanding task, which required long preparation at the expense of other activities such as medical practice. However, it often proved fundamental for the lecturer's career: It was only after having successfully carried out Public Anatomy in 1768 that Galvani obtained an "ordinary lectureship" (i.e., paid) at the University with a stipend of 200 lire, a post that he had not been able to get before.

At the end of the 1760s Galvani had reached most of the goals he had set for himself at the time of his graduation (see Fig. 2.5). He was an anatomist and lecturer at the University, in which he assisted Galeazzi in the course of practical anatomy on the cadaver. In 1764 Galvani had married Galeazzi's daughter Lucia (see Fig. 2.6), and the young couple went to live in his father-in-law's house. Galeazzi became the main sponsor of Galvani's career, playing a major role in his son-in-law's election to the professorship of anatomy at the Institute of Sciences and to the membership as "Benedectine" in the Academy of Sciences, in 1766. Benedectines were a group of 24 members of the Academy (Laura Bassi being the 25th) named after Pope Benedict XIV, who wanted a paid class of academicians to run the institution and make it active and productive. In fact, each Benedictine was to present to the Academy at least one original research dissertation every year. Galvani, who remained Benedictine for more than 30 years, regularly fulfilled this duty, and his dissertations kept in the archives of the Academy represent very important documents to reconstruct his research, including that in the field of electrophysiology (Mesini, 1971, app. IV).

Galvani's role as a professor of anatomy at the Institute consisted of a course of practical anatomy directed not only at students of medicine and surgery but also at artists interested in anatomical drawing. The course was in fact a meeting point between the

FIGURE 2.5. Portrait of Luigi Galvani (1737–1798). This portrait was painted while Galvani was alive and was probably commissioned by the scholar together with that of his wife, Lucia Galeazzi (see Fig. 2.6). Galvani is apparently in his forties and wears the professorial dress of Bologna University teachers. The books on the table and on the shelves are the typical attributes of eighteenth-century men of letters. Note the absence of the frog, which is a constant element of later illustrations of Galvani (private collection).

"Truth and Usefulness"

FIGURE 2.6. Portrait of Lucia Galeazzi (1743–1790), Galvani's wife and daughter of his teacher Gusmano Galeazzi. This late eighteenth-century painting, of a different hand but tightly associated with the portrait of Galvani shown in Figure 2.5, contains typical elements of women portraiture of the period, such as the fine head of hair and the dog (a symbol of marital fidelity). The book opened points to the cultural interests of Lucia and her belonging to the group of eighteenth-century women of letters, who had a very important representative in Bologna in the figure of Laura Bassi, whom the Galvanis certainly knew. In his writings Galvani underlined Lucia's important role in his scientific work both as a laboratory assistant and as a translator and proofreader (private collection).

two academies (of science and of arts) in which the Institute was organized. But it was not performed on cadavers, like the one taught in Galeazzi's house or that carried out at the anatomical theatre; rather, Galvani used the wax models built by Ercole Lelli and those produced later by Giovanni Manzolini and Anna Morandi. The models included some anatomical statues and a number of full-size or magnified representations of human bodily parts, both external and internal, resulting from a combination of anatomical precision and artistic tone. They were kept in the rooms of anatomy of the Institute under the responsibility of Galvani and were one of the main attractions for visitors going to Bologna (Armaroli, 1981; Dacome, 2007; Messbarger, 2010).

Galvani's anatomical activity was differentiated in contents, methods, and addressees. As a university lector, first as Galeazzi's assistant and then as his substitute after Galeazzi's death in 1775, he taught the structure and organization of the human body on the cadaver to students of medicine and surgery. On the occasions of the Public Anatomy, which he performed three more times after 1768, he had to present and to defend his anatomical knowledge in front of a heterogeneous public, who appreciated rhetorical skills and dialectical performance. Last but not least, at the Institute of Sciences he had to teach would-be healers, painters, and sculptors the superficial structure and organization of the body, using wax models. These multifarious activities required many skills and the ability to calibrate contents and methods of anatomy for different publics and in different contexts. What, then, was Galvani's conception of anatomy? And from where did it derive?

2.3. GALVANI'S EARLY ANATOMO-PHYSIOLOGICAL INVESTIGATION

In his inaugural lecture of the 1768 Public Anatomy, Galvani discussed the problem of reproduction and delivery. As he had done in his theses on bones of 1762, he supplemented the description of the structure of the uterus, the placenta, and the other reproductive organs with the examination of their chemical composition and physiological function. For Galvani the purpose of anatomy was indeed to discover the relations between the organization of the human body and its vital functions, as they were established by nature:

> I shall not propose arguments [quaestiones], but I shall present the work of nature as it manifests itself through the observation of the human and animal bodies; neither shall I use verbal ornaments: the works of nature are indeed so clearly illustrated [preclara atque illustria] that any extraneous ornament must be left out. (GMS, cart. IV, plico I, fasc. I)

In this passage Galvani referred to some essential traits of his conception of anatomy: the method of direct observation, the adoption of a comparative approach, and the strong connection between the study of structure and function or—to use a contemporary term—"animal economy." These traits were not only expressed by Galvani in his teaching but were the foundation of his own scientific research.

In the 1760s Galvani focused his research on the anatomy of bones, kidneys, nose, and ear of animals and human beings, presenting his results in some dissertations read at the Academy of Sciences of Bologna. One of these dissertations was then published in the *Commentarii* of the Institute of Sciences in 1767, with the title *De renibus atque ureteribus volatilium* (*On the Kidneys and Ureters of Birds*). The study of birds was a traditional topic of anatomical research, as Galvani himself claimed by quoting important naturalists such as William Harvey, Antonio Vallisneri, and Antonio Maria Valsalva. His aim, however, was not only to describe the structure, composition, and function of the organs of specific animals but also to use this knowledge to understand the human body. Thus, for instance, the understanding "urine secretion" in the kidneys of birds could bring "some light to the problem of urine secretion in quadrupeds" (Galvani, 1967, pp. 65, 77).

The importance given to animal and comparative anatomy by Galvani echoed the view developed a century earlier by Marcello Malpighi. Like Malpighi, Galvani believed that anatomy could not be limited to the observation of cadavers but required the manipulation of animal parts, the use of interventionist methods such as maceration and injections, and vivisection. In one case, he wrote, "I tied the ureters of a live chicken, hoping that urine, once blocked by the ligature, would accumulate in the excretory organs and solidify, so that I could see the structure of the kidneys" (Galvani, 1967, pp. 62–63). To anatomical observation Galvani, like Malpighi, united physiological discourse. For instance, the observation of a progressive and retrograde movement of the ureters in living birds suggested to him that the movement was needed to prevent urine from solidification.

In his anatomical investigation of the 1760s Galvani developed an anatomo-functional approach that already contained all the characteristic elements of his later research on animal electricity. On the one hand, he conceived the study of animals as a means to understand animal economy, including the functioning of the human being; on the other hand, he united the observation of anatomical structures with the investigation of functions, carried out through experimentation on the living. This approach derived from the Malpighian tradition developed in the Institute of Sciences by Galvani's mentors and was reinforced by Galvani's medical and surgical apprenticeship in Bologna hospitals.

In the different places that he frequented as a student and a young graduate—the University, the Institute of Sciences, and the hospitals—Galvani became familiar with the medical and scientific knowledge of the time, but above all he learned a general approach to the investigation of nature. This approach referred to the model of a "new science" represented, albeit with specificities, by great figures like Galileo, Malpighi, and Newton, and had two main rhetorical rallying cries: truth and usefulness. Truth related to the idea that the study of natural phenomena had to be carried out through observation and experiment, eliminating the search for ultimate causes and the use of hypotheses. Usefulness meant that natural investigation should have a practical outcome, which in the case of Galvani consisted of improving medical knowledge and, therefore, the art of healing. As he claimed at the beginning of his most important work, the *De viribus electricitatis in motu musculari*:

> I wish to bring to a degree of usefulness those facts which came to be revealed about nerves and muscles through many experiments involving considerable endeavour, whereby their hidden properties may possibly be revealed and we may be able to treat their ailments with more safety. (GDV, p. 363; translated in Galvani, 1953a, p. 45)

No doubt Galvani adhered to the rhetoric of the scientific Enlightenment, and his work represented an important outcome of the experimental tradition developed in the eighteenth century in the steps of the new science of the seventeenth century. As we shall see, however, the "uses of experiment" could be very different in the individual practice of naturalists like Galvani and Albrecht von Haller, and the rebuttal of conjectures was often more a petition of principle than a coherent choice, as it became clear in the controversy between Galvani and Alessandro Volta.

3
Animal Spirits, Vital Forces, and Electricity

NERVOUS CONDUCTION AND MUSCULAR MOTION IN THE EIGHTEENTH CENTURY

ONE OF THE men of science who best summarized the many explanations of muscular motion that were adopted in the eighteenth century was the "physicist" Alessandro Volta. In his *Memoria prima sull'elettricità animale* (*First Memoir on Animal Electricity*), published in 1792, Volta described the most widespread view in the following terms:

> Those physiologists who had a vague and abstract conception, were satisfied in assuming that *animal spirits*, or nervous fluid, are destined to carry external impressions to the sensorium commune, and to flow—at will's command—through nerves to muscles to produce muscular contractions and the dependent motions; and that these animal spirits, which serve sensations and voluntary motions, are a most subtle, moving and active fluid, whose nature is analogous to light, ether, [or] electric fluid. (VEN, I, p. 21)

This view, which had its roots in ancient medicine, was the general model adopted by most naturalists in the eighteenth century. Some authors, however, refined the model by stressing the affinities between animal spirits and electricity. In Volta's words, "there were [physiologists] who stressed the analogy with *conductors* and supposed that animal spirits did not have the general nature of an ethereal fluid, but the specific one of the electric fluid; they thus concluded that these spirits were identical to the electric fluid" (VEN, I, p. 22).

In this chapter we shall consider the eighteenth-century debate on muscular motion, focusing on the Bologna scientific milieu. Indeed, since the middle of the century

Bolognese scholars were among the protagonists of this debate, and their investigation had a profound impact on Galvani's own approach to muscular physiology.

3.1. THE DEBATE ON HALLERIAN IRRITABILITY

In 1757 Francesco Maria Zanotti, secretary of the Academy of Sciences of Bologna, published in the *Commentarii*—the official publication of the institution—a report on a debate that had profoundly shaken the local scientific community and beyond, so much so that he spoke of "an irritation of all Italy [*totius Italiae irritatio*]" (Zanotti, 1757, pp. 48–49, 55–57). In fact, the debate related to the theory of "sensibility" and "irritability" proposed by the great naturalist and polymath Albrecht von Haller, who was professor of anatomy, surgery, and medicine at Göttingen University, an important politician in Switzerland, and one of the most famous and prolific scholars of the time.

Among Haller's scientific contributions, one of the most significant was his experimental investigation on neuromuscular physiology, which subverted the traditional explanation of sensation and muscular motion, which were two of the fundamental functions of living beings. At the time these two functions were still ascribed to animal spirits, an invisible entity—derived from Galenic medicine—that was secreted by the brain and coursed through the nerves. When flowing from the brain to the muscles, these spirits produced muscular contractions, while when flowing in the opposite direction—from nervous terminations to the brain—they produced sensations. The research carried out in the seventeenth century under the influence of mechanical philosophy criticized the Galenic notion of animal spirits, which came to be considered as occult properties not amenable to scientific explanation. However, naturalists such as William Croone, Marcello Malpighi, Niels Steensen, Thomas Willis, and Francis Glisson were not able to propose valid alternatives to animal spirits, and the physiological processes underlying nervous conduction and muscular contraction remained confined to the area of conjecture.

The most relevant conception of sensation and animal motion in the first half of the eighteenth century was probably that of the Dutchman Herman Boerhaave, teacher of Haller, who defined him as *communis Europae praeceptor* for the great influence he exerted on students and naturalists from all over Europe. In his textbook of medicine (Boerhaave, 1743–1745), which was reissued and translated several times, Boerhaave developed the traditional notion of animal spirits in the light of the seventeenth-century mechanism and Cartesian philosophy. For Boerhaave, the ancient observation that if a nerve was tied or compressed, the part of the body connected to it became insensitive and unable to move, showed the existence of a fluid or "spirit" secreted in the brain, transported by the nerves and responsible for both sensation and muscular motion. Boerhaave thought that this fluid had a material nature even though it did not produce any increase in muscle size (a fact found by Glisson). In fact, the nervous fluid was "imponderable," that is, without a perceptible mass, in a way similar to other natural fluids such as light, electricity, magnetism, and heat. That this fluid circulated in the body, and especially in the nervous system, was supported by the analogy with blood, whose circulation had been established, thanks to the work of William Harvey.

Boerhaave's view of the nervous fluid became very popular and was propagated in many medical textbooks of the eighteenth century. For instance, Jacopo Beccari, Galvani's teacher, taught this notion in his course of medical theory. Instead, Haller soon came to criticize his teacher's view for not being proved by experiments and began a long research aimed at deciding which parts of the animal body were sensitive and which parts had the faculty of contracting. His technique was always the same: He took living animals of different species and age and exposed the part on which he wanted to experiment; waiting for the animal to calm down, he stimulated that part with different instruments or irritating substances, such as heat, spirit of wine, or vitriolic oil. If the animal gave signs of pain or of agitation, the part stimulated was considered as "sensible"; otherwise it was "insensible." If the part being stimulated did contract and shorten, then it was assumed to be "irritable"; otherwise it was not irritable.

Apparently Haller's experimental procedure was simple, but it was in fact rather original and innovative. It developed some fundamental characteristics of the "new science" that emerged in the seventeenth century, such as the active manipulation of the body and the idea of the experimental apparatus, and it embodied new vistas on the scientific method. Haller carried out his experiments on many different animals in order to eliminate the accidental circumstances of phenomena and to find what was constant in a world of nature that appeared variable and difficult to investigate. Haller's "repetitive experimentation" also had an important rhetorical function, aimed at convincing the readers about the accuracy of the procedure and the truthfulness of its results. It was a way to distance himself from those investigators who "do little or no experiments, or who substitute analogies to experiments, affirming that the former have the same force as the latter" (Haller, 1753).

From his experiments Haller concluded that sensibility was a property of nerves because only those parts that had nervous terminations reacted to painful stimuli, while irritability was a property specific to muscular fibers and independent of nerves. In fact, if a nerve was severed so as to deprive the dependent muscle of sensibility, a stimulation of the muscle still produced a contraction. The same happened if the intestines or the heart was isolated from nerves and extracted: Again, an irritation of the part produced the contraction of the muscle. Contrary to what had been claimed before, muscular contraction was therefore a function that did not depend on nervous action, but on an autonomous property intrinsic to muscles.

By ascribing sensation and animal motion to different properties of the body, Haller not only refuted the traditional notion of animal spirits but shook the foundations of a unitary view of the organism, which was a characteristic of both ancient and modern theories. Instead, he related his notion of irritability to the Newtonian idea of gravitation: Like the latter, irritability was an intrinsic force of living matter with no known cause.

Another significant aspect of Haller's notion of irritability lay in its contribution to a new kind of experimental methodology. Being a specific and independent property of a part of the organism that remained when that part was separated from the rest of the body, irritability opened the path to an experimental approach based on the use of animal preparations that could be isolated and manipulated separately. As we shall see, this approach was further developed by Galvani and other great experimenters of the latter part of the eighteenth century, thus contributing to the birth of modern experimental biology.

Haller's research on sensibility and irritability was published in the *Commentarii* of the Academy of Sciences of Göttingen in 1753 and was quickly translated into French and other languages (see Fig. 3.1). A first Italian translation appeared in Rome 2 years later (see Petrini, 1755), a second one was included in a collection of essays entitled *Opuscoli sulla insensitività ed irritabilità halleriana* (*Booklets on Hallerian Insensitivity and Irritability*), edited by the physician Giacinto Bartolomeo Fabri and published in Bologna in 1757 (see Fig. 3.2). Bologna soon became one of the main centers of the debate on "Hallerism," as the new theory was named after his inventor, through the publication of pamphlets and dissertations, public discussions at the Institute of Sciences and at the University, and

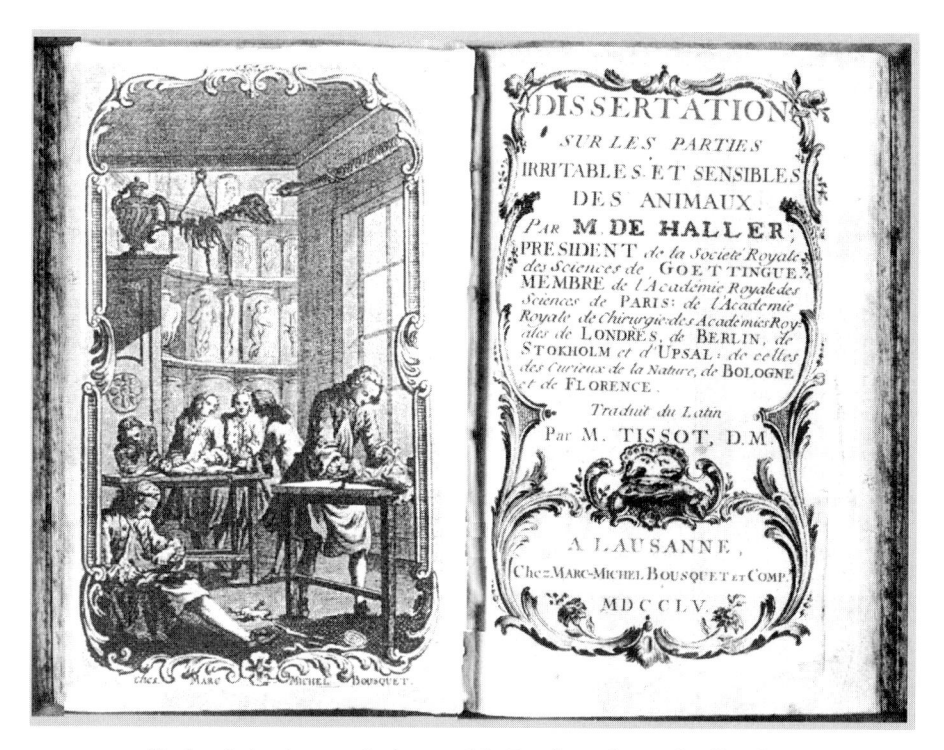

FIGURE 3.1. The frontispiece image and title page of the French translation of Haller's dissertation on irritability (from Haller, 1755).

FIGURE 3.2. Experiments on Hallerian irritability (from the vignette illustration of Fabri, 1757).

collective experimental programs carried out in the private laboratories of Bolognese savants like Laura Bassi and Giuseppe Veratti (Cavazza, 1997a; Dini, 1991; Steinke, 2005).

There were many reasons for the great interest provoked by Haller's theory in Bologna. Its experimental character was well received in a milieu in which experimental philosophy had been successfully introduced in the previous decades. Moreover, the notions of irritability and sensibility promised to have an impact on the understanding of diseases and on the practice of surgery, thus revealing a common ground for natural investigation and the practice of medicine. This common ground was very much sought after by Bolognese scholars, who were for the most part physicians that cultivated natural philosophy with an eye to its useful applications in their professional domain. Jacopo Beccari, for instance, had long been interested in the chemical analysis of foods and drinks for dietary and medical purposes, while Giuseppe Veratti was involved in the study of the medical use of electricity (as we shall see later). Last but not least, the regents of the Academy of Sciences were attracted by the importance of the subject and the fame of Haller, in the opinion that their institution could play a decisive role in such a relevant debate and thus enforce its role in the European cultural context.

The debate on Hallerism involved many prominent naturalists in Bologna, from Pier Paolo Molinelli (professor of surgery at the University and also a well-known figure outside Bologna) to Francesco Algarotti (the author of *Newtonianesimo per le dame*, a work published in 1737 that was translated into various languages—including French, English, and German—and had a great impact in the diffusion of Newtonian philosophy), from Giuseppe Veratti and Laura Bassi to some of Galvani's teachers. However, the main protagonists of the debate were Tommaso Laghi, Marc'Antonio Caldani, and Felice Fontana. Laghi, anatomist of the University and member of the Academy of Sciences, was a prominent figure in the cultural milieu of Bologna and the first to make Haller's research known in the local community. In two dissertations read at the Academy of Sciences in 1756 and 1757 (and both published in 1757), he criticized Haller's experiments and rebutted the notion of irritability. On the one hand, by repeating Haller's experiments he found that several bodily parts, such as tendons and dura mater, which the Swiss naturalist had judged insensible, were in fact sensible. He explained that the difference of the experimental results depended on a serious limitation of Haller's experimental procedure. Indeed, it could happen that the stimulus applied to a part did not provoke any pain in the animal not because that part was insensible, but because the pain provoked by the stimulus was more minor than that provoked by the previous preparation of the part, so that the animal did not react to the second painful stimulus. On the other hand, Laghi criticized Haller's notion of irritability, in particular its independence from the nerves. For the Bolognese anatomist, the observation that the heart separated from its nerves could keep on beating for a while did not prove that muscular contraction depended on an intrinsic property of the muscular fiber, because it could depend, instead, on a portion of animal spirits that had remained in the heart after its evisceration. Moreover, it was possible to make the heart beat again after severing veins and arteries, so as to block the blood flux (which Haller considered fundamental for heart contraction), by stimulating a specific nerve. Laghi thus concluded ironically that "nature seems to attribute to animal spirits more [power] than Haller does" (Laghi 1757a, pp. 115–116).

In his rebuttal of Haller's theory of irritability, Laghi defended the established view of animal spirits, articulated in the "modern" version of the imponderable nervous fluid that derived from Boerhaave, as well as the unitary conception of the organism, which had been reaffirmed in the previous century by Marcello Malpighi, the *numen* of Bolognese medicine. As it happened, Haller's new theory was accepted by two young naturalists, Marc'Antonio Caldani and Felice Fontana, who were less attached to traditional views and saw the support of Hallerism as a way to establish their scientific reputations. If they succeeded in being recognized as the main followers of Haller in Italy, they paid a high price for their success, as we shall see especially in the case of Caldani. What went on in Bologna around Hallerism was also a generation clash, in which personal interests, professional ambitions, and power relations played their role in the name of science.

Caldani and Fontana replied to Laghi's criticism of Haller by accusing their older colleague of neglect with respect to the experimental method: "This savant—Caldani wrote—very well knows how dangerous it is to assert such important facts without having any evidence; this would mean to open the door to all the imaginary structures that the producers of systems need" (Caldani, 1757, p. 465). It was indeed a serious accusation against a person like Laghi, who belonged to the Bolognese experimental tradition and claimed to be a follower of Malpighian rational medicine. But Caldani and Fontana did not limit themselves to criticism. They repeated and varied Haller's experiments, confirming the results obtained by the Swiss naturalist and developing its arguments. In particular, Fontana elaborated on the distinction between "excitant" and "efficient" cause of muscular contraction, proposing a very important analogy with the ignition of gunpowder by a spark:

> A small spark lights a large mass of exploding powder, which force will be prodigious. This spark would not have moved a pebble, while the air contained in an

infinite number of powder grains, by developing its elastic force, turn over rocks. The spark is not the cause of this enormous effect, which greatly surpasses its force; it is only the excitant cause that awakes in the powder the force of an agent, which is confined in the latter. (Fontana, 1757, p. 239)

Because no proportion existed between the excitant cause, or stimulus, of the explosion (i.e., the spark) and its true, efficient cause, Fontana claimed, "the contractile force of the muscle can surpass that of the irritating cause." The idea that the way in which an organism reacts to external stimuli is an expression of its internal force and organization, independent from the force and specific nature of the external influence, is one of the most important novelties introduced by Haller in eighteenth-century physiology and developed by "Hallerians" such as Fontana. From one point of view, this idea would play a fundamental role in the modern understanding of neuromuscular mechanisms (see Chapter 9.2). More important, it influenced the approach developed by Galvani in his electrophysiological experiments and in his interpretation of the role of the electrical stimulus in muscular contraction, as we shall see later.

The most acute and polemic phase of the Bolognese debate on Hallerism lasted some years, until the departure of Caldani, who left Bologna in 1760 to accept a university chair in Padua, and Fontana, who 2 years earlier had gone to Florence to become court physicist and director of the Museum of Physics and Natural History established by the Grand Duke Pietro Leopoldo. Even though, in his report of the events, the secretary of the Academy of Sciences, Zanotti, presented the picture of an austere and polite debate between lovers of truth, Caldani expressed regret because of the "animosity" and personal attacks against him and Fontana (Caldani, 1759; Garelli, 1885, pp. 204–206). In 1762 Germano Azzoguidi, a young supporter of Hallerism, declared that the hostility he had received from "some professors" was "due to [his] adhesion to Haller's irritability."[1] This and other evidence suggests that Haller's theory found many obstacles in establishing itself in the Bologna scientific milieu; nonetheless, it became acceptable some years later, when Galvani began to investigate the problem of animal motion. More important, it offered Galvani new methods and specific questions that could be taken as departing points for a novel exploration of an ancient topic.

3.2. THE STUDY OF ELECTRICITY IN THE EIGHTEENTH CENTURY

During the debate on Haller's theory of sensibility and irritability, which took place in Bologna, Tommaso Laghi did not limit himself to criticizing Haller's experimental results and interpretations, but he also proposed a variant of the traditional idea of animal spirits. For Laghi, "it is in no way contrary to the proper scientific method to conjecture that

---

[1] G. Azzoguidi to L. Caldani, Bologna, February 23, 1762, in BAB, Coll. Autografi, IV, 1091–1123.

the electrical matter—diffused throughout the body by the nervous fluid secreted in the cerebral glands—is so constituted as to flow through the nerves to the senses, fostering motion" (Laghi, 1757b, p. 338; translated in Pera, 1986/1992, p. 58). He observed that one could produce the contraction of a muscle by stimulating its nerve with electricity, even when the nerve was almost dry and all other stimulants were ineffective. It was thus possible that animal spirits had an electrical nature and that this "electric matter" caused in some unknown way both sensation and animal motion.

What was this "electrical matter" referred to by Laghi? More generally, what were the phenomena defined as electrical in the middle of the eighteenth century? It is now time to turn to the study of electricity in Galvani's era in order to understand better the contemporary debate on nervous conduction and muscular motion, as well as Galvani's own electrophysiological research (see Hackmann, 1978; Heilbron, 1979/1999; Home, 1992). At the time electricity was an emerging field, the nature of its phenomena was debated and the definition of electricity itself was uncertain. It was not until the end of the century, thanks also to Galvani's and especially Volta's research, that the concept of electric current was formulated and electrodynamics could be established as a scientific field.

The ancient observation that amber and other substances, when rubbed, attracted small light bodies called for increasing attention in the first half of the eighteenth century. These materials were called "electrics" and their effects came to include repulsion and luminous and sonorous effects such as the spark. Another category of substances, which included metals and most of the fluids, was defined as "nonelectrics" because they did not give electric signs when rubbed but could be electrified if connected with "electrics" or "insulators." It was also discovered that different materials like glass and resin had different responses to electricity: They attracted each other but repelled a body made of the same material. From this observation it was concluded that there existed two different kinds of electricity, called "resinous" and "vitreous."

Around the middle of the eighteenth century electricity was one of the most popular topics among naturalists and a "wonder" that attracted the curiosity of an increasing number of people interested in science. The major scientific institutions in Europe, such as the Royal Society of London and the Académie des Sciences in Paris, discussed the electrical experiments performed by its members or by many other savants scattered over the continent and abroad. Electrical phenomena were included in the natural philosophy courses at many universities, and public demonstrations were offered by itinerant lecturers and instrument makers. In aristocratic salons, "electricians" like Jean-Antoine Nollet entertained a learned public with electrical experiments and divertissements, while a number of textbooks and guides on electricity—describing the apparatus needed to be a self-made electrician—found their way into the publishing market (see Fig. 3.3).

A significant account of the success of electricity in the second half of the eighteenth century is offered by Joseph Priestley, an English naturalist and a great popularizer of

FIGURE 3.3. Electric divertissements in the Enlightenment: the electrified boy (from Nollet, 1746).

science. In the preface to his *History and Present State of Electricity*, published in 1767 and then reissued several times, Priestley wrote:

> Electrical experiments are, of all others, the cleanest, and the most elegant, that the compass of philosophy exhibits. They are performed with the least trouble, there is an amazing variety in them, they furnish the most pleasing and surprising appearances for the entertainment of one's friends, and the expense of instruments may well be supplied, by a proportional deduction from the purchase of books, which are generally read and laid aside, without yielding half the entertainment. (Priestley, 1767, I, p. X)

In Priestley's words, electricity could be managed, and also studied, by anyone with little instruction and small economic resources (see Fig. 3.4).

In fact, things were much more complex, but it is true that this new field exerted a great fascination on the society of the Enlightenment (see Fig. 3.5). Some of the reasons

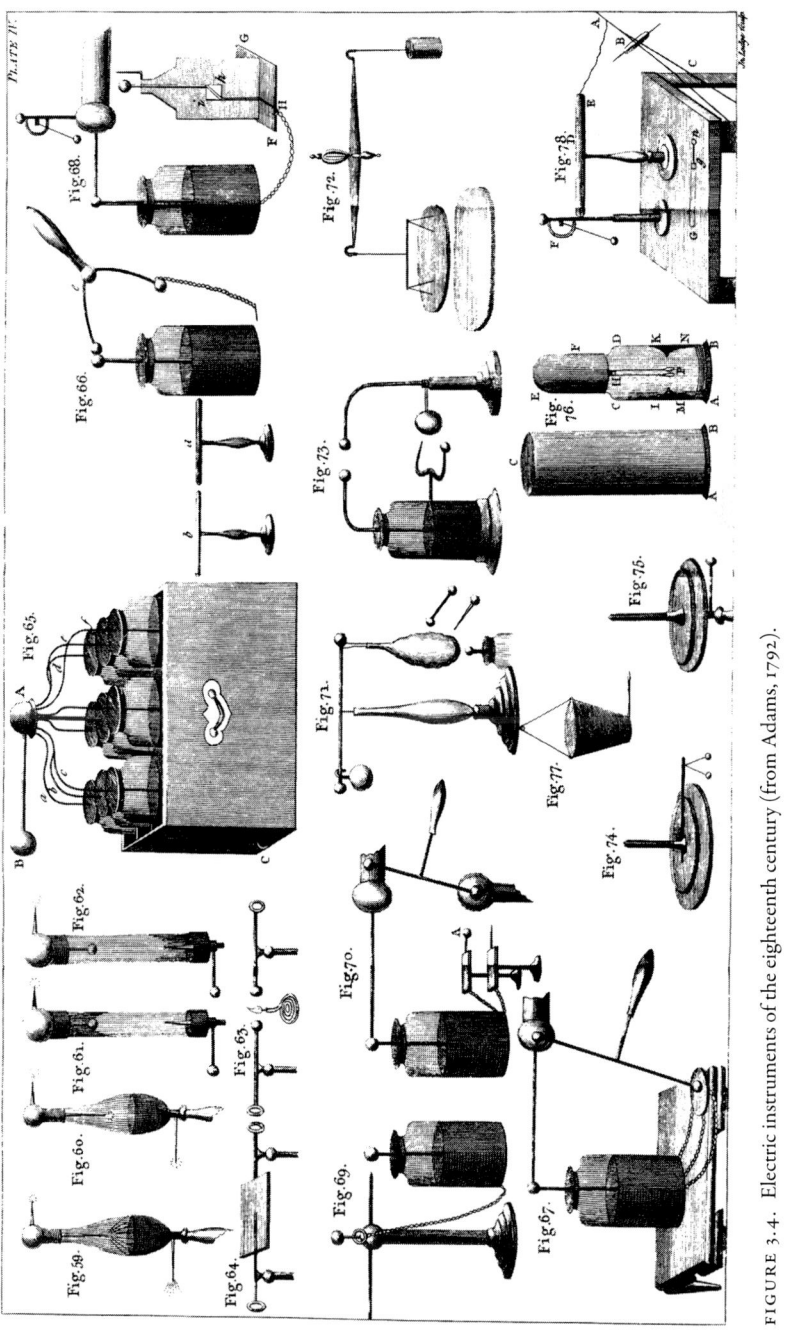

FIGURE 3.4. Electric instruments of the eighteenth century (from Adams, 1792).

FIGURE 3.5. The frontispiece image and title page of *Dell'elettricismo*, an anonymous work (first published in 1746 and attributed to Eusebio Sguario and Christian Xavier Wabst), in the Neapolitan edition of 1747 which incorporates two *Dissertazioni* on the medical applications of electricity (from Sguario and Wabst, 1747).

for this success were described by Priestley himself in a very instructive paragraph of his work:

> If we only consider what it is in objects that makes them capable of exciting that pleasing astonishment, which has such charms for all mankind, we shall not wonder at the eagerness with which persons of both sexes, and of every age and condition, run to see electrical experiments. Here we see the course of nature, to all appearance, entirely reversed, in its most fundamental laws, and by causes seemingly the slightest imaginable. And not only are the greatest effects produced by causes which seem to be inconsiderable, but by those with which they seem to have no connection. Here, contrary to the principle of gravitation, we see bodies attracted, repelled, and held suspended by others, which are seen to have acquired that power by nothing but a very slight friction; while another body, with the very same friction, reverses all its effects. Here we see a piece of cold metal, or even water, or ice, emitting strong sparks of fire, so as to kindle many inflammable substances; and *in vacuo* its light is prodigiously diffused and copious, so as exactly to resemble, what is really is, the lightning of heaven. (p. 548)

One of the most intriguing and marvellous aspects of electricity emerged around 1745. In the context of research on the electrification of water, some naturalists independently discovered that it was not necessary to keep a conducting substance like water isolated in order to make it electric, as the French savant Charles Dufay had argued. If one held a jar with water in one's hand and connected the bottle to an electrical generator by means of a metal hook in order to charge it, and then touched the hook with the other hand, then a tremendous shock was felt in the arm, much stronger than that received if the jar was insulated (see Fig. 3.6). This new apparatus was called the Leyden jar—from the name of the Dutch town where one of its discoverers lived—and it is the first electrical condenser in history. Here is how Priestley described it:

> What can seem more miraculous than to find, that a common glass phial or jar, should, after a little preparation (which, however, leaves no visible effect, whereby it could be distinguished from other phials or jars) be capable of giving a person such a violent sensation, as nothing else in nature can give, and even of destroying animal life; and this shock attended with an explosion like thunder, and a flash like that of lightning? (Priestley, 1767, II, p. 135)

The invention of the Leyden jar contributed to the emergence of a new theory of electricity, proposed by the American man of science and great politician, Benjamin Franklin. Franklin began to study electricity in Philadelphia around 1745, thanks to some electrical apparatus received by an English merchant and member of the Royal Society, and to a report of electrical experiments written by Albrecht von Haller. He soon discovered the

FIGURE 3.6. The Leyden jar (detail from a plate of Nollet, 1746).

power of points to attract and emit electricity, and performed a number of experiments, which he published in his *Experiments and Observations on Electricity* (1751–1754).

Franklin rejected Dufay's theory of two distinct kinds of electricity—vitreous and resinous—and argued for the existence of a single electric fluid, formed by very tiny particles and permeating all natural bodies. Influenced by Newton's concept of gravitation and by Haller's analogy between electricity and fire, Franklin claimed that electric particles repelled each other, while being attracted by those of ordinary matter (Cohen, 1966). All bodies contained a certain amount of electric fluid, which in normal conditions did not manifest itself; however, if this quantity changed by increasing or decreasing, then the body became charged positively (or "plus") or negatively (or "minus") and electricity manifested itself through specific signs such as attractions and repulsions, sparks, and shocks. In the case of the Leyden jar, for instance, if the interior of the jar was positively charged, then the exterior became negative, and the shock felt at the contact of the two surfaces depended on the flow of the electric fluid in search of its equilibrium. Water was no longer necessary to obtain this effect, as it was sufficient for the two superficies to be isolated from each other, like in a glass plate with two metal shields on its opposite faces (see Fig. 3.7). This new instrument, which took the name of "Franklin's square" or "magic square," was added to the Leyden jar among the electrical condensers; it was used also by Galvani in his electrophysiological investigation, as we shall see later.

Franklin's explanation of the phenomena of the Leyden jar made his theory successful and his concept of one single electric fluid largely accepted by "electricians." However, if this theory was useful to explain the processes of charging and discharging, problems

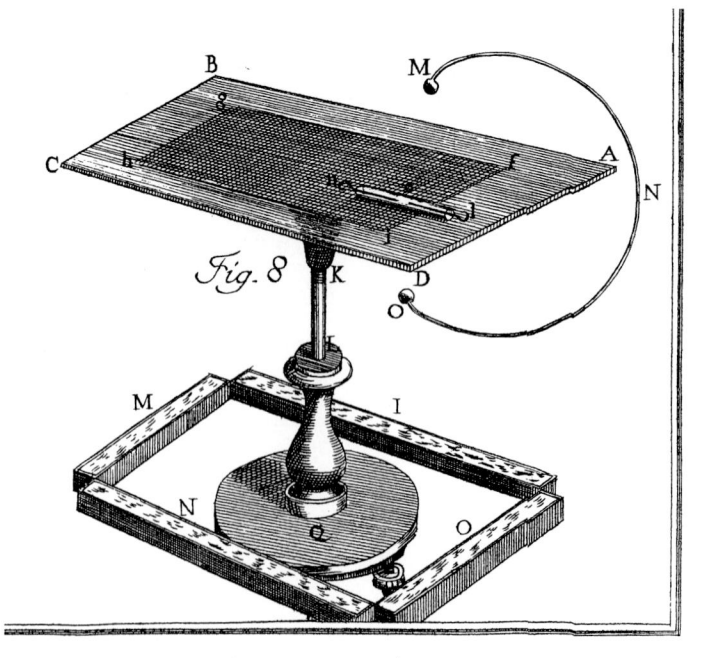

FIGURE 3.7. Franklin's "magic square" (from Beccaria, 1753).

arose with some phenomena of attraction and repulsion, especially in those cases that we now call electrostatic induction. Franklin adopted a pneumatic model to claim, in Newtonian fashion, that in a positively charged body the excess of electricity distributed around its surface like a sort of "electric atmosphere." But the model was not able to explain why two bodies equally charged repelled each other instead of mixing their electric atmospheres, as the pneumatic analogy would have suggested. Neither could it account for phenomena like the electrification of an end of a conductor when a charged body was brought near the opposite end. Equally problematic was the transmission of electricity in a vacuum or even the classification of bodies into insulators and conductors. The fact is that theory had to account for a range of phenomena that were far from being clearly defined and settled, thus making any attempt at systematization precarious and partial.

One important scholar who contributed to the success and development of Franklin's electric theory was Giambattista Beccaria, professor of experimental physics at Turin and a member of the Institute of Sciences of Bologna. Beccaria had friendly relationships with several Bolognese savants, including Laura Bassi and Giuseppe Veratti, and addressed one of his works to Jacopo Beccari, Galvani's mentor. His view of electrical phenomena had a great influence on Galvani's research and was also taken by Alessandro Volta as a departing point for his study of electricity. In 1753 Beccaria published a treatise *Dell'elettricismo artificiale e naturale* (*Of Artificial and Natural Electricity*) in which he proposed a dynamic version of Franklin's theory and a unitary view of electrical phenomena. This and other works of Beccaria's were in Galvani's private library (Bresadola, 1997).

Like Franklin, Beccaria considered electricity as a sort of "fire" or "vapour"—that is a material and imponderable fluid existing in all bodies in a determined amount. This fluid manifested itself only when it was in a condition of imbalance or disequilibrium between two bodies or regions, thus producing its characteristic "signs." The fundamental law of electricity was that "every electric sign is due to the vapour which transfers from one body, in which it is in a larger quantity, to another, in which it is in a smaller quantity, with a vivacity proportional to the difference in quantity between the two bodies" (Beccaria, 1753, p. 17). Instead of Franklin's pneumatic model, Beccaria adopted a hydrodynamic model, based on the flow of a fluid, which tended to stay in equilibrium or to regain it once it was lost for some reason. It was in the re-establishment of this equilibrium, which implied a real flow or circulation of the fluid, that electrical phenomena occurred.

Beccaria's theory was a refinement of Franklin's while keeping the latter's main concepts, including that of electrical atmosphere. Other naturalists, however, criticized some tenets of the one-fluid theory and proposed an explanation of the phenomena of induction in terms of the action of forces and "influences" instead of material atmospheres. In the second half of the century, Tiberio Cavallo, a Neapolitan "electrician" who had settled in London, confirmed the ambiguous state of electrical studies in his *Complete Treatise on Electricity*, a work that underwent numerous editions and was translated into many languages. About the distinction between insulators and conductors, for instance,

Cavallo wrote that "there have been several conjectures offered, but, except one probable hypothesis, there is nothing as yet ascertained" (Cavallo, 1786, vol. 1, p. 123). Even less established was the understanding of induction:

> In respect to the place occupied by the electric fluid superinduced on a body, it has been thought, by several ingenious persons, that, when a body is electrified, all the superfluous fluid, or all the deficiency of it, in case the body is electrified negatively, resides as a kind of atmosphere all around the body [...]. But to this assertion it is answered by others, that if the electricity communicated to a body did reside round it like an atmosphere, it should certainly repel the air contiguous to that body; but this is not the result of experiments. (Cavallo, 1786, vol. 1, pp. 130–131)

If electricity could offer great possibilities for new discoveries, so that "were the thing possible, to entertain such a man as Sir Isaac [Newton] for a few hours with its principal experiments"—as Joseph Priestley underlined (Priestley, 1767, vol. 1, p. XV)—it was also an uncertain terrain, difficult to explore also for one who had all practical and conceptual means at his disposal.

3.3. "Artificial" electricity, "natural" electricity, and their role in the human body

In his work on electricity Giambattista Beccaria distinguished two main sorts of electricity: "artificial" and "natural." The former included those electric phenomena produced by instruments such as the Leyden jar or the electric machine like the one used by Beccaria himself (see Fig. 3.8). This was formed by a glass cylinder that could rotate around a wooden frame and by another long brass cylinder—called the primary conductor of the machine—with a pointed end which was located very close to the first cylinder. During rotation, the glass cylinder rubbed against leather or some other insulating substance, thus producing a charge of electricity, which was then picked up by the prime conductor and became usable by the experimenter. Other generators used glass disks instead of cylinders, like the one used by Galvani in his laboratory for his electrophysiological experimentation (see Fig. 1.3 in Chapter 1). In addition to electric machines and condensers, the apparatus of the eighteenth-century electrician included metal rods, demonstrating devices, and all sorts of conducting or insulating substances.

"Natural" electricity was instead that associated with natural phenomena, especially weather conditions. In his work Franklin observed that the effects of lightning were analogous to those produced by electrical experiments, and thus suggested that lightning had an electrical nature. He also designed an experiment to prove his idea, which was performed in France in 1752 and then replicated in several other places, including Bologna. The experiment consisted in raising a long iron, pointed and isolated rod during

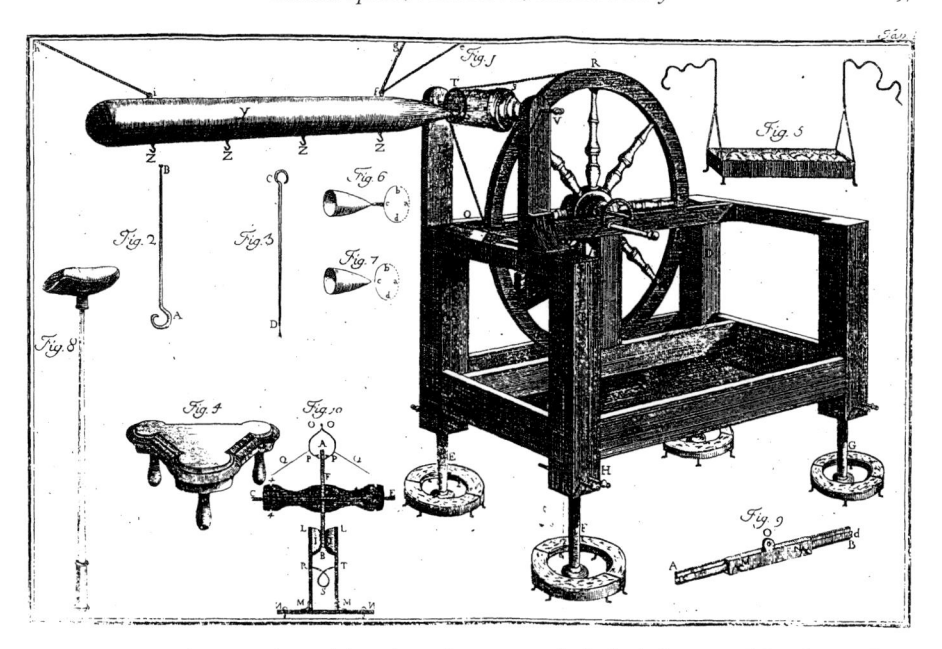

FIGURE 3.8. An electric machine of the eighteenth century in the "cylinder" version of Giambattista Beccaria (from Beccaria, 1753).

a stormy weather, and then in bringing an isolated conductor near the rod: One could thus obtain sparks and shocks similar to those produced by electrical instruments. This experiment convinced most naturalists that Franklin's idea was sound and that electricity was diffused in the atmosphere, as well as in all bodies. As we shall see, during his electrophysiological research Galvani carried out experiments on "atmospheric electricity" and devoted a chapter of his memoir on animal electricity to it.

Around the middle of the eighteenth century other cases of "natural" electricity were revealed or conjectured. One was the shock produced by fish like the torpedo and the eel of Guyana (see later); another was the mechanism of vital functions such as sensation and animal motion. Tommaso Laghi in Bologna was far from being alone in suggesting that the organism contained an electric matter which flew in the nerves to produce sensations and muscular contractions. A similar idea had been proposed some decades earlier by the famous English naturalist Stephen Hales, who explicitly referred to some Quaeries in Newton's *Opticks* in order to suggest that animal motion was the effect of "a vibrating electric virtue" which "can be conveyed and freely act with considerable energy along the surface of animal fibres, and therefore on the nerves" (Hales, 1733, II, p. 57). Hales became a sort of authority for those who supported a neuro-electric view and the analogy between nervous fluid and electricity. Giambattista Beccaria, for instance, quoted Hales in his 1753 textbook on electricity, in which he devoted some chapters to the effects of natural and artificial electricity on plants and animals. He observed that electricity favored perspiration, nutrition, and the growth of plants, while in animals it expanded bodily fluids, increased heartbeat, and penetrated into "nervous and muscular parts of the animal, causing them to dilate and shorten" (Beccaria, 1753, p. 124).

Beccaria supported these conclusions with quotations from other authors and with original experiments. In one case he took a living chick, separated a muscle from the body except for its tendons and nerves and tied each tendon with a brass wire. Then he connected the two wires to a Franklin's square and produced a spark on the muscle of the chick: "In the same moment that the spark was produced, the chick extended his leg with great impetus" (p. 129). For Beccaria, this and other observations, such as the sparks produced by rubbing a cat's or a dog's fur, suggested the existence of a "natural electricity in living beings." In particular, the electric vapor could be involved in sensation and muscular motion, as Beccaria suggested in an often-quoted paragraph of his work:

> The further experiments and discoveries in electricity—of which Newton had seen only the principle—seem to reinforce the great Philosopher's doubts. The speed with which the electrical vapour moves, changes direction, stops and races forth again seems consistent with the speed and changes in animal sensations and motions. The singular ease of its travel—in general, through electrical bodies by communication, and in particular, through the nervous and muscular parts of animals—is consistent with the ease with which the mutations induced in organs by various objects are conveyed to the seat of sentience; it is also consistent with the agility with which other motions correspondingly ensue in the body. And the contractions and dilatations caused in the muscles by an electrical spark or electrical shock are arguments, perhaps even decisive ones, for the above-mentioned conjecture. (Beccaria 1753, pp. 126–127; translated in Pera, 1986/1992, p. 58)

Beccaria was very cautious in claiming the electric nature of animal spirits and the existence of natural electricity in animals, but his work undoubtedly contributed to making this view acceptable or at least amenable to scientific investigation. In the 1750s the topic of sensation and muscular motion was very popular thanks to Haller's research on irritability and sensibility, and in many places the relationships between electricity and life was explored from a practical point of view. Indeed, the effects of electricity in favoring perspiration and evacuation, in accelerating heartbeat, and in moving arms—all included in Beccaria's work—were initially proved by electricians on their own bodies or those of volunteers, but they soon gave the idea of using electricity for therapeutic purposes. Medical electricity—as the new field was called—emerged around the middle of the eighteenth century, thanks to the initiatives of a number of people with different background, social status, and professional affiliation, and soon became a widespread and debated practice (see Fig. 3.9).

In 1747 the Venetian polymath Gianfrancesco Pivati sent a letter to Francesco Maria Zanotti, secretary of the Academy of Sciences of Bologna, in which he described a new therapeutic method based on electricity. He had been successful in treating several

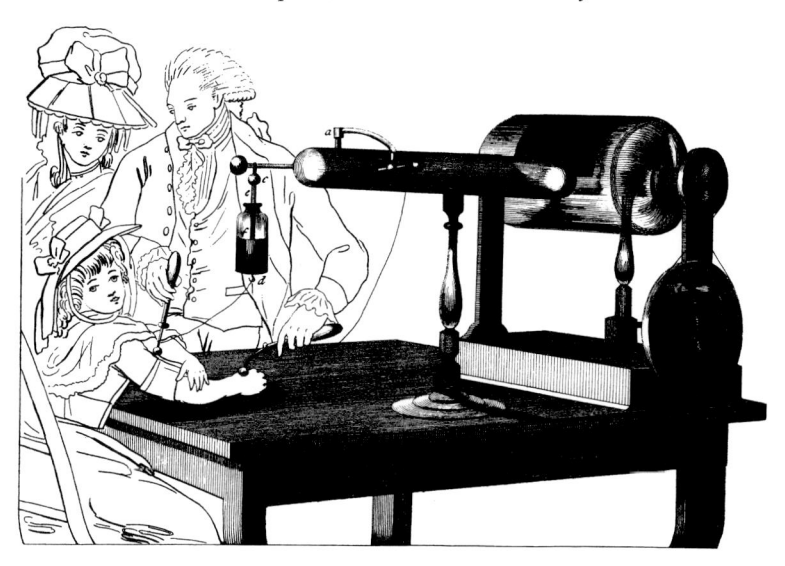

FIGURE 3.9. A séance of "medical electricity" (from Adams, 1785).

patients, thanks to the administration of remedies put inside glass tubes, which were then electrified and applied to the patient. Zanotti and the other regents of the Academy were impressed by Pivati's "medicated tubes" and thus decided to test this method and, more generally, the use of electricity in therapy. They assigned this task to Giuseppe Veratti; he was both a physician and a naturalist interested in electrical studies and could make use of the electrical equipment in the laboratory created in their house by his wife, Laura Bassi. One year later Veratti published a book of *Osservazioni fisico-mediche intorno alla elettricità* (*Physico-Medical Observations on Electricity*), which represented an important episode in the early history of medical electricity and had a great influence on Galvani's investigation (Bertucci, 2007).

In his book, Veratti was much more cautious about the therapeutic efficacy of electricity than Pivati. In the foreword he claimed that "many attempts are still to be done in such a vast field to define what kinds of diseases can be treated with this remedy, and what kinds cannot" (Veratti, 1748, unnumbered page). More important, Veratti underlined the need to study more deeply the effects of electricity on the living body before attempting to use electricity in medicine. This statement was in line with the view of "rational medicine" supported by Marcello Malpighi and widely accepted by Bolognese savants. It is important to note here that the same view would be adopted by Galvani in his research project three decades later, when he decided to investigate the mechanism of muscular motion.

In spite of his caution, Veratti argued that medical electricity could be useful in the treatment of some diseases, which depended on the thickening or accumulation of bodily fluids in one part, or which affected nerves and muscles such as deafness and paralysis. The reason lay in the "subtlety and energy" of the electric fluid and in its efficacy in penetrating into the body, "accelerating notably the flow of fluids, and augmenting the insensible perspiration." These characteristics made the electric fluid similar to animal spirits or nervous fluid, which had to be equally subtle and active to produce sensation and muscular motion.

Veratti's work on medical electricity, as well as Beccaria's views on natural electricity, had a great influence on Bolognese scientific milieu, and especially on the debate over Hallerism in the 1750s. During the replication of Haller's experiments on sensibility and irritability, Leopoldo Marc'Antonio Caldani and Felice Fontana asked Veratti and Laura Bassi whether they could use their laboratory to carry out electrical experiments on animals, a new method not described by the Swiss physiologist. The two young naturalists electrically stimulated the heart, intestines, and nerves of some frogs, observing that "all the parts of frogs, either live or dead, do contract with electric sparks" and that the electrical stimulation of crural nerves produced the contraction of the inferior limbs even when any other mechanical or chemical stimulus was ineffective. In some experiments they dissected a frog, leaving only the inferior limbs attached to their crural nerves, a preparation very similar to the one later adopted by Galvani (Caldani, 1756).

It is worth noting that in Caldani and Fontana's electrical experiments, as well as in that of Beccaria on the living chick, the stimulus usually consisted in a spark produced by the electrical instruments of the time, which gave small quantities of electricity at very high voltages (10,000 volts or more). It was thus an instantaneous discharge, inappropriate for giving a prolonged flow of electricity, which in fact became easy to obtain only after Volta's invention of the battery. Therefore, in these experiments it was not at all evident that electricity must form a discharge circuit in order to produce its physiological effects. At the beginning of his research, Galvani also used instantaneous and sparking stimuli on the frog, but he then realized the importance of applying electricity through a circuit, and he explored with great care the role of each element of the circuit (animal tissues, metals, and other conductors) in the action of electricity. This change of view constituted a breakthrough not only in Galvani's investigation of animal electricity but also in the study of electrical phenomena in general (see Chapter 5.2).

From his and Fontana's experiments with electrical stimuli, Caldani concluded that electricity was the most efficient stimulus to produce muscular contractions and that nerves were the most conducting matter in the animal body (Caldani, 1756, p. 332). However, he did not agree with Tommaso Laghi that the electric fluid was the true cause of animal motion, but he explained these experimental results in Hallerian terms, claiming that electricity was the most powerful exciting cause of irritability, which in turn remained the true cause of muscular contraction. Together with Fontana, he also raised some objections to Laghi's neuro-electric theory, arguing that this theory contradicted the laws of electricity. Taking Franklin's one-fluid theory of electricity in the version developed by Giambattista Beccaria as their frame of reference, the two Hallerians underlined that electricity could produce its effects only if an electrical imbalance occurred between the bodies involved. But how could any imbalance exist between nerves and muscles if these animal tissues were both conducting substances, as the experiments seemed to suggest? If, on the other hand, nerves were found to be insulators, how could the electric fluid leave them and reach the muscle so as to produce its contraction?

These objections, based on the electric properties of animal tissues, took their force from the common understanding of electrical phenomena and from showing the contradiction of two alternatives hypothesis (nerves conductors—nerves insulators). They were very convincing, and in fact no satisfactory solution to them was proposed until Galvani's research. In the eight tomes of his monumental *Elementa physiologiae corporis humani* (*Elements of Physiology of the Human Body*) published in the period 1757–1766, Haller reiterated Caldani and Fontana's objections to the neuro-electric view of animal motion, adding a third problem, which would have great importance in physiology until the nineteenth century (see Chapter 9). He observed that a ligature of a nerve blocked both sensation and muscular contraction but did not block the passage of electrical fluid: Indeed, the electrical stimulation of the same nerve at the site opposite to the muscle produced the contractions. Therefore, animal spirits and electric fluid must be different. Moreover, Haller underlined that the characteristics of electricity, especially that of diffusing through conducting matter, was contrary to the localized and precise control needed for motor functions. As an example, he observed that, if electric signals were electrical, it would be impossible to move only the big toe at will, because the electric fluid would pass from one nerve to the near ones, making us move also the other toes of a foot (Haller, 1762, p. 380).

In the second half of the eighteenth century the objections against the neuro-electric theory went hand in hand with the success of Haller's theory of sensibility and irritability. In his *Traité des nerfs et de leurs maladies* (*Treatise on Nerves and Their Diseases*), published in the period 1778–1780 and soon translated into several languages, Samuel Tissot made both these points, but he also invited "enlightened physicists, well versed in experiments" to make new experiments on electricity and nerves (Tissot, 1778–1780, vol. 1, pp. 217–218). Shortly after, Galvani, who had Tissot's treatise and all other relevant books of physiology in his private library, responded to this invitation and began his research program in electrophysiology.

In fact, Galvani had taken an interest in irritability and the problem of muscular motion since at least the early 1770s. In 1772 he read a dissertation at the Academy of Sciences of Bologna, *On Hallerian Irritability*, followed in the next 4 years by memoirs on frogs' nerves and the motion of the heart. Unfortunately, he did not publish these works, and the manuscripts are not kept in the archives of the Bolognese institution. However, we have the text of a lecture given by Galvani during his Public Anatomy of 1772, in which he spoke of animal spirits as fundamental to the life of the organism and disagreed with Haller's explanation of the decrease of the heart's irritability with age. In this period Galvani carried out a number of experiments on the heart, using frogs as his preferred animal and developing various preparations of them. During this research he found that it was possible to stop the heartbeat by damaging the spinal cord, thus questioning Haller's idea of the independence of the heart motion from the nervous system. He also irritated the heart with various stimulants in order to find out their effects on pulse. Although it is not clear what conclusions Galvani reached from these experiments in vivo, they certainly increased his ability to operate on the animal body and his awareness of the importance of experimental preparations and procedures.

In 1780 Galvani was, for the third time, in charge of the Public Anatomy in the anatomical theatre of Bologna. A witness to the function reported that Galvani wanted to explain "everything with electricity and fixed air," the latter being one of the new gases discovered in the previous years. The last lecture of that Anatomy was devoted to the causes of death, which Galvani located in the extinction of "that most noble electric fluid on which motion, sensation, blood circulation, life itself seemed to depend." The professor of anatomy suggested that "death comes when blood ceases to circulate and to produce the electric fluid by friction in the brain and the nerves" and called that idea "plausible, if not true" (Galvani, 1966, pp. 134–137). Some months later Galvani began a long series of electrical experiments on muscular motion, one of the fundamental functions that characterized life in the animal world.

Galvani's idea of the role of electricity in vital phenomena was very similar to that expressed by Laghi and other naturalists some decades earlier, but in the early 1780s it acquired new force especially thanks to the research done on some very peculiar fishes, known since antiquity for their extraordinary properties and which had become protagonists of the scientific debate in the previous years. These fishes included the torpedo, which lives in the Mediterranean and in other seas; a particular catfish of the African lakes and rivers; and a particular type of fish that was discovered in the rivers of South America (particularly Guyana and Surinam) characterized by an elongated body, which resembled that of an eel (but belonging to the genus of knifefish not to the eels).

3.4. ELECTRIC FISH

A common property of these fishes is the ability of giving a strong shock, capable of painfully numbing humans and animals. The effect is particularly strong in the case of the so-called eel of the Surinam (also referred to as the "trembling eel" because of the tremulous consequences of the shock on the experimenter's arm). Torpedoes and African catfish had been known to Western scholars since the classical period and were given the generic names νάρκη (in Greek), *torpedo* (in Latin), and *ra'ad* (in Arabic), while the shocking eel of South America came to the attention of Westerners only after the discovery of America (Finger & Piccolino, 2011; Piccolino, 2003a). The effect of shocking fishes had called for the attention of classic scholars mainly because for the impossibility of accounting for its mechanism on the basis of known physical and philosophical conceptions of the age. It was one of the phenomena consigned to the province of the "un-nameable" (or "hard-to-define") qualities, eventually referred to in the Renaissance as "occult properties." Until the dawn of the modern era, they were also considered by physicians and magicians alike for the possible therapeutic or enchanting action of the shock as well as for ointments prepared from their bodies (particularly from the liver).

It was only, however, with the scientific revolution of Galileo and Newton that these peculiar fishes became the subject of a great experimental attention. This had the purpose of interpreting what seemed more akin to the prodigious and "supernatural" in the light of the new scientific paradigms. In the second half of the seventeenth century, many studies in Italy (by the second generation of Galileo's followers) were concerned with this aim, leading to a mechanical explanation of the nature of the shock. This was advocated particularly by the great naturalist and physician Francesco Redi (1671) and his pupil Stefano Lorenzini, who (in 1678) was the first to publish a book entirely devoted to the torpedo fish (see Fig. 3.10). The mechanical explanation became the reference hypothesis, especially after it was endorsed in France by the celebrated naturalist and member of the Académie des Sciences, René-Antoine Ferchault de Réaumur. In 1714, after studying its shock in his native town La Rochelle (on the Atlantic Coast of the Poitou), Réaumur devoted a famous article to the study of torpedo (see Fig. 3.11).

Progress of electrical science was hastened after the invention of the Leyden jar made feasible the comparison of the shock produced by the fish with that caused by artificial electricity. Thereafter voyagers, adventurers, missionaries, academicians, and amateur scientists

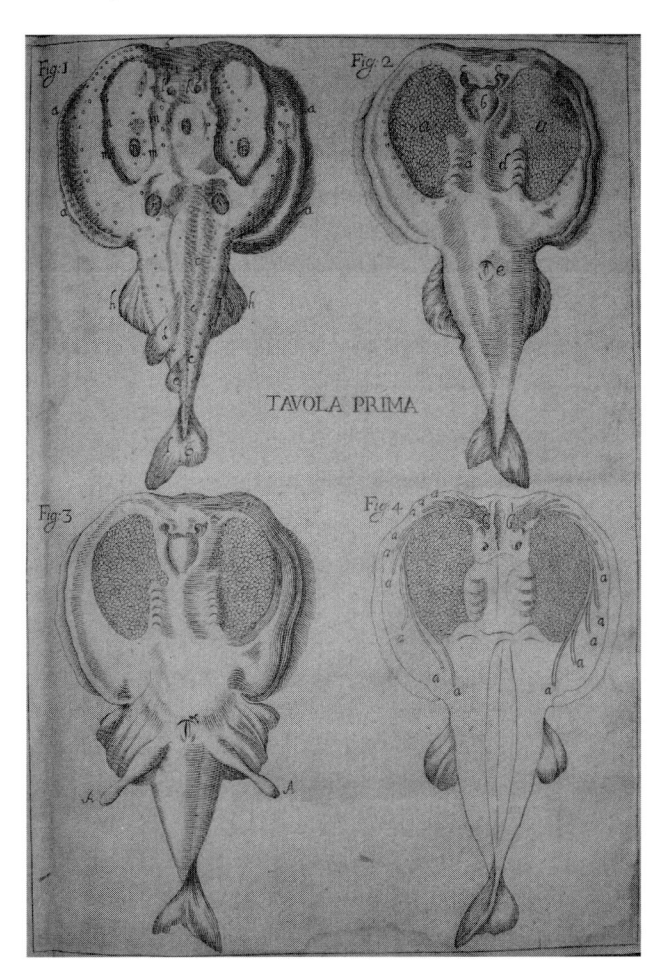

FIGURE. 3.10. A plate with anatomical preparations of torpedo (from Lorenzini, 1678).

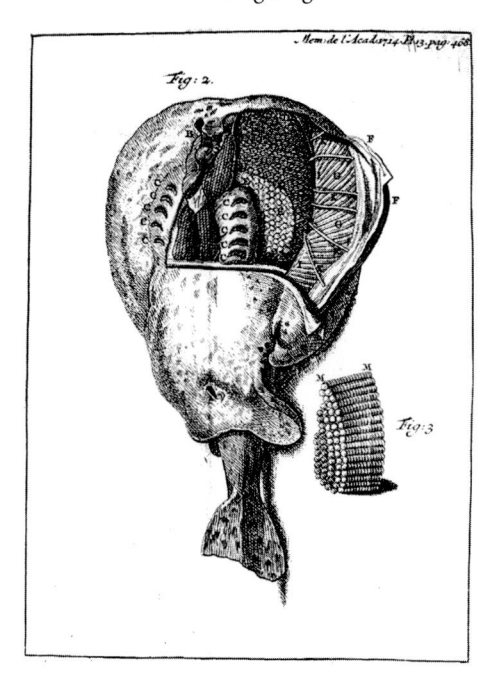

FIGURE 3.11. An anatomical preparation of a torpedo with the detail of the electric organ (from Réaumur, 1714).

started to notice the similarity between the two effects and to assume that the fish shock could also involve electricity (Finger & Piccolino, 2011; Piccolino, Finger, & Barbara, 2011). Around the middle of the century, it was the shocking eel of South America that took the center stage of attention for naturalists and laymen. This was so especially, but not exclusively, for Dutchmen, who then owned the marsh regions of Surinam, in which these fish were abundant (Koehler, Finger, & Piccolino, 2009; see Finger & Piccolino, 2011 for a review). In 1769, after that the electric hypothesis of the eel shock had been reiterated by the American physician (and double agent in the war between the mainland and the American colonies) Edward Bancroft, the amateur scientist, John Walsh, got into the field and investigated the shock of the torpedo during a scientific trip made in the summer of 1772 to La Rochelle (where Réaumur conducted his experiments).

After an extremely intense series of studies, conducted partially in collaboration with local literati (and with the assistance of his nephew Arthur Fowke and his secretary David Davies), Walsh concluded that the torpedo shock was electrical (Piccolino, 2003a). This was especially the case because the effects of the fish were transmitted through metals (and other electric conductive matters) and blocked by insulating bodies. The discovery was first announced in a letter to Benjamin Franklin written on June 12, 1772 (which would become the starting text for a communication presented on Walsh's behalf by Franklin to the Royal Society of London on 1773). In the letter Walsh expressed his conviction:

> that the effect of the Torpedo appears to be absolutely Electrical, by forming its circuit through the same conductors with Electricity, for instance metals, Animals

and moist Substances; and by being intercepted by the same nonconductors, for instance Glass, and Sealing wax. (Walsh, 1772, p. 71 verso)

Walsh's article of 1773 was published together with a companion paper devoted to anatomical investigations performed by the famous Scottish surgeon John Hunter on some torpedoes brought by Walsh from La Rochelle to London "in brandy" (see Fig. 3.12). Hunter (1773) showed that the particular organs responsible for the shock (referred to by Walsh as "electric organs") were formed by the stacking, one above the other, of numerous flat disks or "prisms," according to an arrangement that would be, about 30 years later, an important source of inspiration for Volta's invention of the electric battery (see Chapter 9).

Despite numerous attempts in La Rochelle, Walsh was not able to produce an electric spark from the shock of the torpedo, nor could he measure the fish's electricity with the common devices of the age. As he remarked in his journal of the experiments, the apparent impossibility of producing a spark (and of having other evident electric signs) could not be discounted simply by saying that the fish's electricity was weak. This was because the shock produced by the torpedo was strong and more intense than that produced by the discharge of ordinary electric jars (which, on their side, produced powerful sparks). The problem was tackled by Henry Cavendish, one of the most brilliant English physicists of his age, who collaborated with Walsh and Hunter in London. Cavendish considered the difficulty from both a theoretical and practical point of view,

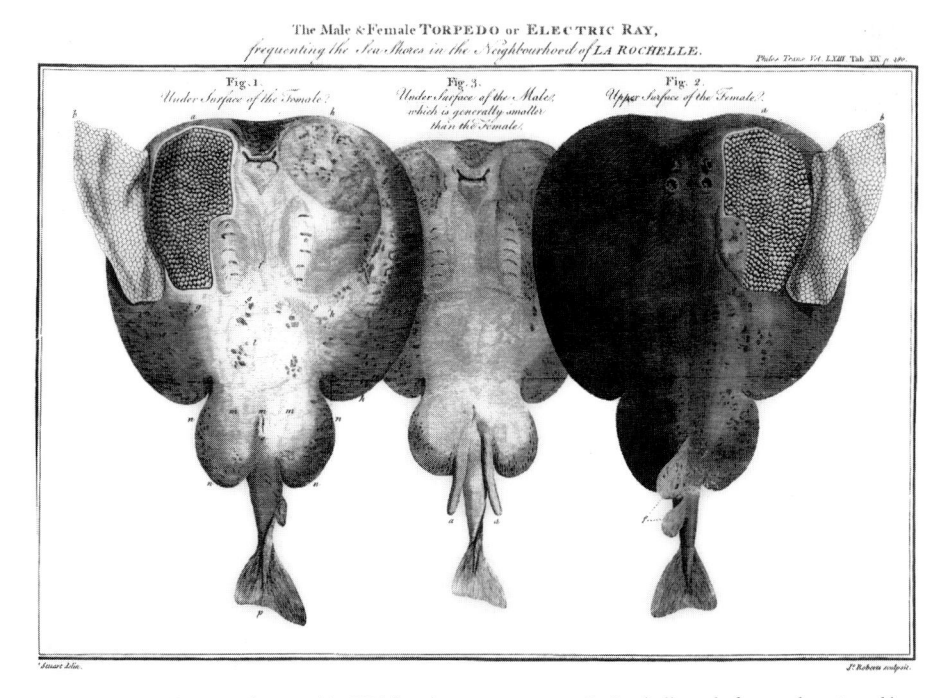

FIGURE 3.12. The torpedoes used by Walsh in his experiments at La Rochelle and afterward sectioned by Hunter in his anatomical research (from Hunter, 1773).

and constructed a model of the torpedo (called the "artificial torpedo") capable of imitating the effects of the real fish in both the strength of the physiological effects and the absence of typical electrical signs (Cavendish, 1776; see Maxwell, 1879). Cavendish's artificial torpedo was similar in the shape to the fish and was powdered by a large assembly of Leyden jars connected "in parallel" (according to modern terminology) and charged at a relatively low "degree of electrification" (a parameter corresponding to electric "tension" or "potential" in modern physics). According to Cavendish, the effects of an electric device depended in different ways on two parameters, the quantity of charge and "degree of electrification." Spark production depended exclusively on the degree of electrification and this explained why a small Leyden jar, charged at a high degree, easily produced visible sparks. However, the physiological effects (like muscle contractions and the sensation of shock) depended on the combination of the two parameters. The fish organs (that Cavendish assimilated to an assembly of small, square capacitors) were supposed to be charged to a relatively low degree of electrification but to contain an enormous quantity of charge because of the huge combined surface of the disks making up their columns (and acting as capacitors).

The other problem considered by Cavendish was the difficulty of determining how the torpedo could direct its discharge to relative distant objects (prey or predator) despite the electric shunt caused by the seawater layer immediately surrounding its body. He succeeded in overcoming this difficulty by showing that, contrary to the assumptions of electric science in his day, water and water solutions were much worse conductors than metals. Because of this, the fish's shock could be transmitted at distances (as it happened with his artificial torpedo), although with progressive attenuation according to a spatial diffusion process that was portrayed in a famous figure which represents the first graphical illustration of an electric field (see Fig. 3.13). As we shall see in Chapter 8, Cavendish's interpretations of electric fish were well known to Volta, who was, however, convinced that the electric battery he invented was a better model for the fish's discharge.

Returning to Walsh, he succeeded in producing a spark from the shock of an electric fish in 1776, working with trembling eels imported from Guyana. This achievement is accounted for by the larger potential of the shock of this fish (ca. 500 volts) compared to the torpedo which is about 50 volts (see Piccolino & Bresadola, 2002; Finger & Piccolino, 2011). These experiments, which were largely demonstrated by Walsh to his colleagues and friends, were never published by its author. Nevertheless, news of them circulated rapidly through the many pathways of communication existing in the Republic of the Letters of the era (journals, private letters, reports presented to academies, and books). In particular, Walsh's eel experiments became widely known in Italy, where an extract of an article dedicated to them by the French academician Jean-Baptiste Leroy was published in a widely circulated journal (Le Roy, 1776).

Walsh's experiments on electric fish were of enormous importance in convincing even the most sceptical about the possibility that electricity could play a role "in animal economy." All previous objections, and particularly those previously raised by Haller and his

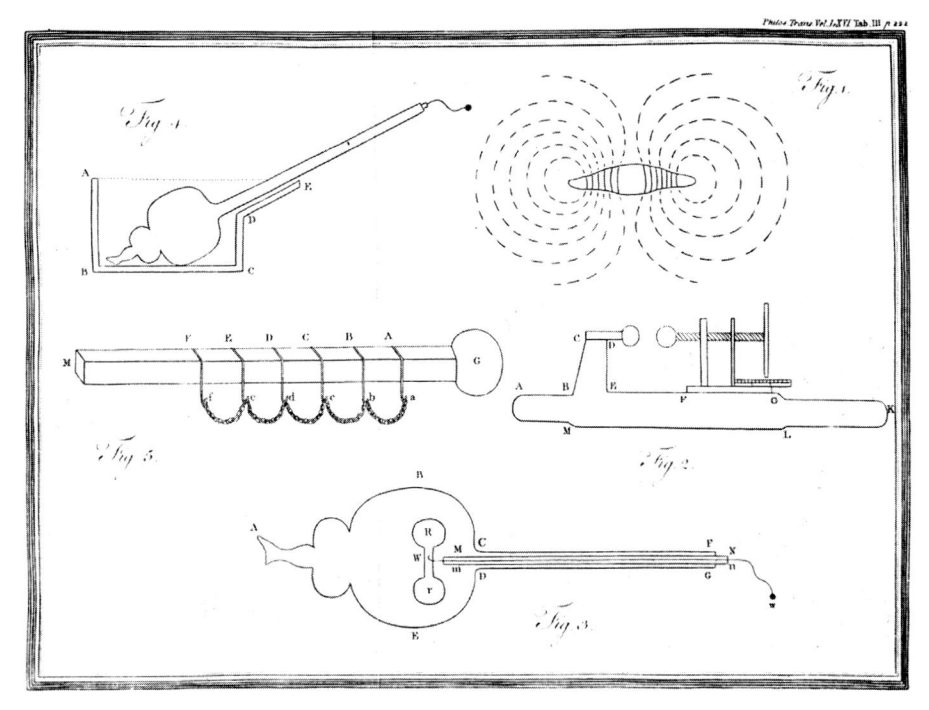

FIGURE 3.13. Cavendish's "artificial torpedo" with the diagram of the spatial diffusion of the electric shock in water (from Cavendish, 1776).

followers, despite their soundness, were thus undermined. In the case of electric fish, not only the animal body appeared to be unsuited to the physiological agency of electricity (because of its electric conductivity), but the liquid habitat also made the possibility unlikely. And still, the shock was definitely electrical, and the fish was even capable of producing an electric spark, the landmark of electricity for eighteenth-century science. The force of experiment had knocked down any logical argument, according to one of the *regulae philosophandi* expressed by Newton in his *Principia*:

> In experimental philosophy, propositions collected from the phenomena by induction, are to be deemed (notwithstanding contrary hypotheses) either exactly or very nearly true, till other phenomena occur by which they may be rendered either more accurate, or liable to exception. (Newton, 1726, p. 389)

After Walsh's experiments, even one of the strongest followers of Haller (and one of the fiercest opponents of the electrical nature of "animal spirits"), Felice Fontana, changed his mind (as seems to be the case from a passage included in the French edition of his treatise on the venom of the viper published in 1781 in Florence):

> In short, not only the mechanism of muscle motion is unknown, but we cannot even imagine anything capable of accounting for it. It seems that we are forced to

make recourse to some other principle, perhaps to something analogous to electricity, if not to ordinary electricity itself. The electric Gymnotus and the torpedo make the thing at least possible, if not probable. One could believe that this principle follows the most ordinary laws of electricity. It might be even more modified in nerves than it is in the torpedo and the Gymnoti. Nerves would be the organs devoted to conducting this fluid, and possibly also to exciting it. But all remains to be done. (Fontana, 1781, II, pp. 244–245)

As a matter of fact, at the moment Fontana's treatise was published, something in the direction he indicated had been already done. Indeed, as we shall see in the next chapter, at the end of the previous year Galvani had started annotating the results of a series of experiments that would lead him, after about 10 years of intense experimental research, to advocate publicly for the electrical nature of animal spirits.

4

Artificial Electricity, the Spark, and the Nervous Fluid
GALVANI'S EARLY RESEARCH ON MUSCULAR MOTION

THE STUDY OF electric fishes and Walsh's demonstration of the electrical nature of their shock were considered by many naturalists as strong evidence of a neuro-electric view of animal motion. In a treatise on nerves published in 1778, the Swiss physician Daniel de La Roche suggested a "strong analogy," if not identity, between the electric and the nervous fluids, referring to the observation made on the torpedo and the eel of Surinam made by "Mr. Walsh and some other physicists." La Roche was among the first authors to spread news about Walsh's research, as well as Hunter's observations on the electric organ of the torpedo, which appeared to the Scottish anatomist very rich in nerves, whose function was that of "producing, picking up and directing" the electric fluid (La Roche, 1778, vol. 2, pp. 300–304).

La Roche's treatise belonged to Galvani's private library, and it was probably an important source of knowledge about electric fish for the Bolognese anatomist (Bresadola, 1997). As we have seen in Chapter 3, a "Hallerian" such as Felice Fontana, who had formerly refuted the identity between animal spirits and electric fluid, changed his opinion in the light of electric fish research, suggesting in 1781 a neuro-electric explanation of muscular motion. However, he invited his readers to base scientific theories not on analogies but on experiments, and if electric fish made the analogy between electricity and the mechanism of muscular motion "possible, if not probable [...] everything yet remains to be done." Fontana then explained:

> We must first assure ourselves by firm experiments, whether the electrical principle has really its site in the contracting muscles; we must determine the laws that this

fluid observes in the animal body; and after all it will yet remain to be known what excites this principle, and how it is excited. (Fontana, 1781, vol. 2, p. 245; translated with revisions from Pera, 1986/1992, p. 60)

It is worth noting that Fontana's words represented very clearly the research program Galvani had begun some months earlier. This means that experimental research on the effects and role of electricity in muscular motion was considered desirable, and possible, by one of the main physiologists of the time. That it was not Fontana, nor anyone else, who carried out this research, but Galvani, depended on several reasons, which we shall explore in this chapter and the next. In fact, we shall reconstruct Galvani's investigations during the 1780s, which resulted in the publication of his *De viribus electricitatis in motu musculari* (*On the Forces of Electricity in Muscular Motion*), the memoir in which Galvani described his discovery of animal electricity.

In what follows Galvani's investigative pathway is reconstructed in unprecedented detail, thanks to the surviving laboratory notes and other documents kept in the archives of the Academy of Sciences of Bologna. A fresh look at his laboratory allows us to clarify his approach to the study of the living beings and the detailed development of his experimental activity reveals difficulties, blind alleys, and interruptions, as well as successes, accumulation of data, and turning points. From all this there emerges a new image of Galvani's scientific practice, which shows some important differences with those offered in other historical studies or by Galvani himself. As Lawrence Holmes has argued, this sort of account of segments of investigative pathways "show[s] how individuals manage their ordinary work, how they think about the events of the day, how they plan their next steps in response to what just happened, whether foreseen or unexpected" (Holmes, 2004, p. 171). Though the resulting story is unique, local, and essentially tied to the personality of the scientist investigated—Galvani in our case—the reconstruction of investigative pathways can contribute to throw light on "the fine structure of scientific creativity"—to quote one of Holmes's most important papers (Holmes, 1981).

Most historians have followed the reconstruction of Galvani's research offered in *De viribus*. Indeed, in his memoir the Bolognese anatomist explicitly aimed at reporting his experiments in chronological order, starting from the effects of "artificial" electricity (that produced by electrical instruments) on muscular motion, passing on to consider the effects of "natural" electricity (lightning and other weather manifestations), and, finally, concluding with the discovery of the role of conducting arcs in producing muscular contractions. This research path was supposedly marked by some crucial experiments, defined by historians as the "first" and the "second" experiment, which revealed a relevant role of chance in Galvani's discovery (Dibner, 1952; Fulton, 1926, 1966; Hoff, 1936; Pera, 1986/1992).[1]

---

[1] Pera takes into consideration some of Galvani's experiments not described in *De viribus*. See also Innes Williams (2000).

A first, crucial point in Galvani's investigation—so the story goes—was the observation of contractions at a distance, when he observed the movements of a frog's leg coincident with a spark being extracted from the electrical machine and with no material connection between the animal and the instrument. This observation was followed, some 5 years later, by a second crucial experiment, that is, the possibility of producing muscular contractions by connecting the nerve and muscle of a "prepared" frog through a metal arc. Galvani described these two discoveries in the *De viribus*, while other "crucial experiments" were performed during the controversy with Alessandro Volta in the 1790s. In particular, the "third" experiment consisted of connecting the nerve and the muscle directly, without using any metal or other conducting material, while a variation of this was the direct contact between the crural nerves of two different frogs' legs.

This is in brief the development of Galvani's electrophysiological investigation, described on the basis of what Galvani himself reported in his published writings. However, if we consider the publication as part of scientific practice (being neither simply a faithful representation of that practice nor a mere rhetorical instrument of persuasion) and integrate it into all the historical material available, including unpublished documents, there emerges a different picture. In the following pages, we shall devote great attention to experiments not reported in *De viribus* but fundamental to understanding his laboratory practice, to ideas that Galvani did not develop in public but which are important to understand the genesis of his discovery of animal electricity, and, moreover, to research programs that left few traces in publications but had a crucial role in his investigative pathway. In this way we hope to contribute to dissolve some stereotypes regarding Galvani's science, such as the chance character of his discoveries and his ignorance of physical knowledge, and to give the reader the taste of participating in the activity of a creative man like Galvani.

4.1 THE BEGINNING OF ELECTROPHYSIOLOGICAL EXPERIMENTATION

Though Galvani had studied the physiology of nerves and muscles in the 1770s (see Chapter 3), the first surviving laboratory notes on electrophysiological experiments date from November 1780. The recent discoveries on electric fish and Galvani's interest in the neuro-electric theory of vital functions, testified by his discussion of these ideas during the Public Anatomy lectures early that year, were probably among the reasons that induced him to design a research program focused on the study of the role of electricity in muscular motion. This program was clearly delineated by Galvani some time later in an unpublished draft:

> As several anatomists have thought that the electric fluid either enters into the composition of that very subtle fluid which is considered, and not without reason, to flow through the nerves, or it is that same nervous fluid, so I decided to carry out

some experiments on the nerves with the electric fluid, in the hope that they could disclose the truth or at least contribute to throw some light on the obscurity of the phenomena of nerves. (GOS, p. 123)

In the next 10 years, until the publication of *De viribus electricitatis in motu musculari* at the beginning of 1792, Galvani performed a huge number of experiments, to which he devoted up to 15 days a month. It was a very demanding activity, which had to be reconciled with Galvani's professional duties as professor of anatomy at the University, professor of anatomy (and since 1782 of obstetrics) at the Institute of Sciences, practising physician, and member of the medical authority of Bolognese territory. The possibility of carrying out the experiments at any time he could spare from his professional duties may have played a role in his decision to establish a laboratory in his house, instead of using the facilities offered by the equipment hosted in the rooms of the Institute. A domestic laboratory was a convenient facility and defined a space under Galvani's complete control, who could thus decide who was allowed to take part in his research and what information could leave the private space. This could be an important matter, as Galvani had realized some years earlier when he had been involved in a polemic about the priority of some anatomical discoveries.

A home laboratory was not infrequent in eighteenth-century Bologna. We have seen that a famous laboratory existed in the house of Laura Bassi and Giuseppe Veratti, where courses in experimental physics had been offered to students and curious people, and research on various topics had been performed, including the replication and extension of Hallerian investigation on irritability and sensibility (see Chapter 3). Moreover, in Bologna it was common for university professors to use their homes to give lectures to students, a famous example being that of Giovanni Antonio Galli, who had formed a collection of obstetrical models to be used in his "private" course. Galvani's teacher and father-in-law Gusmano Galeazzi also had taught his course of practical anatomy on the cadaver at home, and Galvani did the same when he took the place of Galeazzi as professor of that discipline. It is probable that Galvani had used the same spaces and anatomical instruments when he performed his early investigation on the kidneys, ears, heart, and other animal parts.

A laboratory for electrical experiments on animals was quite a different matter, however. Figure 4.1, a watercolor of Galvani's laboratory—then published in a slightly changed form in the *De viribus* (see Fig. 1.3 in Chapter 1)—shows an electrical machine, a Leyden jar, and other sorts of electrical equipment together with apparatuses invented by Galvani himself, located in two adjoining rooms. From Galvani's laboratory notebooks we know that he also had measuring instruments, a pneumatic pump and several chemical apparatuses, and other objects appear in the inventory of Galvani's collection of scientific instruments, which was acquired at the beginning of the twentieth century by Henry Wellcome for his museum. This collection, which includes also instruments from a later era like some batteries, is now kept in stock at the Science Museum in London, where it

FIGURE 4.1. Original watercolor image of the first plate of Galvani's *De viribus* (AASB, Fondo Galvani, cart. VI, plico 15,©).

is possible to find an electric machine and some Leyden jars very similar to those represented in the image of the laboratory (see Bresadola, 2011a).

Galvani's home laboratory was well equipped with many instruments for natural- philosophical investigation, and especially with the complete apparatus of the "electrician" of the time, as it was represented and described in the works of authors such as Giambattista Beccaria and Tiberio Cavallo. Galvani had these and many other books on electricity in his private library, probably located in the same rooms of his laboratory so that they could be used as sources of knowledge, methods, and ideas. Of course, all these materials were very costly and Galvani could only afford them after establishing his professional position in the academic and medical world and probably also thanks to the inheritance he received from his parents, who had died in the late 1770s. Financial aspects did thus play a role in making Galvani's electrophysiological research possible at a certain moment in his scientific life.

Although the laboratory was a domestic space, it was not frequented by Galvani alone. The performance of many experiments needed the participation of at least one assistant, as when the operation of the electrical machine and the manipulation of the animal were done at the same time but at some distance. Some important observations were made by collaborators or friends, as Galvani reported in his writings, and witnesses were called for experiments considered particularly relevant. Among these frequenters of Galvani's laboratory, the most assiduous for many years was his wife Lucia, Gusmano Galeazzi's daughter, whom Galvani had married in 1764 when she was 20 years old (see Fig. 2.6 in Chapter 2). Lucia had received the typical education of high-class women in eighteenth-century Bologna: Instructed in historical and religious matters, she learned Italian and Latin so well as to be able to revise all her husband's writings. She also took an interest in scientific topics and animated the "conversazioni" that took place in

Bologna's salons of aristocratic and other notable families. Though very different from contemporary fellow citizens such as Laura Bassi, who struggled for a public career in science, Lucia Galeazzi had an important role in the scientific activity of her husband (Bresadola, 2011a)

Other important collaborators of Galvani's were two nephews, Camillo Galvani and Giovanni Aldini—the latter was also one of the main promoters of animal electricity in the 1790s—but there were also people with no family connections. In an unpublished note of 1786, for instance, Galvani referred to the "most erudite Spanish Rialpus" as the author of a very important observation made in the laboratory. Raymond Rialp (Latinized as Rialpus) was a Jesuit who abandoned Spain after the suppression of his order and found refuge in Bologna. In *De viribus*, he is quoted as the person who formed with Galvani the "electrical chain" that allowed the flow of animal electricity between nerve and muscle to produce muscular contractions (see Fig. 4.2).

But what did Galvani and his collaborators do exactly in the domestic laboratory?

From the first experiment reported in laboratory notebooks, dated November 6, 1780, Galvani used frogs "prepared in the usual manner." This phrase has led some historians

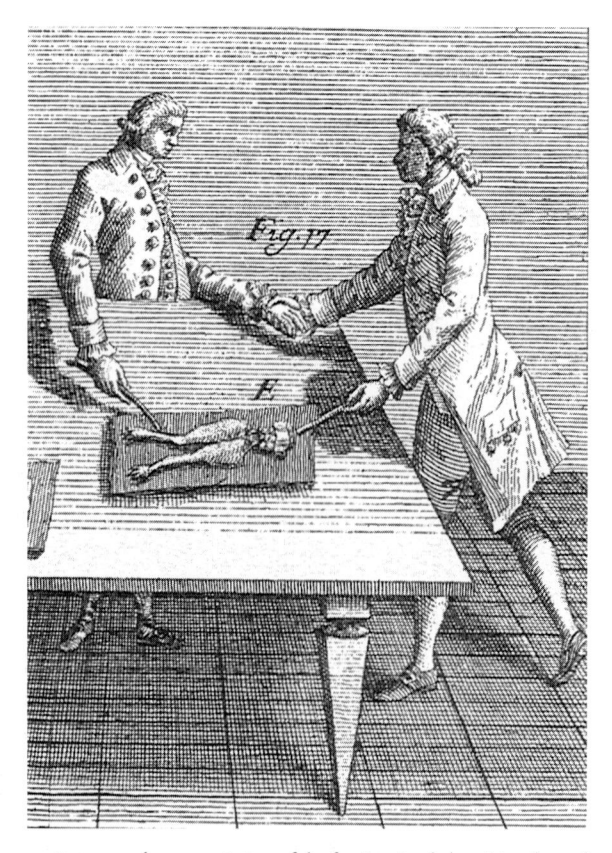

FIGURE 4.2. The experiment on the transmission of the frog's animal electricity through a "human chain" (detail from plate IV of Galvani, 1791).

to believe that Galvani had made electrophysiological experiments before and that we have lost their traces. It is possible, however, that here Galvani referred to a preparation developed during his previous research on the heart and other animal parts without the use of electricity. More important, a similar preparation had been developed by Leopoldo Marc'Antonio Caldani and Felice Fontana during their electrical experiments aimed at testing Haller's theory of irritability. However, though Galvani's frog preparation was not entirely original, it was he who made this preparation one of the most famous and repeatedly used experimental objects in the history of life sciences. It is worth noting that it was still used in 1930s by Alan Hodgkin in the early stage of the research which led him to give an experimental demonstration of the existence of animal electricity, somehow thus concluding the path opened at the end of the eighteenth century by Galvani (see Chapter 9).

Galvani's choice of the frog also had precedents. In the seventeenth century this animal had been used by Marcello Malpighi in his research on lungs, which had led him to the discovery of capillaries and the clarification of the alveolar structure of this organ. Malpighi had compared the killing of frogs he had done for the needs of his scientific investigation to the epic massacre of this animal perpetuated by mice as described in the pseudo Homeric *Batrachomyomachia*. Another great anatomist and microscopist of the seventeenth century, Jan Swammerdam, also made great use of frogs in his research, which included the demonstration of muscular contractions performed in front of the Tuscan Court. More recently, frogs had been largely used by Lazzaro Spallanzani, an important naturalist and correspondent of Galvani, in his research on reproduction. The ubiquitous presence of the frog on the experimental table of early modern and modern scientists has made this animal a real "martyr of science" (du Bois-Reymond, 1852; Holmes, 1993).

There were several reasons that made the frog very suitable for studying muscular motion: Its nerves are easily located and separated, muscular contractions are evident and, more important, these contractions last some time after the animal's death. It was Galvani who discovered that the animal could manifest the motions of its limbs, if stimulated with electricity, as long as 44 hours after being killed and prepared (GME, p. 267). This characteristic was very important for Galvani, who decided to focus his research on voluntary motions, that is those "motions that in the living animal are dependent on the soul [*animo*]" but could also be produced by "a mechanical cause applied to the nerves." Contrary to sensation, which "can be felt [by the subject] but it is totally occult to the observer," muscular motion "is manifest to the eyes of the observer, who thus can, through experiments and observations on the action of the nerves, do violence to nature—so to speak—and force it to show some of the secrets it contains" (GOS, p. 124). As we shall see, Galvani's distinction between sensation and muscular motion would be an important issue in the controversy with Alessandro Volta, who based his criticism of Galvani's theory, and his own alternative theory, on experiments performed on sensations such as taste.

The problem with voluntary motions was that the action of the soul, or will, was for Galvani "beyond the limits of human sight and understanding" and thus could not be subjected to experimentation nor could be taken under control by the observer. Hence, this explains the need to use dead animals, in which "the action of the soul had ceased," and to choose a "mechanical cause," which was still effective in these conditions. Caldani and Fontana had shown that it was possible to produce muscular contractions in a dead frog, using electricity as the stimulating agent. These were just the conditions required by Galvani, who consequently developed a particular preparation of the frog, which he described in the following terms (see Fig. 4.3):

> The frogs being cut transversally below their upper limbs, skinned and disembowelled, I left only their inferior limbs joined together, with their long crural nerves inserted [in them]. These latter were either left loose and free, or attached to the spinal cord, which in turn was either left intact in its vertebral canal or artificially extracted from it and partly or wholly separated. (GOS, p. 125; translated with revisions from Pera, 1986/1992, p. 65)

In the *De viribus* Galvani underlined the fundamental role of this preparation in the experimental outcome. As he wrote, "the phenomena we have described occurred

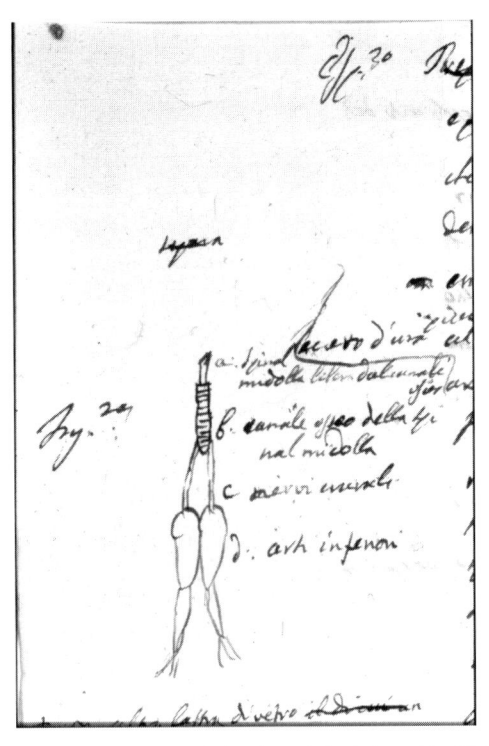

FIGURE 4.3. The "prepared frog" in an autograph sketch made by Galvani (AASB, Fondo Galvani, cart. II, plico II,°).

only if the animals had been prepared for experimentation in the manner we have mentioned above; otherwise the contractions failed to take place" (GDV, p. 374; translated in Galvani, 1953a, p. 56). The dependence of observation on the experimenter's apparatus had emerged as a new tenet, and a debated issue, of scientific knowledge since Galileo's telescope. In the case of Galvani's research, the choice of a different animal preparation, made for instance by Volta, did in fact produce different results and supported different interpretations of the phenomena of muscular contractions, as we shall see later.

The choice of dead animals distinguished Galvani's experimentation from Haller's, which was based instead on the stimulation of living tissues. In this regard it appeared to be problematic since doubts arose about the validity of inferences on a fundamental function of the living being, such as muscular motion, derived from experiments performed on dead bodies. This point was indeed made by some of his contemporaries. Although Galvani did not explicitly tackle this question, there were several reasons behind his choice. His idea that animal motion could be understood by studying the action of "mechanical causes" was rooted in the mechanical view of vital phenomena which emerged in the seventeenth century, thanks to the work of naturalists such as Marcello Malpighi. In a work published posthumously, which is a sort of manifesto of his "rational" approach to medicine and a fundamental text in the history of life sciences, Malpighi made a comparison between the human body and a clock or a mill:

> I know that the way our souls use the body in operating is ineffable. Yet it is certain that in the operations of growth, sensation, and motion the soul is forced to act in conformity with the machine on which it is acting, just as a clock or a mill is moved in the same way by a pendulum of lead or stone, or by an animal, or a man; indeed, if an angel moved it, he would produce the same motion with changes of position as the animals or other agents do. Hence, even though I did not know how the angel operates, if on the other hand I did know the precise structure of the mill, I would understand this motion and action, and if the mill were out of order, I would try to repair the wheels or the damage to their structure without bothering to investigate how the angel moving them operated. (Malpighi 1697, pp. 111–112; translated in Adelmann, 1966, vol. 1, p. 571)

Although, as we shall see, Galvani's concept of mechanism was more complex than that of Malpighi, the two Bolognese naturalists shared the idea that it was possible to study some fundamental vital functions, such as animal motion, independently from the immaterial principle on which they ultimately depended. Moreover, they were convinced that the investigation of these functions in the animal body could offer the key to the understanding of the same functions in the human body, a view that gave a central role to comparative anatomy and physiology in the life sciences. This conviction implied not only that the mechanism involved in sensation and muscular motion was similar in the animal and the human bodies but also that a similar immaterial principle directed these functions in both animals and humans. In fact, Galvani's choice of dead frogs for his investigations was based on the idea that voluntary motions depended on the animal's will which, like the human will, could not be subjected to scientific experimentation. This idea was not only quite different from Descartes's notion of the "beast-machine," but it could be problematic from the point of view of traditional Christian thought, which denied that animals had an immortal soul. However, it circulated among some Catholic religious orders and circles active in the first half of the eighteenth century in several places, including Bologna. For these "enlightened Catholics" animals do not possess a spiritual soul, but they have a sort of "animal soul," which controls sensory perception and acts of volition like voluntary motions (Mazzotti, 2007). Galvani was educated in a Catholic environment that was opened to these ideas, but Volta—who had a less conservative education—believed that animals have a soul (Bresadola, 2011a; Pancaldi, 2003).

On November 6, 1780, Galvani experimented on the effects of an intense electrical discharge on the spinal cord, crural nerves, and muscles of different frogs. As we know from his laboratory notes, Galvani worked as follows: First, he placed a dead prepared frog on an electrically charged Franklin square, where the metal coatings applied to the inferior and superior surfaces were, respectively, positively and negatively charged. Then he connected the inferior coating or "armature" of the Franklin square and the part of the frog on which he intended to experiment with a metal arc, in order to release an electric discharge on the latter. Finally, he observed whether the frog's limbs contracted. In fact, this is one of the standard procedures that Galvani adopted for a long time and that was based on an artificial stimulation of some parts of the frog's body. Following this method, an artificial electrical discharge, produced by an instrument, was directed to different parts of the frog's body in order to register its effects on the muscular activity of the inferior limbs. From the earlier report, it is also clear that Galvani started his research using "sparking" electricity and only afterward he adopted stimulation techniques involving less transient electric flows.

The experiments performed by Galvani on November 6, as well as in the following months, were variations of experiments conducted by other authors in the previous decades, including those carried out in Bologna by Caldani and Fontana in the middle of the eighteenth century (see Chapter 3). In particular, Giuseppe Veratti had performed some experiments very similar to Galvani's during the 1750s, with the aim of studying the effects of electricity on the body of both humans and other animals (Veratti, 1752, 1755). Scientific literature was full of reports of both humans and animals who had been struck by thunder and, when they had survived, showed evident impairments especially in motor functions. In the attempt of artificially reproducing the effects of thunder, Veratti decided to use Franklin's square, as in this instrument "sometimes is contained a quantity of electric vapor sufficient enough for generating a *little thunder*" so that it was supposed to produce the maximum discharge possible at that time. As a result of his experiments, Veratti found that, if it was directed to either muscles or nerves, the discharge of the square could provoke an irreparable structural damage, by making the animal unable to produce movements or detect sensations.[2]

While the similarity between the instruments and the methods adopted by Veratti and Galvani suggests a direct influence of the former author on the latter, the scope of Galvani's investigation was surely more ambitious than Veratti's. Galvani was not interested just in studying the dangerous effects of electricity. Rather, he was interested in understanding how electricity could influence muscular motion, in order to make such a physiological mechanism clear. This theoretical objective was consistent with the idea of rational medicine that played a pivotal role in Galvani's research. It would later inspire the view of the relationships between physiology and medical practice developed by the French physiologist Claude Bernard in his *Introduction àl'étude de la médecine expérimentale*.

In the first stage of his experimental research, Galvani considered some problems concerning the neuro-electric conception of muscular motion that had been noticed by the Hallerians. More precisely, he focused on the two main questions raised by Haller, Caldani, and Fontana in the debate with Laghi and other supporters of the identity between animal spirits and the electric fluid (see Chapter 3): the effects of nerve ligature and the conductive—or insulating—properties of nerves. Galvani was also very interested in another question usually touched in the debate about irritability and neuro-electric conception, namely the problem of determining whether the electric stimulus acted principally on nerves or on muscles. Additionally, since he soon observed that "very little electric fluid—which is very far to produce the least electric sign—is sufficient to produce the contractions" (GME, p. 243) he designed some experiments aimed at discovering the minimum force of an electric stimulus, namely the lowest quantity of electricity capable of eliciting muscular contractions.

From this point of view, Galvani was in line with the general interest that contemporary naturalists devoted to the so-called weak electricity. Indeed, several "electricians" of the time, like Bennet, Henly, and Cavallo in England; Saussure in Switzerland; and Vassalli and Volta in Italy, were interested in studying the "lowest electric degrees," as documented by the construction of progressively more and more accurate and sensitive electrometers. At the basis of such an interest, there was the increasing awareness, in natural philosophers of the eighteenth century, that electricity was not involved only in the case of thunder and sparks, or in the sensational artificial events provoked by electric instruments, but also in a lot of other minor natural phenomena. A special interest was provided by studies aimed at detecting the level of electricity present in the atmosphere both on stormy days and on clear days. At least in part, this interest was motivated by the idea that a relation between atmospheric electricity and climate variations could be reasonably hypothesized. Finally, a great interest in these phenomena came from the medical

---

[2] On this subject Veratti read two dissertations at the Academy of Sciences of Bologna in 1769 and 1770 (AASB, Tit. IV, Sez. I).

community, since doctors could use the measurement of atmospheric electricity to provide a "physical" basis for the influence of the atmosphere on human health, which had been considered a central aspect of medicine since antiquity.

In 1780, when Galvani was starting his electrical experiments on frogs, Volta invented his "*condensatore*," an instrument that allowed the detection of very weak electricity, thanks to the variation of electric capacity due to the removal of the one of two metal plates from the other which was connected to ground. In 1782, Volta used this instrument to perform several experiments in Paris—some of which with Laplace and Lavoisier—to demonstrate that water evaporation, as well as some other physical and chemical phenomena were accompanied by the production of electricity.[3] In the following years, the scientist from Como associated the *condensatore* to a "straw electrometer," thus constituting a "straw micro-electrometer," also named *elettroscopio condensatore*—an instrument able to detect electric potentials lower than 1 volt. Thanks to this instrument, in 1792 he was able to obtain a physical measure of the minimum electricity necessary to elicit contractions in a frog and then, in 1796, a measure of the electricity generated by the contact between different metals (see Chapter 6.2).

Let us now return to Galvani and his first experiments in late 1780. After 2 months of experimental practice, the investigation had made important progress. Although Galvani focused his attention on problems that had been explored before him, his research presented a systematic structure that had been absent in previous works. Moreover, by inventing a series of original experimental arrangements, he had by that time already acquired a great familiarity with manipulating experimental preparations and in using electrical instruments. Surely, these technical arrangements concerning the frog, its parts, and the electric source represent some of the most interesting elements in the first stage of Galvani's experimental work. Indeed, Galvani soon realized that the way in which instruments, as well as animal preparations, were arranged on the working table was fundamental for the phenomena that could be observed.

When designing an experiment, he was faced with a series of complex decisions: He had to choose which part of the frog to use and how to prepare it, its best position on the table, the right electric instrument, and how to produce the electrical stimulus or where and how the stimulus should be applied. In addition, he had to determine the correct disposition for the metal hooks used to provide the electrical stimulus. Every single decision resulted in a different experimental arrangement, thus giving rise to a new experiment. Such procedures, systematic as well as open to further experimental indications, were shared with other experimenters—like Lavoisier in his early chemical research—and presented some aspects that were central to the subsequent experimental science (Bresadola, 2003).

---

[3] In the presence of flammable or explosive material, the production of electricity by evaporation of water or expansion of gas (first proved by these experiments) can cause serious accidents. Primo Levi (1965) alludes to this possibility in one of the chapters of his *L'altrui mestiere* (Levi, 1985).

If it is true, on the one hand, that Galvani was inspired by Veratti's research and by Caldani's and Fontana's experiments on the electrical stimulation of muscular contractions, on the other hand, the variety and richness of his experimental dispositions constituted a relevant novelty in the eighteenth-century research framework. From this point of view, Galvani's experimental approach, based on constantly changing experimental details, is in evident contrast with the relative monotony of the methods adopted by Haller in his research on irritability, where the changes were concerned only with either the part of the animal selected for stimulation or the agent used in the stimulation. On the contrary, Galvani was very capable in combining preparation techniques—which he derived from the anatomical-physiological tradition originated in Bologna between the seventeenth and eighteenth centuries—and technical and experimental variations typical of contemporary research on electricity. We can say that he introduced a frog into a "room for experiments" that, even visually, was more similar to the laboratory of an "electrician" than to a classic laboratory of anatomy (see Fig. 4.1). This attitude was certainly influenced by the climate of general interdisciplinary interaction that was animating the Institute of Sciences of Bologna, where scholars of physics, chemistry, anatomy, and medicine efficiently worked together (see Chapter 2).

Galvani's investigative approach consisted of applying the instruments, the concepts, and the laws of electricity to the study of muscular motion, and this represented a fundamental aspect of his entire research activity. Indeed, the very same approach can be found also in his subsequent research concerning chemistry and physics, where he used the technical and conceptual tools of the newborn "pneumatic physics" to study the effects of "airs" on vital functions. Even under this perspective, Galvani cannot be considered a supporter of an exclusively physiological or "electro-biological" approach to natural phenomena (Pera, 1986/1992). First of all, because, as we have already suggested earlier, by following this line of thought one could risk using some disciplinary categories that were not available at the time of Galvani but also because in such a way a fundamental aspect of his work would be neglected: his unitary vision of natural phenomena.

At the end of 1780, Galvani decided to put down on paper the results he had obtained so far. On Christmas Day he thus planned a Treatise on the "nervous force"; he wrote the beginning of the first chapter and formulated some laws derived from his experimental work. We shall return later to the reasons why he preferred the term "nervous force" to that of "animal spirits." For the time being, it will suffice to say that by nervous force he meant a "very active principle existing in the nerves," which, by acting on muscles, was supposed to produce muscular contractions. This definition was followed by a list of rules derived from the examination of this principle applied "first to cold-blood animals and to muscles related to the Will," that is, in the conditions assumed in the experiments performed in November and December. In this project, every rule would be treated in an independent section of his general work. These rules, in turn, were followed by the formulation of some laws:

> Law 1: The muscular contraction produced by nerve irritation is proportional both to the minimum parts of the nerve moved by the stimulus and to the force by which they are moved.
> Law 2: Independently on the irritating cause, the irritation is almost uniquely local, that is, it spreads very poorly—if any—beyond the place of application.
> Law 3: The communication, as well as the propagation to the muscle, of either the action [of the nervous force] or of the induced motion is dependent only on the nerve. (GMS, cart. III, plico II aa, fasc. I)

These laws were generalizations of the results obtained in the first 2 months of experiments. Accordingly, their validity was supposed to be general rather than limited to the electric stimulus. In particular, the first two laws were derived from the experiments involving nerve ligature, while the third one synthesized a series of experiments performed in December. By elaborating these laws, Galvani brought an original contribution to the comprehension of the relation between stimulus, nervous force, and muscular contraction, and tried to provide the basis for a quantitative analysis of the phenomenon. Moreover, it should be noted that these laws were phenomenological in nature, in the sense that they were aimed at reducing phenomena to general rules, without referring to a specific theoretical model. The terms used by Galvani, like "irritation" or "irritating cause," while in line with the Hallerian terminology, did not imply that the concept of "irritability" was distinct from the concept of "sensibility," as well as from that of nervous fluid or force. Indeed, as we shall see later, his early experiments brought Galvani to conclude that the electrical stimulus was primarily directed to the nerve rather than to the muscle. Such laws were thus acceptable both to those who identified the efficient cause of muscular motion in a fluid flowing in nerves and to those—like Haller and his followers—who considered it as a specific property of the muscle. Their meaning was analogous to that of electrical laws described in contemporary handbooks like Cavallo's *Treatise*, where the first part was devoted to explaining "such natural laws concerning electricity, as by innumerable experiments, have been found uniformly true, and are independent on any hypothesis" (Cavallo, 1786, vol. 1, p. VI).

Since the beginning of his experimental investigation Galvani had used wires and other metallic objects that were inserted in different parts of animal preparation, in order to properly direct the electric stimulus. More specifically, he was interested in establishing whether the stimulus was more efficient if applied to the nerve or to the muscle. In line with contemporary terminology, he classified these objects as "conductors" of either the nerve or the muscle, respectively. It should be noted that this fact started a long and complex relation between the animal preparation and metals, and that this relation will become very important in both Galvani's subsequent experiments and in the controversy following the publication of his *De viribus*. Moreover, the use of a physical terminology in the context of animal preparations had a great influence on the conceptual development of Galvani's research and, more specifically, on the idea that a frog could be considered as an "animal Leyden jar."

Though he had obtained some important and original results in his experiments, or maybe exactly for this reason, Galvani soon realized it was necessary for him to continue further his research in this direction. He thus interrupted the writing of the treatise he had planned the previous Christmas and on December 29, 1780 he returned to his laboratory. He placed a frog on a "large glass sheet" and connected it to the electrostatic machine. Then, turning the disk of the machine 15 revolutions, he observed muscular contractions at every turn, exactly as he did in his previous experiments. When the disk was stopped, the contractions ceased. Conversely, Galvani obtained 15, and "no more," contractions by touching the frog's shinbones with a metal object. In this experiment it seemed that electricity, after being communicated to the frog by the conductor of the machine, was discharged in small stages at each contact between the metal object and the shinbones, thus producing a contraction. However, when Galvani disconnected the frog from the conductor, contractions ceased. Finally, once that the contact was re-established and the sheet of glass reheated—in order to dry and insulate it—by replicating the experiment, he obtained the contractions at the touch of the animal's shinbones again.

Galvani derived two important "corollaries" from these results. The first corollary regarded the electric properties of nerves and muscles. According to this corollary, nerves were considered not very efficient in conducting electricity; in Galvani's words, "the conductor does not easily discharge through the nerves; there always remains a little [electricity], which flows from nerves to muscles at every touch. This shows how difficult the flow [of electricity] through the nerves is" (GME, pp. 242–243). Indeed, when Galvani touched the frog's shinbones, the previously accumulated electricity was not discharged in a single shot, as it would be if the nerves were good conductors. This result could suggest an answer to one of the Hallerian objections to the neuro-electric hypothesis. Indeed, it showed that nerves could hold back the electric fluid, but not in a sufficient quantity for preventing it from reaching the muscles and from provoking contractions. Galvani focused his attention on the idea that an electric fluid could be "pinned" or "tied" in nerves in a later experiment, dated January 17, 1781, where the concept of an intrinsic electricity to the animal was introduced.

The second corollary of December 29 can be considered as a sort of fourth law about the action of the nervous force, in addition to the three laws formulated some days before. This corollary stated that the contraction was the minimum sign of the presence of electricity in a condition of imbalance; in Galvani's words, "very little electric fluid, which is far from giving any electric sign, is able to excite the contractions" (GME, p. 243). As Galvani himself will afterward remark, this was a true "discovery" that he had made more than 10 years before the publication of his *De viribus*. When, at the beginning of 1792, Volta read Galvani's treatise, he was particularly struck by the frog's capacity of revealing any electricity that was "so weak, that we cannot feel any shock from it nor make it sensible to the most delicate electrometer" (VEN, I, p. 25); indeed, this was one of the most interesting aspects of the work that induced the physicist from Como to replicate Galvani's experiments. From an historical point of view, this fact shows the importance of Galvani's initial experiments, which he did not include in *De viribus* and which, for this reason, have always been underestimated by historians. In fact, these experiments are fundamental to understanding Galvani's experimental practice and to reconstruct the path that led him to his famous "first experiment" at the end of January 1781. As we shall see, Galvani returned several times to reason about the high sensibility of the frog to electric stimulation. Indeed, he repeatedly tried to measure the minimum "quantity of electric fluid which, applied to the nerves, is sufficient to excite the contractions of the muscles."

On December 30 and again on January 10, 1781, Galvani focused on the flow of electricity in the system composed of the frog, the electrostatic machine, and the Franklin square. More precisely, he returned to an experimental disposition that he had adopted earlier and varied the connections between the frog and the instruments in several ways (see Fig. 4.4). As a result, he elaborated the following: "Reflection—It is the frog, then, that contains the electric vapour which, as soon as the conductor is touched, flows from the frog to the inferior surface [of the square]" (GME, p. 246).

Though this explanation referred to all the phenomena he had observed with such an experimental arrangement, the following week Galvani re-examined the question. According to the methodological approach of his research, in fact, he was never satisfied with those results that were obtained only once, but found it necessary to replicate the experiment under different conditions, in order to clearly pinpoint that the obtained result was not due to particular conditions. During his investigation Galvani often underlined the need to replicate an experiment to eliminate "the abnormalities and inconstancies" of a phenomenon, to "have a better knowledge of the reason" of a certain given result, or to clarify experiments previously classified as "uncertain and treacherous." Moreover, he often introduced a control experiment in the series of experimental

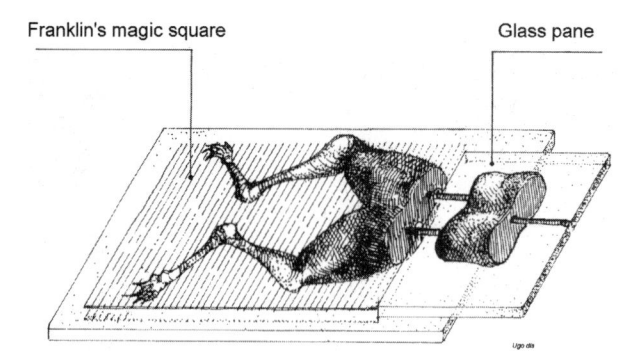

FIGURE 4.4. Scheme of Galvani's experiment of December 9, 1780, designed to determine the insulating or conducting property of nerves. The prime conductor of the electrical machine was connected to the spinal cord and a spark was extracted from it (sketch by Ugo Sorrentino).

variations, in order to clearly understand to what extent a given phenomenon depended on a particular experimental condition. Such an attitude is evident in several passages of his works, both in his papers and in his laboratory notebooks. In a laboratory note dated December 29, 1780, for instance, he wrote the following "warning": "In order to solve [the problem], two frogs are to be used: in one the spark of the [Franklin's] square must be extracted from the muscle, but not in the other. The same has to be done for the nerves" (GME, p. 244).

Years later, in 1795, during a brief journey along the Adriatic coast, Galvani conducted some experiments on torpedoes. In these experiments, after having demonstrated that the "electric body [i.e., the electric organ of the animal]"—if separated from the animal and with all the nerves cut—"neither produced the least shock, nor induced the least contraction," he used the other—intact—organ as a control element. On May 17, he noted in his *Taccuino* (notebook) that "the other electric body, located in the animal in its natural condition, produced many shocks, and excited many contractions that progressively diminished" (Galvani, 1937b, fol. 32). On the basis of observations like this, Galvani came to demonstrate the importance of the nerves in electric organs for the genesis of the torpedo's shock (see Chapter 7.3). In this respect, it is worth noting that in many cases control experiments—designed in response to either personal doubts or to external criticisms—marked some of the highest moments of Galvani's research. As we shall see, this was the case for the experiments about the "contractions without metal" that Galvani conducted in 1794 in response to the objections by Volta who, at that time, "attributed everything to metals." And this was also the case for the experiments performed in 1797, once again in response to Volta's criticism, this time concerning the possibility of producing an electrical disequilibrium by making two heterogeneous conductors—not necessarily metallic—touch each other. Indeed, in 1797, Galvani showed that this heterogeneity—that was central for Volta's theoretical framework—was not a proper cause of the contractions, as it was possible to produce them by directly connecting two nerves, that is, two animal tissues of the same kind. It should also be noted that Galvani's discovery of muscular contractions produced by metal arcs without the use of "artificial electricity" produced by the electrical machines or without "stormy atmospheric electricity" was again the product of a very fortunate control experiment. In 1786, while he was trying to replicate some experiments designed to study the "power of daily and quiet electricity," Galvani indeed discovered that a simple contact between a nerve and a muscle, via a metal conductor, elicited contractions. In these experiments he intended to evaluate, through an experiment of "negative control," the role of the atmosphere in the production of such contractions (see Chapter 5.2).

Let us now end this long digression on Galvani's methodology and return to his investigative pathway. On January 17, 1781, Galvani designed a new experimental arrangement that consisted of separating the frog from Franklin's square and of connecting both the frog and the square to the conductor of the electrostatic machine (see Fig. 4.5). After turning the machine's disk in order to charge the square and to transmit positive electricity to

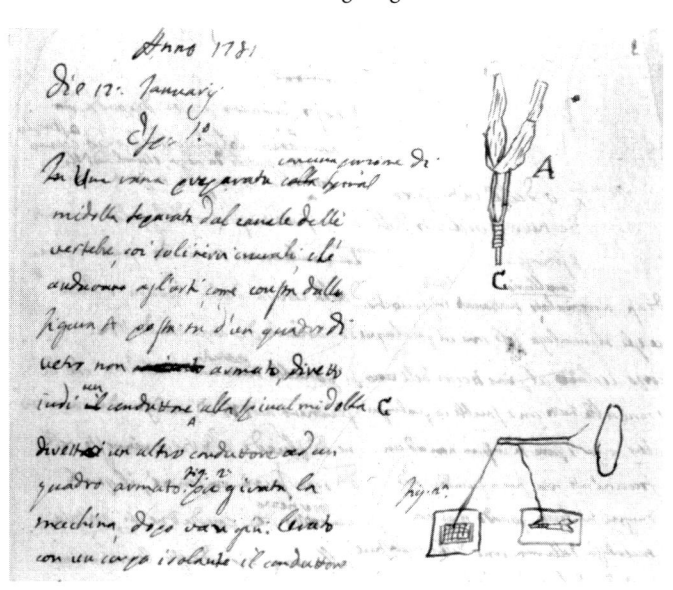

FIGURE 4.5. Galvani's laboratory notes on the experiment of January 17, 1781; the autograph sketch shows the electric machine connected to the frog and Franklin square (AASB, Fondo Galvani, cart. II, plico II,©).

the frog, he connected the inferior (uncharged) armature of the square and the animal's spinal cord through a conducting arc. At each contact, and for 15 consecutive times, he obtained the legs' contractions. At first, Galvani considered this result a confirmation of the "reflection" elaborated the previous week, namely that at each contact between the frog and the square a quantity of the electricity accumulated in the former was transmitted to the inferior square, and that such an electric flow excited the contractions of the inferior limbs. However, at a later time Galvani corrected this explanation, by rewriting it in his laboratory notebook as follows:

> This phenomenon seems to depend either on the extrinsic electric fluid accumulated, held and confined by the frog's nerves and spinal cord, or on that intrinsic to nerves awakened by the extrinsic one, which flows to the lower coating of the armed [i.e., Franklin] square, considerably discharged by the charge of the upper one and by the touching of conducting bodies. (GME, p. 248)

In other words, the frog's electricity that made muscles contract could be due either to the electricity of the machine accumulated in the frog or to the intrinsic electricity of the frog set in motion by the electrical stimulus produced by the electrical machine. It was the first time since the beginning of his experiments that Galvani realized that he had found some evidence to support the neuro-electric conception of animal motion he had proposed in his anatomical function of 1780—so much so that he recorded it in his notebook. Even more significantly, Galvani elaborated such an explanation while reflecting on the outcome of the experiment, and in particular on the new arrangement he had adopted. In fact, the problem being investigated—that is, the circulation of electricity between the electric machine, the Franklin square, and the frog—was the same as that of January 10. What was new was the arrangement of the experimental objects, since the frog and the Franklin square were only connected indirectly through the conductor of the electrical machine. A simplification of previous ones, the new setup allowed Galvani to see more clearly the relationships between the animal and the electrical instruments, and the circulation of electricity among them. In this case, the experiment was not designed as a test of the neuro-electric hypothesis; rather, the hypothesis of an intrinsic electricity was a sort of afterthought suggested by the experimental arrangement (Bresadola, 2003).[4]

The experiment of January 17 is important not only because it offered some evidence for the existence of a frog's intrinsic electricity but also because it marked a change in the scope of Galvani's investigation. Whereas up to that point Galvani had focused on the laws of muscular contractions at a phenomenological level, he now entered the more difficult and hazardous area of theoretical reasoning about the cause of the phenomenon. In fact, he became more cautious and adopted an open-minded attitude toward the experimental outcomes, taking into account other possible causes of the phenomenon under observation and systematically submitting all these explanations to experimental testing. During these attempts, Galvani fully realized the difficulties of the investigation he was carrying out, as when he observed that "in all the experiments carried out during this year, as well as in many others, a great irregularity and inconstancy has been observed not only when different frogs were used, but also with the same ones" (GME, p. 252).

Galvani's frustration about such experimental variability—the fact that the very same experiment conducted on different animals or even on the same animal often gave different results—is well known even today by everyone working on the experimental study of the animal organism. However, Galvani was convinced that such variability was due to a difficulty in completely managing the conditions of the experiment, rather than to an essential characteristic of the animal's preparation. Once the fundamental circumstances of a certain phenomenon are understood—the reasoning went—it should be sufficient to reproduce these conditions in order to obtain the phenomenon in question. This general framework based on a sort of "experimental determinism," which would be fully developed only during the nineteenth century, is clearly evident in different passages from Galvani's laboratory notebook. For instance, in the experiments performed on December 13, 1781, in which he observed the production of muscular contractions "at every turn of the disk [of the electric machine]," Galvani noted: "Warning—It would be convenient to find out the circumstance on which these contractions depend, as it could give great light" (GME, p. 321).

---

[4] The change in the interpretation of this experiment has not been noticed by historians because they have relied on the published edition of Galvani's laboratory notebooks, which reports only the final version of Galvani's note without any sign of the changes he had made in the manuscript.

In this "warning" it is easy to note that Galvani attributed a positive value to this apparent variability; indeed, it could sometimes give valuable cues that a careful experimenter should notice and use as a starting point for further research. The idea that variability of the results depended on the experimental situation, and not on nature's indeterminacy, is also to be found in *De viribus* when Galvani discussed some reasons for the "inconstancy and variation" of animal electricity. More precisely, here he was dealing with the experiments on metal arcs, in which contractions "differed in accordance with the various seasons of the year and especially with atmospheric conditions" (GDV, p. 393; translated in Galvani, 1953a, p. 71). However, such a variability of results could be explained—and thus discarded—by following the indications offered in the experimental report and thanks to practice and experience. Galvani was confident of the truth of his results, and he concluded his argument with a significant statement based on the confidence and on the pride typical of a careful experimenter:

> We purposely mention these facts so that no one who replicates our experiments will either deceive himself or think that we have been deceived in assessing the strength of contractions and electricity. For, if one carries out these same experiments many times, he will discover time and again the phenomena which we have revealed through trial and experience. (p. 394/p. 72)

Some years later, in one of the most heated moments of the controversy on animal electricity, Galvani returned to this question. More precisely, in 1797, he reacted to some of Volta's criticisms about the doubtful validity of the experiments on the direct contact between nerves and muscles. After underlining that in these experiments "the effect has not occurred only a few times out of a hundred," Galvani wrote:

> I really hope that Mr. Volta, who is such an accurate philosopher, will not doubt about the truth of a fact and of an experiment only because it has not occurred a few times; otherwise in physics one should not admit any fact deduced from the experiments. He knows better than me that there are so many and tiny circumstances, often unknown, on which the outcome of an experiment depends, that it is too easy that one of them may be absent, even though all diligence is used; and therefore the phenomenon, always constant in itself, sometimes cannot take place. (GEA, p. 6)

4.2 A "PROBLEMATIC" TURN: THE OBSERVATION OF CONTRACTIONS AT A DISTANCE

The variability of the experimental outcomes found in the early stage of his electrophysiological research did not dissuade Galvani from his research project, drive him to assume conjectural shortcuts, or modify his experimental approach to phenomena. This attitude received a first and important reward on January 26, 1781. In the initial experiments on this day Galvani varied the disposition adopted on January 17. Once again, he did not obtain a very clear set of results. In particular, some results seemed to be in contrast to the idea that contractions were the minimum sign of an unbalanced electricity, thus constituting an argument for relating the muscular contractions to the electricity of the electrostatic machine rather than to an electricity internal to the frog itself. However, the results of the sixth experiment brought it all up for discussion again. In this experiment, a prepared frog was placed on a glass panel, which, in turn, was at a short distance from the electrostatic machine. Although no physical connection was established between the frog and the machine, something happened:

> When my wife or someone else brought a finger close to the conductor [of the machine] and elicited sparks [from it], and at the same time I rubbed the crural nerves or spinal cord with an anatomical knife provided with a bone handle—or even if [the knife] was just brought close to the said spinal cord or nerves—then there were contractions even though no conductor was applied to the glass where the frog lay. (GME, p. 245)

In his laboratory notebook, Galvani reported that the phenomenon "was consistent, and is doubtless marvellous." In *De viribus* he wrote that he had been informed of the phenomenon while he was involved in totally different research and that from that moment on he was "inflamed" (*incensus*) with an incredible desire and curiosity (*incredibili studio et cupiditate*) to conduct the experiment personally, in order to shed new light on its mysterious nature (*quod occultum in re esset in lucem proferendi*) (GDV, p. 364; see Fig. 4.6). Galvani was undoubtedly used to reporting in full detail the emotions derived from his observations, with terms like "singular" and "marvellous" recurring in his texts. However, the frog's contraction provoked by the distant electric spark appeared to him as a really

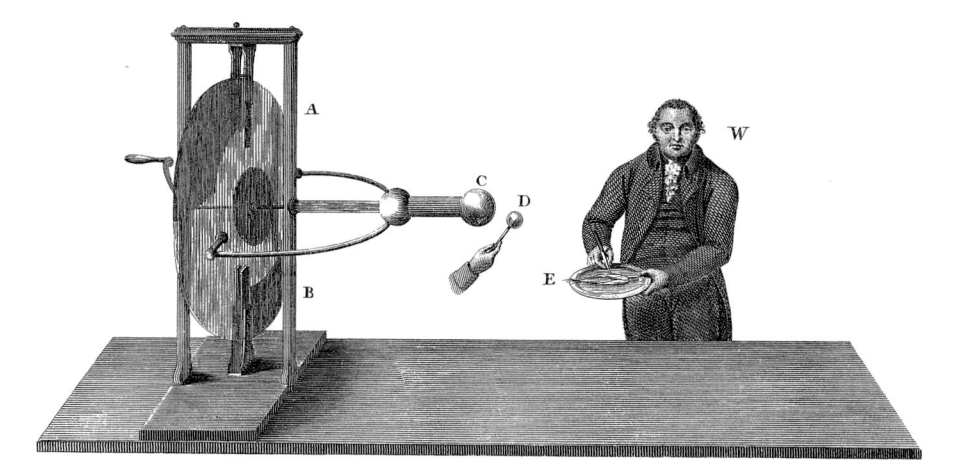

FIGURE 4.6. Early nineteenth-century illustration of Galvani's experiment on the contractions at a distance of frog's legs, the so-called first experiment (from Wilkinson, 1804).

spectacular phenomenon and worthy of the greatest interest. In fact, all electrophysiological experiments performed by Galvani from this date to the other relevant date of September 20, 1786, consisted of creative variations of the experiment on the "distant spark" with the aim of revealing the mechanism of the phenomenon.

But, where did Galvani's wonder and curiosity, when observing the contractions at a distance, come from? First of all, it was an unexpected result in the light of the facts established in his previous experiments: The occurrence of contractions when the animal was unconnected and at some distance from the electric machine conflicted with the idea that it was the electricity of the machine which flowed to the frog and produced the contractions. Moreover, the experimental conditions were totally new: Up to then the electric stimulus was applied directly to the frog, whereas in this experiment the animal was physically disconnected from the electric source and the stimulus—that is, the spark produced by the machine conductor—was delivered only indirectly. It must have been really surprising, indeed "marvellous," to see the frog's legs move exactly at the same time as a spark came out from the electric machine. Furthermore, none of the existing hypotheses was able to explain unequivocally the observed phenomenon, so much so that Galvani wrote that the "cause" of the contractions was "unknown." The question about the cause, on which Galvani had focussed his attention in previous days, was then reopened as a consequence of a new experimental outcome.

In the "warning" that immediately followed the report of the experiment on contractions at a distance, Galvani explicitly recognized that the action at a distance between the electric source and the frog was one of the most important aspects of the phenomenon. Indeed, he advanced the possibility that an analogous mechanism could be at the basis of those contractions that had been observed in other frogs placed on the glass panel of the Franklin square "when nerves were touched either with the conductor or with some other [means], and no connection whatever was established between the machine and the glass" (GME, p. 254).

Since the sixth experiment performed on January 26, 1781 was described at the beginning of *De viribus*, it is also known as Galvani's "first experiment." Especially in general works on the history of physiology and of science, it is considered a typical chance observation which represented a fortunate start to Galvani's electrophysiological research (see Chapter 1). Though the description contained in *De viribus* may suggest a similar interpretation, the reconstruction of the previous experimental pathway sheds a different light on this experiment. As we have seen, if on the one hand this result was unexpected and controversial, on the other hand it derived from an experimental disposition that was a variation of previous experimental arrangements. In other words, only a scientist who was already involved in experimenting on the role of electricity in the mechanism of animal motion with an approach based on the systematic variation of experimental conditions could discover a phenomenon like contractions at a distance. And, in fact, it was Galvani who first observed this phenomenon and who derived from it some important consequences. To sum up, if the observation of contractions at a distance was partly the product of chance, one must add that in this case chance favored an already qualified mind that was ready to take the opportunity due to external motives (a laboratory with electric instruments, and a frog prepared for the study of muscular contractions) and internal reasons (the effort of an investigator who examined the experimental results on the basis of new hypotheses as well as his attitude to note every new detail that could provide cues for solving an open problem).

Several of the readers of *De viribus*—both contemporary and subsequent—did not agree with Galvani's explanation of his "first experiment" in terms of electricity intrinsic to the frog, which in turn was elicited by the electricity produced by the electric machine. As we have noted in Chapter 1, in his *Memoria seconda sull'elettricità animale* (*Second Memoir on Animal Electricity*)—published some months after *De viribus*—Volta suggested an alternative explanation for this phenomenon and accused Galvani of ignoring some fundamental laws of electricity, in particular that related to "the action of the electric atmospheres." In Volta's opinion, if Galvani had been aware of this law, he would have identified the cause of the phenomenon with the electricity of the electrostatic machine rather than with electricity intrinsic to the frog. Moreover, he would have related the experiment to the phenomenon known as "return stroke" (VEN, I, pp. 46–47). According to this phenomenon, described for the first time by Charles Stanhope (Lord Mahon) in 1779, a body—the frog, in Galvani's case—can be struck by an electric discharge if an electrified distant object—for example, an electric cloud or a conductor of an electrostatic machine—were suddenly discharged. The modern explanation of the "return stroke" is based on the concept of electrostatic induction and is quite close to the explanation provided by Volta in agreement with a nonmaterial conception of electric atmospheres that gained great success among the "electricians" in the latter years of the eighteenth century (see Chapter 3.2). According to this conception, a charged body, by acting on conductors situated within its field of influence—or, in modern terms, in its electric field—attracts opposite charges and repels charges of the same value. When the electrified body is suddenly discharged, electric charges reflow into nearby conductors in order to re-establish the equilibrium according to the new conditions of the electric field. The electric current thus generated rapidly passes through these conductors or through other bodies, eventually provoking secondary sparks.

The effect of the "return stroke" can be particularly strong if the objects influenced by the electric field are connected either with another conductor or with the ground, because in these cases the "return" current can be very intense. This provides a good reason for the fact that in Galvani's experiment the phenomenon occurred especially when someone touched the crural nerves of the frog with the metallic part of the "knife" while it did not take place if the knife was held by the bone handle. Thus, in *De viribus* Galvani wrote: "Since the scalpel had a bone handle, we found that when this handle was held in the hand, no movements were produced at the discharge of the spark. They did occur, however, when the fingers touched the metallic blade or the iron nails that secured the blade of the instrument" (GDV, p. 364; translated in Galvani, 1953a, p. 47).

In his laboratory notebook this sort of "control experiment" is reported on January 31, when he found that "if a glass or an old bone cylinder is applied to the nerves or spinal cord, no motion occurs, while if one applies or touches [them] with any conducting body—metal, finger or whatever—the phenomenon and the contractions occur" (GME, p. 255).

That Galvani's observation of contractions at a distance could be explained by the "return stroke" theory was not only claimed by Volta but also by François Arago in the nineteenth century. According to Arago, if this phenomenon "had presented itself to some good physicist, familiar with the properties of electric fluid, it would have scarcely attracted his attention" (Arago, 1854; see also our Chapter 1). Such a claim has been substantially repeated by many modern historians, including Ierome Bernard Cohen, who, in his introduction to the most relevant English translation of *De viribus*, has underlined that Galvani's "knowledge of physics, even in some of the most elementary aspects of electricity, appears rudimentary" (Cohen, 1953, p. 30).[5] There is no doubt that, even though Galvani used the electrical instruments and concepts in his study of animal motion, he was not involved in the cutting-edge research about physical electricity, nor did he probably know Stanhope's explanation of the "return stroke" at the time of the first performances of his experiment. Indeed, the treatise where Stanhope described this phenomenon for the first time was originally published in English (a language which Galvani did not know and which was not the main language of scientific communication in the eighteenth century) in 1779, that is, less than 2 years before Galvani's observations, and it was translated into French first in 1781.

Galvani's ignorance about the return stroke probably contributed to making his astonishment even greater when he observed the contractions at a distance. It must be said, however, that in that period the study of electric phenomena was not at all a well-established body of knowledge and the agreement on the most fundamental aspects of that study, including electrostatic induction or the action of "electric atmospheres"—to use Volta's words—was not univocal even among the most important "electricians." For example, in his textbook on electricity Tiberio Cavallo expounded different conceptions about these atmospheres, from their corpuscular interpretation—according to which they were fluids and material "effluvia"—to the idea of spheres of influence of a charged body defended by Volta, without taking a position in the debate. At the end of the eighteenth century, an influential scientist like Alexander von Humboldt was still a supporter of an essentially material conception of electric atmospheres (Humboldt, 1797; see Pera, 1986/1992; and Finger, Piccolino, & Stahnisch, 2013a).

---

[5] While Cohen endorses the old explanation of the contractions at distance observed by Galvani's as the outcome of a return stroke phenomenon according the eighteenth century views (i.e., electrostatic induction), a somewhat different view about the interpretation of the phenomenon was expressed by Quirino Majorana an Italian physicist, who gave important contributions to the development of the wireless telegraphy. According to Majorana, the phenomenon involved electromagnetic mechanism and thus "it contained the germs of modern Marconi's telegraphy" (Majorana, 1937, p. 54).

More important, what Galvani observed in his laboratory on January 26, 1781 really was extraordinary. It was the first time that a signal (a piece of information) was transmitted from a source (the electrostatic machine) to a receiver (the frog) without the contribution of an intermediate conductor. In other words, this was the first time that a communication was realized through the elusive and somewhat immaterial medium that ancient physicians named "ether" (and that still retains the same name, at least in ordinary language). From this point of view, the explanation of the phenomenon on the basis of the "return stroke" is a stereotype established by a long tradition that makes it difficult nowadays to appreciate the novelty and the importance of the phenomenon itself, as it was observed by Galvani and his assistants on January 26, 1781. No doubt, the return stroke played a role in the phenomenon of the contractions at a distance produced by the spark elicited from the machine, though it is plausible that other mechanisms of electromagnetic induction were involved, at least in certain experimental conditions (Majorana, 1937: Sarkar et al., 2006, p. 258; Susskind, 1964). However, the return stroke was only a partial, in some way secondary, aspect of the complex and astonishing phenomenon that attracted Galvani's attention. In the case of a return stroke, as Stanhope originally described it, the energy moved by the spark's discharge was the direct cause of the observed effects. Such effects are evident, even dramatic, only if the energies involved are extremely high in intensity, as it happens when the charged body is a big cloud and the "spark" is lightning. Indeed, with his theory Stanhope was able to explain why some great accidents could happen in places located under a great cloud after a thunderstorm had exploded at one extremity of the cloud itself, even if this thunder was very distant from the provoked event (see Fig. 4.7).

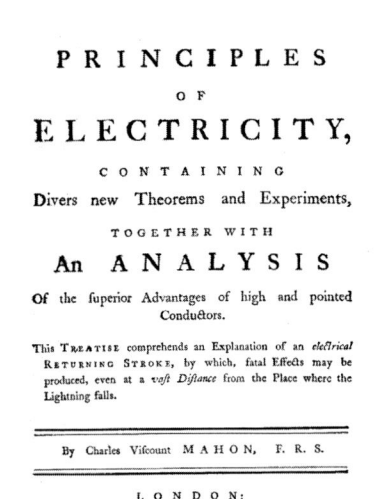

FIGURE 4.7. Title page and plate IV of Stanhope's *Principles of Electricity* with, at the bottom, the illustration of the *return stroke* phenomenon (from Stanhope, 1779).

However, when the energies involved were small—as surely happened in the laboratories of eighteenth-century "electricians" and particularly in Galvani's experiments—the "energetic" effects provoked directly by the discharge were lower and consisted mainly of light or sound phenomena (like sparks, "flakes," "pencils," "electric lines," "auras," "globes," and "stars"), which usually exhausted at a small distance from the discharged conductor. This was because the electric instruments, and especially the electric machine, were able to produce electric potentials higher than 10,000 volts, but they involved only a low quantity of charge. Moreover, even if the phenomena associated with the discharge of the machine were well visible, they were quite unimportant with regard to their intensity, so that the experimenter could touch the conductor of the machine with a hand without particular risks. At worst, he might feel an unpleasant, but tolerable, sensation limited to his finger or hand, or even to his arm or shoulder, as for instance if he had used large Leyden jars or had touched the conductor with wet hands. For example, in his textbook Cavallo warned neophytes not to manage charged "electric jars" (Leyden jars) without precautions, "for an unexpected shock, though not very strong, may occasion several disagreeable accidents" (Cavallo, 1786, vol. 1, p. 186). Only in very exceptional circumstances could electrical experiments be really dangerous. This had been the case of George Wilhelm Richmann, a member of the Russian Academy of Science, who had been killed by a thunderbolt while studying the phenomenon of electricity with a long metal cable in stormy weather.

Despite these cases, no one gave great importance to the possible harmful effects of electricity; on the contrary, the effects of electrical discharges on the body of the experimenter were considered important clues for the understanding of these phenomena. Moreover, the application of sparks and shocks produced by electrical instruments to the human body was used as a therapeutic method by many physicians and healers who practiced medical electricity, and it was often used for pleasure and entertainment in the *salons* of the aristocracy (see Chapter 3.3).

When one was at some distance from the electrical instrument, nothing important happened—no particular sensations were felt, nor tremors. Neither did objects on experimental tables suddenly move after a spark's discharge, nor did they change temperature, nor were they affected by other relevant physical-chemical modifications, which could imply a violent external agent. Such absence of effects at a distance was due to the fact that low energies were involved in the discharges, and also because the action of electricity rapidly decreased with the distance from the machine's conductor. Moreover, in eighteenth-century laboratories there was nothing in the "ordinary" conductors capable of specifically reacting to electric influences of low energy; that is, there were no electrical "amplifiers."

We have said earlier that no particularly interesting events occurred away from the machine. This is not completely true. For the sake of precision, we should say that *nothing worthy of interest* took place until January 26, 1781, when Galvani—or one of his assistants—observed that "suddenly all the muscles of the limbs were seen to contract so that they appeared to have fallen into violent tonic convulsions" without any apparent cause that could explain the phenomenon, apart from the fact that another person "lightly applied the point of a scalpel to the inner crural nerves" of a prepared frog (GDV, p. 364; translated in Galvani, 1953a, p. 47; see Fig. 4.8). From what Galvani reported in *De viribus*, the spark that exploded exactly when the experiment was performed must not have been very strong, as it did not illuminate the room or produce an audible sound. In fact, it was not the person who was touching the frog who noticed the simultaneous discharge of the spark, but another assistant, who "thought he observed that this phenomenon occurred when a spark was discharged from the conductor of the electrical machine" (ibid.). As Galvani soon realized, only the frog with its contractions clearly manifested an action external to the discharge of a spark at a distance in this experiment. No other object, among those placed on the laboratory bench, was either moved like the frog, or showed cues of the involvement of a violent energy—or "force," in the terminology of the time—which struck it suddenly.

Subsequently, Galvani carried out specific experiments to prove that the electric influence acting at a distance in the experimental conditions of his "first observation" was very low. Indeed, he showed that the spark's discharge neither produced any evident attractive—as well as repulsive—phenomena on light bodies, nor could it be detected by "very sensible electrometers." These outcomes were in agreement with the fact that

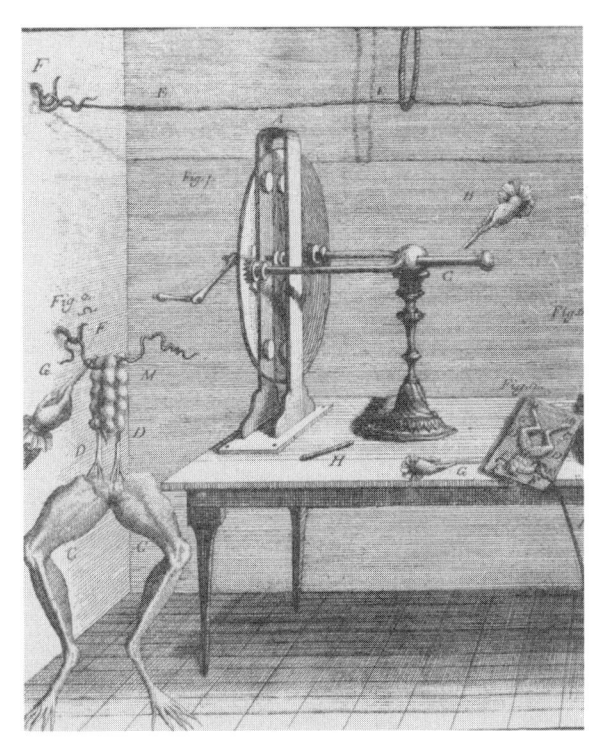

FIGURE 4.8. Detail of plate I of *De viribus* showing Galvani's experiment on the contractions at a distance (from Galvani, 1791).

previous experiments had shown that the frog was highly sensitive to electric stimulation. Five days after observing the contractions at a distance for the first time, Galvani noted some characteristics of the phenomenon that made the idea of the simple influence of an electric atmosphere hardly sustainable. On January 31, 1781, he reported in his laboratory notebook that:

> The phenomenon [of the contractions] similarly occurs at the first spark as well as after many, but it does not seem that from a single spark one can expect such a [big] atmosphere; and even so, contractions should increase in proportion to the atmosphere, i.e., in proportion to the turns of the machine's [disk], but the phenomenon is almost the same at the first spark as at the hundredth; therefore it is not the electric atmosphere of the conductor or of the disk [of the electric machine] that is the author of the phenomenon. The phenomenon occurs at the first spark even at a distance of six feet, but the electrical atmosphere of a spark cannot diffuse to such a distance; therefore the said atmosphere is not the author [of the phenomenon]. (GME, p. 256)

Although here Galvani seems to develop his reflections within the conceptual framework of "electric atmospheres," it is quite clear that even in these early observations he realized that in the case of the contractions at a distance the frog reacted to external influences of small intensity. Moreover, no relation of simple proportionality existed between the cause (the electric force, or "atmosphere" moved by the spark) and its effect (the contractions). These two aspects were largely independent of the theoretical assumptions involved in the explanation of action at a distance. If one assumed that these contractions were a simple consequence of the external action of the spark, then they could be elicited only at short distances from the machine's conductor. But Galvani noted that they were observable "even at a distance of 6 feet" (about 2 meters). It should be noted here that Stanhope explicitly claimed that the return stroke was detectable in the laboratory only if the distance between the conductor and the object was very small (no more than "20 thumbs," which corresponded to about 50 centimeters) (Stanhope, 1779, p. 80). Moreover, it was reasonable to suppose that there was a proportion between the strength of the contractions and the intensity of the electric "force," so that contractions would increase in intensity "according to the machine's turns" and decrease by continuing to elicit sparks from the conductor so as to discharge it. But, as we have seen, in Galvani's case the experiments produced the opposite results.

To understand the characteristics of the prepared frog's response to the electricity associated with the spark's discharge, which caused Galvani's wonder, a brief description of the basic aspects of the physiology of nervous fiber is required (we shall return to it in more detail in Chapters 9 and 10). A nervous signal typically consists of a rapid (less than 1 millisecond duration) variation of electric potentials transmitted along the nerve fiber. Even if the fiber has a very high longitudinal electrical resistance, the signal is transmitted without losing amplitude, as it would instead occur if transmitted along an ordinary electric cable of comparable resistance. This is an aspect of a complex physiological mechanism—based on the local stimulating action of the signal on nearby points of the fiber—which constantly regenerates the signal during transmission. To be efficient, such a mechanism must rely on a fiber's membrane, which has a high degree of sensitivity to electricity. As a matter of fact, this membrane is provided with an amplifying mechanism that can produce electrical signals of maximum magnitude in response to weak stimuli, as soon as the intensity of the stimulus exceeds a minimum value called a "threshold." It has been calculated that during this process the stimulus is amplified approximately 100,000 times. Such an amplification process is typically "nonlinear," in the sense that it is based on an "all-or-none" mechanism—in which the signal can assume only two values, namely either the maximum or zero—rather than on a simple proportion between stimulus and response. In this respect, the nerve signal is very similar to those digital signals that are now at the basis of modern telecommunication.

These characteristics of nerve conduction explain why the nerves of a prepared frog were so sensitive to an external electric influence and also why there was no proportion between stimulus and response—as Galvani observed in his experiments of January 1781. In Galvani's experiment of January 26, in particular, the strictly electric amplification involved in the generation—and propagation—of the signal in the muscle or nerve fiber was accompanied by another, "mechanical" amplification, namely the contraction of the frog's muscle brought about as a consequence of the eventual excitation of the muscle fibers. Moreover, it should be noted that exposed nerve tissue—like that in Galvani's "prepared frog"—was particularly appropriate to show the effects of external electrical influences, as a great quantity of the energy generated in the animal preparation by electrostatic induction passed preferentially through the nerve fibers instead of the surrounding tissues. Finally, the great efficacy of a spark in producing the contractions—an aspect on which Galvani repeatedly focused his attention—depends on the fact that the nerve membrane is somehow "tuned" to respond to electric stimuli which consist of rapid and impulsive variations of electric potentials, while it responds very little (or does not respond at all) to electric stimuli which remain constant or change slowly over time.

That Galvani was ready to note such surprising aspects of the electrical influence of a spark on the frog's limbs did not depend on that "happy ignorance" of physical phenomena underlined by Arago. There existed, in the eighteenth-century physiological tradition, some conceptual frameworks to which Galvani could refer and that appeared to be more suitable to explain the complexity of a living machine than any physical explanation based on a direct cause-effect relation. One of these frameworks was Haller's theory of irritability, as developed by scholars such as Felice Fontana, who had shown the absence of a simple proportionality between muscular contraction and the stimulating agent (be it mechanical, chemical, or electrical). Fontana had efficiently represented this fact by an analogy with gunpowder, in which a small spark could produce the explosion of a great mass of gunpowder (see Chapter 3.1). In the experiments performed in late January 1781,

Galvani was thus prepared to detect the lack of proportion between the low intensity of the stimulus (i.e., the discharge of a feeble spark at a distance) and the strength of muscular contractions—described in *De viribus* as "violent tonic convulsions." Therefore, he theorized that the electrical force elicited by the sudden spark acted by affecting the muscle's irritability.

In Haller's opinion, irritability was a property of muscles, so that it was reasonable to suppose that the stimulus was efficient only if directed to the muscles. However, Galvani's experiments showed that to obtain any effect it was necessary to direct the spark's action on the nerves by touching them with a conductor exactly when the electrostatic machine was discharged. On the contrary, no contractions were observed if the spark's action was directed to the muscles. In the fourth corollary of January 31, Galvani thus noted that: "How can one believe that the muscle contracts due to its irritation if the contraction occurs when [electricity] exits from the nerve and not when it enters the muscle? What about irritability?" (GME, p. 257).

Though he was referring to the theory of irritability, here Galvani pinpointed that some results of his experiments were inconsistent with the conclusions predicted by the Hallerian framework. In such a way, he showed his independence from the doctrine of the Swiss physiologist. If irritability was at play in these phenomena, it must be a property of the nerves rather than of the muscles, and this was in contrast with Haller's principles. It should be noted that at this time Galvani preferred to use the term "nervous force," instead of muscle irritability, to indicate the physiological principle at the basis of contractile responses to both internal and external stimuli.

In Galvani's view, the spark experiment clearly highlighted that the animal response to an instantaneous electrical stimulus was the expression of an internal organization and force rather than a simple effect of the strength of the external electrical stimulus. In such a way, the notion of a fluid intrinsic to the animal could be reintroduced, as in the first corollary of the experiment of January 31, where Galvani spoke of "a very subtle fluid that exists in nerves" as the "true author of such a phenomenon [i.e., the muscle contraction]" (GME, p. 257). The spark set this intrinsic fluid in motion by a sudden and instantaneous action, whose characteristics were analogous to those of the discharge of the electrical machine. In this corollary, Galvani used terms like "hit," "vibration," or "impulse" to describe the aforementioned action, and these terms were also often used in the description of subsequent experiments. Furthermore, they somehow constituted the leitmotif of an unpublished memoir that Galvani started to write in November 1782. It is thus since the early stage of his electrophysiological research that Galvani became aware of one of the fundamental aspects of the mechanism of production and transmission of the nerve signal, namely the instantaneous and rapid nature of the electrical stimulus.

It is interesting to note that, in the first corollary of January 31, Galvani did not argue for the electrical nature of the fluid existing in the nerves. However, in the following corollary, he advanced this hypothesis on the basis of the observation that contractions were not produced "if a glass or an old bone cylinder is applied to the nerves or spinal cord, [. . .]

while if one applies or touches [them] with any conducting body—metal, finger or whatever—the phenomenon and the contractions occur" (GME, p. 255). In the partition of hypotheses or elaborations about a single phenomenon into different corollaries, we can see Galvani's tendency to separate each logical step in the development of his ideas. This way of reasoning allowed him to keep the various research programs coming out from his work somehow independent from each other. More specifically, the idea that the contraction agent was an internal fluid was a hypothesis that could maintain its validity even when the possibility would emerge that this might be nonelectrical in nature (as it would actually happen in Galvani's subsequent investigation).

4.3 GALVANI'S *SAGGIO SULLA FORZA NERVEA* OF 1782

All the experiments performed by Galvani from late January 1781 to February 14, 1783 consisted of systematic variations of the same experimental arrangements in order to understand the action of the spark and of similar instantaneous electrical effects. In many of these experiments Galvani used conductors of different lengths that, in turn, were connected to different parts of the animal; moreover, from February 10, 1781 on, he introduced "for the sake of convenience [...the use of] a little machine formed by two flasks joined at the mouth," an instrument he used constantly from then on (see Fig. 4.9). The idea of this two-flask apparatus was developed out of an experimental procedure adopted on February 4, when he put the frog in a glass flask firmly sealed with fat. In the experimental setting adopted that day, a long metal wire connected the frog to a different

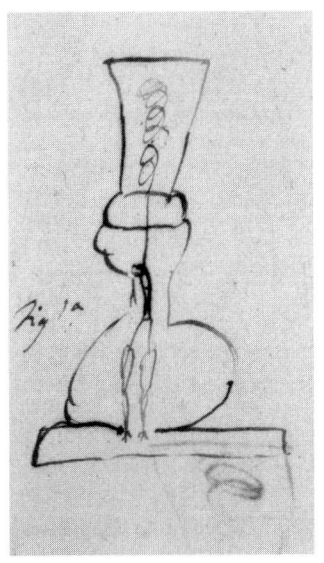

FIGURE 4.9. Illustration of apparatuses invented by Galvani for his electrophysiological experiments (autograph sketch from Galvani's notebooks, and detail from the original watercolor of plate I of *De viribus*, AASB, Fondo Galvani,©).

room, where the electric machine was located, by passing through a long glass pipe fixed in the fat. While an experimenter touched the wire attached to the frog, a spark was elicited from the machine. Evidently, this experiment was aimed at testing whether the spark's effect could be explained in terms of "material atmospheres" or electric corpuscular effluvia driven from the machine to the frog through the conductor, or in terms of mere mechanical actions related to the spark's discharge (GME, pp. 262–263).

Since the problem of the propagation of electrical effects in a vacuum was a central topic in the study of electricity in the eighteenth century, it is clear why Galvani was so interested in performing this kind of experiment (see Chapter 3.3). Indeed, both the frog and the conducting wire were hermetically sealed in their glass containers, while the far extremity of the conductor—as well as the experimenter—were located in a separate place. The observation that "beautiful and great motions" of the frog occurred at the discharge of the spark excluded any material or mechanical influence in the production of contractions. The same experimental arrangement was replicated in the experiment performed on February 7, when Galvani decided to fill the flask that contained the frog with water. A very interesting observation coming out from such an arrangement was that the animal preparation preserved the "nervous force" so long that the experiment was prolonged until February 9 and contractions were visible 44 hours after the beginning of the experiment (very likely as a consequence of the aqueous milieu used to preserve the vitality of the animal tissues).

The day after this experiment, Galvani introduced the two-flask apparatus for the first time. This piece of equipment provided him with the possibility of varying the experimental conditions in several new ways. By isolating the frog from the surrounding environment, in fact, this machine allowed him to test the possibility of a purely material influence on the spark's effect. Usually the frog was placed in the lower flask while the metal wire—typically connected to the nerves or to the spinal cord—passed through the mouth and ended in the upper flask. In the lower flask some metal pellets (which functioned as conductors of the muscles) might be inserted or it might be filled with water or some other fluids (like olive oil) to test how different materials (conducting or insulating) influenced the production of the frog's muscular contractions.

With the sole interruption of summertime, Galvani carried out his research until February 1782. He then abandoned his laboratory for many months, being occupied by academic duties (as we shall see later). He resumed his experiments in December of that year and continued to work until next February. Notwithstanding the high number of experiments performed and the information gained, Galvani was not satisfied with the results he had obtained. In fact, he was not able to derive any clear indication from these experiments either about the mechanism of the action of the spark's artificial electricity or about a possible role of an electric intrinsic fluid in muscular contractions, which was surely a more relevant problem for him. He soon realized the difficulty of explaining the contractions produced by the spark. The results obtained from the experiments performed in February and March 1781 were more likely to raise objections to the neuro-electric hypothesis than to corroborate the idea of an intrinsic electric fluid. On March 31 he formulated three corollaries that explicitly contradicted the conclusions about the existence and role of animal electricity he had reached 2 months earlier. In particular, in the third experiment performed that day he electrically stimulated a muscle isolated from the rest of the animal. As a result, he noted that "the fibre did not shorten at all, even if a spark was elicited from it." That implied that "the passage of the electric fluid trough the fibres [...] does not make them shorter, [...] even though such a fluid would be located in nerves. [...] Hence the fluid inhabitant in the nerves cannot be the electric fluid, that is, the animal spirits are not the electric fluid" (GME, pp. 286–287).

In modern terms, such a result can be explained on the basis of the different electrical excitability of the muscle and the nerve: A muscle requires longer and more intense stimulation to be activated, and this explains why a direct application of a sparking electricity to it is relatively ineffective. As a consequence of his previous research, Galvani had abandoned the idea that the action of electrical stimuli in producing muscular contractions could be explained in terms of Hallerian concepts. Now he felt that the neuro-electric hypothesis also needed to be rejected, as the idea that electricity shortened the muscular fiber was not confirmed by the experiments.

We do not know what kind of mood Galvani was in at this point. If he was, in all probability, frustrated, this did not prevent him from continuing his experiments. On April 10 he was again in his laboratory, and after less than a month he proposed the hypothesis of an "internal electric nervous fluid" once more. A new experimental campaign kept him busy during the winter of 1781/1782, and then he decided to put down on paper the problems he considered solved and those which needed further investigation. He re-examined his experimental notes and grouped them according to the problems or the topics investigated during his research, he compared his results with the physical and medical knowledge of the time, and he outlined some treatises, as well as some dissertations, about nerve and muscle physiology. Among these notes—which are kept in the archives of the Bologna Academy of Science—there is a *Saggio della forza nervea e sue relazioni colla elettricità*(*Essay on the Nervous Force and Its Relations With Electricity*), dated November 25, 1782, which Galvani left unfinished and unpublished. In this essay the Bolognese physician summarized what he had observed in the last 2 years and tried to clarify the theoretical picture that had emerged in the first stage of his electrophysiological research.

A key concept of Galvani's essay is that of "nervous force," a term he had used to define the principle responsible for muscular contractions since the beginning of his experimental investigations. If the term "force" was derived from the Newtonian tradition and was thus typical of the standard vocabulary of natural philosophy, the expression "nervous force" was introduced into the physiological literature to provide an alternative to the idea of animal spirits. Probably, Galvani was inspired by two texts that had been published some years before. The first was published in Bologna by Germano Azzoguidi, a young physician of the same age as Galvani; the second was by Leopoldo Caldani, one of the most important Italian supporters of Haller (see Chapter 3). Although Azzoguidi and Caldani remained within the framework of Haller's theory, which considered irritability as a force or property intrinsic to the muscular fibers, they were more radical than the Swiss physiologist in their criticism of the notion of animal spirits. In his *Institutiones physiologicae*—first published in 1773 and re-edited several times in the following years—Caldani examined six fundamental aspects of the theory of animal spirits and rejected each of them on the basis of their purely conjectural nature, or of some contrary evidence. For instance, the observation that nerve compression caused the related muscle to cease its motion could not be considered as proof of the existence of a nervous fluid because the same effect could be obtained in another way, namely by legating veins or arteries. Nevertheless, nobody identified the efficient cause of such motion in the "influence of venous and arterial blood." Caldani could thus conclude:

> Though by bringing these and several other reasons I do not intend to claim that animal spirits must be completely eliminated from the theories of physicians, or to mock those who fight for them, I just want to raise a question: let them forget their convictions for a moment and imagine that no one has never thought that a fluid flowing through the nerves was the governor of sensation and motion. In this case, let them judge if it would be possible to derive from the experiments just one thing: that ligatures, sections and any other damage to the nerves or to the brain destroy the integrity of these parts and above all the very condition without which sensation and muscular motion cannot be produced. (Caldani, 1778, p. 171)

For Caldani, the theory of animal spirits was therefore a mental construction neither confirmed by experiments nor necessary to account for the phenomena observed. While the nerves were still supposed to play a role in muscular motion, especially in voluntary motion, the proper nature of such nervous force remained mysterious. Indeed, as Caldani wrote, we "cannot understand—with our human mind—how and through which means the will acts on the nerves." In his *Institutiones medicae*—published in 1775—Azzoguidi further articulated the concept of *vis nervea*. Sensation and irritability were both mutations that needed "some force inherent and proper to the nerves" which activates them: "It is necessary that a force which holds and sustains the said mutation exists in the nerves; this force we called nervous." Such nervous force worked as a stimulus to activate the irritability of voluntary muscles, but its proper nature, as well as the mechanism underlying its functioning, remained "inexplicable" (Azzoguidi, 1775, vol. 1, pp. 145–146).

As noted by the Italian historian of science Walter Bernardi, the concept of nervous force elaborated by Caldani and Azzoguidi allowed a link between sensibility and irritability to be re-established after the rejection of the theory of animal spirits. Moreover, in this way nerves kept their role in determining muscular motion, even if such a role was limited to a stimulating action (Bernardi, 1992). Finally, Caldani and Azzoguidi chose not to define the proper nature of the nervous force, exactly as Newton had done with gravitational attraction and Haller with irritability. Since Galvani had failed to identify its nature as well, he focused his attention on the properties and the effects of the electrical stimulus on muscular contractions. It is therefore clear why, in this period, Galvani preferred to use a term like "nervous force" instead of terms like "animal spirits" or "nervous fluid," which were less neutral in the debate concerning "essences" and "natures."

In the first part of his 1782 essay, Galvani focused on the characteristics of the electrical stimulus needed to elicit contractions of the frog preparation. If, on the one hand, in the previously performed experiments he had obtained contractions in different ways, on the other hand he had noticed that the stimulus was particularly efficient when it was accompanied by a spark's discharge. This spark was not just that produced by the sudden discharge of a machine distant from the frog—note that the distance could be considerable, as in the experiment performed on December 16, 1781 when the frog was "30 feet away from the machine." It could also be a spark extracted from a frog placed very close to the machine's conductor, if not directly connected to it, or that produced by a machine and fortuitously directed to the frog's nerves and muscles. So, the central problem consisted of determining why the spark was so efficient if compared to all the other applications of the electrical fluid produced by an electrostatic machine. More specifically, Galvani was struck by the observation that, although a direct connection between the frog and the charged machine allowed the flow of a great quantity of electric fluid through animal tissues, it was usually inefficient if it was not accompanied by a spark's discharge or by some other manifestation of "sparking" electricity—like the "pencil," the "star," or the "fire globe"—that produced a short "effort," "hit," "impulse," or "vibration" in the flow of the electric fluid (GOS, pp. 126–132).

Galvani had already observed that a short and impulsive stimulus was much more efficient than a prolonged one during his previous experimental investigation and had reported this observation several times in his laboratory notes. After noticing this peculiarity of the electrical stimulus during the initial observations of contractions at a distance (in the experiments of late January and early February, 1781), he had focused particularly on that on February 8, when he noticed that "these motions and contractions are not always and absolutely proportional to the quantity of the electric fluid, but only when [the electric fluid] produces a particular motion, perhaps a tremor or vibration" (GME, p. 267). He then returned to the same issue during the experiments performed in April and May and, once again, in those carried out during the winter of 1781. In his 1782 essay, he noted some relevant observations aimed at explaining the difference between stimuli of different nature on the basis of the physiological characteristics of the mechanism of muscular contractions rather than in agreement with the physical parameters of the stimulus. He described an experiment in which the frog was placed on the upper surface of a Franklin square—either "armed" or "disarmed" (i.e., covered with metal coatings or not)—and received the electric fluid from the conductor of the electric machine "on its spinal cord." Presumably, the lower surface of the square was, instead, connected to the ground. In such an experimental arrangement, contractions occurred "not a few times" during the first turns of the machine's disk and then progressively ceased, even if the machine was still in function. Galvani observed that "the ceasing of the phenomenon probably depended on some alteration induced in the spinal cord by the continuous torrent of the electric fluid" and then wrote:

> Once contractions had ceased at the turning of the machine's [disk], if the machine and the frog were left at rest for some time, they re-occurred though fewer and less powerful, as if either the alteration induced by the [electric] torrent in the spinal cord faded with the rest, or the nervous muscular force rekindled itself thanks to the rest. (GOS, p. 134)

In this case, Galvani again underlined some important aspects of the mechanism involved in the action of the electric stimulus on the excitable tissues of the organism, which would be completely clarified by modern electrophysiology (see Chapters 9 and 10). He also reported some conditions in which contractions occurred when electrical stimuli were so feeble as to be incapable of producing any spark or luminous phenomenon. More precisely, he dedicated the third chapter of his essay to the examination of the "[smallest] quantity of the electric fluid applied to the nerves which is sufficient to excite contractions in the muscles." In this investigation, he explicitly referred to the experiments performed between December 1780 and January 1781, when he had obtained contractions with a "small" quantity of electricity, that is, "without spark and aura or light" or "without spark, attraction of light bodies, without aura or other electric sign." More precisely, he quoted the experiments performed with the Franklin square and the Leyden jar, in which contractions had repeatedly occurred at each discharge even though no spark or light was observed and the quantity of electricity in the instrument progressively decreased until it became extremely small (GOS, p. 140). As we have noted earlier, the great sensitivity of the prepared frog to the electrical stimulus was one of the first important results obtained by Galvani in his electrophysiological investigations.

Another relevant aspect of the contractions induced in the prepared frog by artificial electricity—already observed by Galvani in his initial experiments and confirmed in the 1782 essay—regarded the fact that a stimulus applied to the nerves was more efficient than a stimulus applied to the muscles. This observation—as Galvani had underlined during his experimental investigation—was at odds with the Hallerian theory of irritability, which located contractility in the muscles and not in the nerves. For instance, although in an experiment performed on February 3, 1781, Galvani had obtained some contractions at the spark's discharge also by touching the muscles, he added that "for these contractions [to occur] an impetus of the electric fluid coming from the nerves was needed when the muscles were touched." On May 8 of the same year, he observed that: "It has always been seen that the electrical fluid firstly given to the muscles, that is, when it has to pass through the muscles to get to the nerves and spinal cord, is less active, as it cannot avoid losing some of its force in passing through them" (GME, p. 301).

In his essay Galvani pointed out some other characteristics of the stimulus application that made it particularly efficient in eliciting contractions. For instance, he noted that an electrical stimulus, with or without a spark, was more efficient when its application was made in the direction of the nerve's axis rather than when applied perpendicularly to the same axis. This observation, which is easy to explain on the basis of modern electrophysiology, shows the extraordinary experimental ability and observational acumen that Galvani developed in his investigations. Moreover, it was interpreted by Galvani as evidence of the fact that electricity acted on the whole nerve and not only at the point where the stimulus was applied. By supporting the view of a flow of electricity inside animal tissues, this conclusion allowed him to relate more intimately the influence of the electrical agent on the internal agent directly involved in muscular contractions, thus offering a solution to one of the most complex problems faced by Galvani during his experiments. In fact, he had long been uncertain whether the "vibration," "hit," or "impulse" associated with the spark was a mere mechanical event that acted only on the surface of excitable tissues whose action was quite different from that of the internal agent of contractions; or contributed instead to the intimate mechanism involved in muscular motions. If the model developed after the observation of contractions at a distance implied that the electrical stimulus did not act directly on the frog but through air or some other fluids interposed between the animal and the electrical machine, later he changed his mind. For instance, on February 21, 1781 he reported that "the frog's conductor is what gives [electricity] to the frog and consequently the [electric] vapour of the frog's conductor enters the nerves" (GME, p. 277). Again, on May 6 he noted that "this experiment may show that it is the fluid portion of the nerve that conducts and ties the internal nervous electric principle or fluid, or that conducts also the external [electricity]" (GME, pp. 299–300).

In his 1782 essay, Galvani did not take an explicit position about the nature of the agent responsible for muscular contractions. Even though during his experimental research he had been often close to concluding that such agent—"a very subtle fluid existing in the nerves"—was electrical in nature, this hypothesis had often been contradicted by his experimental results. Moreover, he probably realized that, even after 2 years of intense research activity, he was not able to elaborate a plausible model to explain how this hypothetical nervous agent produced the contractions. He thus preferred to define this agent as a "nervous force"—as reported in the title of the essay—or as a "muscular nervous force" and limited his analyses to the description of its phenomenological aspects, like the way it was excited by external electric stimuli or its characteristic response to them. In this regard, the action of the external agent consisted of "exciting" this internal force, which was "the real author of the contractions."

In the final chapter of his essay, Galvani focused on the lack of proportionality between stimulus and response, further developing Fontana's arguments and getting close to the idea of "threshold" and to the concept of "all-or-none" mechanism, which are fundamental aspects of the process of generation of the nerve signal as revealed by modern electrophysiology. The same consideration was suggested by those experiments on conductors connected to the nerves that Galvani had carried out in the initial phase of his research. For instance, on April 10, 1781 he reported that "the nerve cannot receive more than a certain amount [of electric fluid], so that, though the conductors of the electric fluid to the nerve increase, its amount does not and so do the contractions" (GME, p. 289).

In *De viribus* Galvani preferred to emphasize the importance of the experiments on metal arcs, from which he inferred the explanatory model of the animal Leyden jar, rather than the description of the phenomenological aspects of the muscular response to the external electrical stimulus. However, in the first part of his work—devoted to the effects of "artificial" electricity—he analyzed the stimulus-response relation by taking into account the experimental circumstances that had characterized his initial research about 10 years earlier. More precisely, he focused on the stimulus intensity, on the length and disposition of metal conductors applied to the animal, and on the distance between the animal preparation and the electrical source. With a clear reference to the 1782 essay, Galvani exposed the laws that defined the complexity of such phenomenon in the following way:

> Muscular contractions of this sort, then, seemed to us within certain limits to stand in a direct proportion [*ratio*] to the strength of the spark and of the animal as well as to the length of the conductors, particularly the nerve-conductors. They seemed to stand in an inverse proportion, however, to the distances from the conductor of the electrical machine [...]
>
> I have said, to be sure, that a direct proportion seemed to me to be maintained in the contractions, but only within certain limits. It has been found, for example, that if, after fixing a certain length of a nerve-conductor which is sufficient to bring about contractions, you shorten it, then the contractions do not diminish, but disappear. If, however, you lengthen it, the contractions in fact grow stronger, but only up to a certain point, beyond which—however far you extend the nerve-conductor—the contractions are augmented imperceptibly, if at all. The same can be said about the other elements of the proportion I have set forth. (GDV, pp. 368–369; translated in Galvani, 1953a, p. 51, with revision)

To appreciate fully the value of Galvani's statements, we should compare them with modern physiological notions developed between nineteenth and twentieth centuries, thanks to the research initiated by Henry Bowditch and Francis Gotch, and culminated in the discoveries of Alan Hodgkin (see Chapter 9). A single nerve (or muscle) fiber produces a maximum response when the stimulus exceeds a minimum value, called a "threshold," while it does not produce any response to stimuli with an "under-threshold" intensity. This is the so-called all-or-none law. However, since different fibers (nervous or muscular) have different thresholds, there is a relative proportion—a *ratio* in Galvanian terms—between the intensity of the stimulus and the physiological response, but this happens "within certain limits," that is, within the interval between the thresholds of the most excitable fibers and the thresholds of the least excitable ones. If the intensity of the stimulus is increased above the thresholds of the least excitable fibers, the contractions will remain constant, that is, they "do not augment." On the other hand, if the intensity of the stimulus decreases below the threshold of the most excitable fibers, "contractions do not diminish, but disappear altogether," as Galvani explicitly stated.

In concluding the reconstruction of the first phase of Galvani's research we are now in a better position to understand that the phenomenon of contractions at a distance—observed by Galvani at the beginning of 1781—implied much more than a simple action at a distance explicable in terms of the influence of nonmaterial electrical atmospheres or, more specifically, in terms of the theory of the return stroke, as Volta and many others have claimed after him.

5
A "Fortunate" Discovery
GALVANI'S THEORY OF ANIMAL ELECTRICITY

THE *SAGGIO DELLA forza nervea*, dated November, 1782, embodied Galvani's highest achievement in the first phase of his research, which was devoted to the analysis of the role played by artificial electricity on muscular contractions. Although he continued his experimental activities until 1783, Galvani did not make significant progress until 1786 when he started his experiments concerning "atmospheric" electricity. In fact, from the analysis of the available documents, it transpires that Galvani dedicated this period to other commitments as well as to other research projects, which were at least partially related to his interests in electrophysiology. They would also play a great role in providing a new impetus for Galvani in his research on the relationship between electricity and muscle motion. One of the reasons that forced Galvani to abandon temporarily his "room for experiments" was that in February 1782 he was elected professor of obstetrics at the Institute of Sciences of Bologna, in succession to Giovanni Antonio Galli, one of his former teachers. It was a demanding task, since it implied the preparation of lessons that were addressed not only to students of the faculty of medicine but also—and maybe most of all—to midwives. Galvani's notes are still available and show his familiarity with this field of knowledge (Galvani, 1965, 2009).

However, this activity was not the sole element that diverted Galvani from electrophysiological experiments. In fact, he had become aware of the difficulties in identifying the agent of the "nervous force" in an electrical fluid, so he turned his attention to other fluids that were considered interesting by contemporary natural philosophers. In particular, as we can see in a series of unpublished manuscripts kept in the archives of the

Academy of Sciences of Bologna, he focused on the study of the chemical and physical aspects of the organism. This focus was driven by his interest in the physiological mechanisms of the human body and by the research projects on pneumatics that were drawing increasing attention in the scientific community of the time. Not only were these research fields interesting in and of themselves but they also suggested new directions of research in the field of electrophysiology. In 1786, he returned to his electrophysiological experiments and began to study the effects of "atmospheric" electricity on muscular motion. By working in this direction, he discovered the function of metal arcs in producing contractions in the frog's muscles. This discovery represented a turning point in Galvani's scientific investigative path and led him to develop the model of the muscle as an animal Leyden jar. Between 1786 and 1791, Galvani elaborated several versions of his theory of nerve conduction and muscular contraction, which varied with the progress of the experiments performed during this period as well as on reflections concerning both electric fish and tourmaline. The final version of his theory of animal electricity was formulated in 1791 and published for the first time in *De viribus electricitatis in motu musculari*, a major contribution in the history of science.

5.1 THE STUDY OF "AIRS" IN RELATION TO THE LIVING ORGANISM

Since the 1770s the study of the so-called airs—the term used at the time for gases—had become a highly debated topic in the scientific community (Abbri, 1984; Crosland, 1963; Golinski, 1992). The research had been started at the beginning of the eighteenth century by Stephen Hales, who was the first to measure arterial pressure in animals and to whom we owe one of the first formulations of the neuroelectric hypothesis (see Chapter 3.2). After this initial contribution from Hales, who demonstrated the role of air in the composition of bodies, the study of airs received a decisive impulse from the discovery of various airs provided with different and peculiar characteristics. For example, in just a few years, between 1760 and 1770, it became possible to isolate "inflammable air" (hydrogen), "phlogisticated air" (nitrogen), "fixed air" (carbon dioxide), "swamp air" (methane) and "dephlogisticated air" (oxygen). In 1775, Joseph Priestley, one of the major discoverers of new gases, wrote:

> I have seen abundant reason, since the publication of my former volume of *Observations on different kinds of air*, to applaud myself for the little delay I made in putting it to the press; the consequence having been that, instead of the experiments being prosecuted by myself only, or a few others, the subject has now gained almost universal attention among philosophers, in every part of Europe. [...] and this branch of science, of which nothing, in a manner, was known till very lately indeed, now bids fair to be farther advanced than any other in the whole compass of natural philosophy. (Priestley, 1774–1777, vol. 2, p. V)

Priestley's "prophecy" would come true quite soon. Just a few years later, in fact, Antoine Laurent Lavoisier would elaborate a totally new explanation of combustion and of respiration, based on different principles from the English natural philosopher's, thus initiating the chemical revolution. Galvani was well informed about this new "branch of science," as Priestley had named it: Besides being in close contact with Giuseppe Veratti, who was the leading expert in this field in Bologna, in his library he had the works of authors like Hales, Torbern Bergman, and Carl Wilhelm Scheele, as well as the *Opuscoli scelti sulle scienze e sulle arti* (published until 1778 under the title of *Scelta di opuscoli interessanti*), one of the main journals of the time, where many recent works on airs were published (as well as excerpts of articles that first appeared in other scientific publications in Italy and elsewhere). Besides, Galvani knew and frequently cited the research performed by Priestley, by Volta and by the other natural philosophers involved in this field. As in the case of electricity, Galvani's interest in airs was fed by a great familiarity with the body of knowledge and experimental practice of the time. But, most of all, it developed in a particularly fertile ground of the Bologna scientific community, where chemical research had always been connected with the study of the "animal economy" and of medicine. We are referring, for example, to the research that led Jacopo Beccari, one of Galvani's masters, to identify gluten—considered up to then an animal substance—in wheat, or to Veratti's studies on "airs contained in the bodies [...] of animals" and on blood (Angelini, 1993; Beach, 1961).

On May 2, 1783 Galvani was at the Academy of Sciences of Bologna to read a dissertation that recalled Veratti's research in its title: *Su de' principi volatili cavati insieme coll'aria fissa da varie parti solide, e fluide di vari animali* (*On Volatile Principles Extracted With Fixed Air From Various Solid and Fluid Parts of Several Animals*). On November 27 he read a second dissertation that probably was the continuation of the one given in May, titled *Sopra l'aria infiammabile delle parti animali* (*On the Inflammable Air of Animal Parts*). Neither was published nor are they to be found in the Academy Archive. However, among Galvani's manuscripts there is an incomplete version of a dissertation that may be connected with this specific research, where he declared:

> In the meantime we diligently examined the above mentioned principles [i.e., the various airs] with the aim of finding their quality and quantity in the various fluid and solid parts [of animals], an examination—as far as we know—that nobody has done before. (GMS, cart. I, plico I, fasc. a)

In this passage, not only did Galvani focus the target of his research on physiological chemistry, but he also highlighted its originality. While this research was profoundly inspired by Veratti, it was not a simple continuation of the work of his elder colleague. Before the members of the Academy of Sciences, among whom Veratti was probably present, Galvani added:

> We have decided to communicate to you this new branch of physiology, though in a rough way, as not to be anticipated by anyone. The delay in doing so might

have led you to know it through the efforts of some foreigner, instead of those of a colleague of yours. (GMS, cart. I, plico I, fasc. a)

Galvani was fully aware of the originality of his work as well as of the fact that the same theoretical topic also interested other researchers. In this sense, Galvani was in line with a cutting-edge field of research that had been inaugurated by Lavoisier and by his theory of respiration. Although these researches of Galvani, performed between 1783 and 1786, are very interesting—both from a scientific and a methodological point of view—they cannot be described in full detail here (see Seligardi, 1997, 2002; Holmes, 1999). Accordingly, we will focus on some aspects that are closely related to his electrophysiological research. Similar to the way he had previously applied the existing notions concerning electricity to the study of muscular motion, now Galvani applied the instruments, the methods and the concepts of pneumatics to the study of physiology. His aim was to study the gases contained in animal tissues and organs in order to understand their role in body composition and functions. In particular, he was looking for a fluid with abundant energy but different from electricity, which could account for the forces that were at the basis of the fundamental physiological processes. For this reason, he was particularly attracted by "inflammable air" (hydrogen), which was discovered around 1766 by Henry Cavendish and so called because it flared up and exploded when it was mixed with ordinary air and ignited.

Galvani found that inflammable air was present in all animal parts—solid and fluid—and that it was particularly abundant in nerves. This observation, together with the "great speed, subtlety, and pureness" of this gas when compared to other gases, led him to suppose that this "inflammable principle" was "wholly or partly that fluid in the nerves called nervous fluid, or animal spirits, [which is] the author of sense and motion" (GMS, cart. I, plico I, fasc. a). The nervous fluid, in fact, had to be very thin in order to flow through nerves, and very fast in order to account for the quickness of sensations and of movements. Furthermore, muscular contraction could be described as a sort of little explosion, analogous to what happened when a lit candle was placed close to a container full of inflammable air. In his chemical-physical research, then, Galvani was not forgetting the main objective that had kept him busy in the previous years; indeed, he was trying to understand the muscular mechanism by approaching it in ways complementary to the electrophysiological framework.

The experiments performed on the "inflammable animal air" led Galvani to suppose that such gas could somehow be absorbed by foods and by digestion and could be transported throughout the organism in order to exert its functions, among which was the production of muscular contractions. However, the source or the means that caused the ignition of the inflammable principle was still to be identified. On the basis of Volta's investigations, which had demonstrated that an electric spark could flare up and produce an explosion in inflammable air, Galvani conjectured that an analogous role was played by electricity, in particular, by the electricity present in the atmosphere and acting on the body (GMS, cart. I, plico I, fasc. a). It is interesting to note that shortly after making these observations, Galvani focused again on the effects of electricity on muscular motion, starting a new series of experiments dealing with the effects of "atmospheric" electricity.

The chemical-physical research on "animal airs" had put in place a new principle—inflammable gas—which could be identified with nervous fluid and was closely related to electric fluid. The presence of this principle, especially in nerves, will be used by Galvani in the *De viribus* to solve the well-known Hallerian objection regarding the electric properties of nerves. In his memoir Galvani reported the objection in the following terms:

> For either the nerves are of an idioelectric [i.e., non-conductive] nature, as many affirm, and for this reason they could not function as conductors, or else they are conductive; and how can it then happen that they retain the animal electric fluid within them so that it is not diffused and spread to adjacent parts, with, undoubtedly, a great diminution of muscular contractions? (GDV, pp. 398–399; translated with revision from Galvani, 1953a, p. 76)

The presence of inflammable gas was evidence of a substance of an oily nature, and thus electrically insulating, which entered in the composition of nerves. Galvani underlined the fact that he had been able to extract from nerves a great quantity of such gas, that when lit "showed a capacity for emitting a brighter, clearer, and most lasting flame than the inflammable air drawn from other parts is wont to produce." And then he added:

> On the other hand this idioelectric substance of the nerves, which seems to have the function of preventing the electric nerve fluid from being dispersed with great consequent damage, will not prevent this fluid, coursing through the inner conducting substance of the nerves, from leaving these same nerves, when needed, in order to produce contractions. (GDV, p. 399; translated with revision from Galvani, 1953a, p. 76)

The research on gases allowed Galvani to solve one of the main problems in the investigation of the role of electricity in animal motion and a major objection to the neuroelectric theory. Besides, his view that nerves are constituted by a conducting internal matter surrounded by an insulating sheath represents an important novelty from the previous tradition, based on the 1,000-year-old notion of nerves as hollow tubes in which animal spirits flowed (see Chapter 3). A similar view would play a central role in the explanation of the passive aspects of electrical conduction in the nerve fiber, as developed especially by Lord Kelvin, who in 1885 formulated the so-called cable equation to describe the electrical transmission along a cable surrounded by an insulating sheath.

Actually, in the course of his experiments, Galvani had obtained contrasting results on the conducting or insulating nature of nerves. With his model of the conducting nucleus surrounded by an insulating sheath which he had developed after the discovery of inflammable gas in nerves, Galvani could not only respond to the Hallerians' objections, but he could also account for the apparent incongruence of his experimental results. Moreover, he could overcome another important objection to the neuro-electric hypothesis, regarding the effects of nerve ligature. According to Galvani, the ligature stopped the transmission of the intrinsic electrical fluid responsible for the nervous signal, but not the passage of the external electricity applied to the nerve, because this propagated through the wet tissues surrounding the nerve. As the intrinsic electrical fluid flowed exclusively inside the nervous fibers, however, the ligature slowed down or stopped this flow by increasing the contacts between the insulating parts of the fiber.

All the different research paths that Galvani was following in the first half of the 1780s found an explicit point of intersection in a series of studies in which electricity, airs, and animal economy were treated together. On April 6, 1786 Galvani read a dissertation at the Academy of Sciences of Bologna entitled *Dell'accordo e delle differenze tra la respirazione, la fiamma ed il fiocco elettrico uscente dal conduttore acuminato di una bottiglia di Leida* (*The Agreement and the Differences Between Respiration, Inflammation, and the Electric Flake Issuing From the Point of the Conductor of a Charged Leyden Jar*). In the dissertation, as can be seen from the title, Galvani aimed at establishing some analogies between the phenomenon of respiration—one of the most debated topics of the time—the properties of the flame and the electrical signs manifested by the Leyden jar. In particular, Galvani conceived respiration as a process that "contaminates the air and renders it inappropriate for maintaining it [the respiration] and for life's continuation," as it "corrupts [the air] with acid, phlogiston and above all with watery vapours" (GOS, p. 154). Although this conception would become outdated shortly afterward with the affirmation of Lavoisier's theory, it cannot be considered it as old-fashioned with respect to the debate of the time, as it referred to Priestley's ideas and to the debate enlivened in Italy by outstanding scientists like Felice Fontana, Alessandro Volta, and Marsilio Landriani (Holmes, 1999).

In his experiments, Galvani had observed that the "electric flake [*fiocco*]" visible at the point of the conductor of a charged Leyden jar behaved, in respect to atmospheric air, similarly to fire and to respiration. For example, when a charged Leyden jar was placed "under a glass recipient [...], the flake [...] grows weaker and weaker until any light extinguishes itself," in the same manner as what happened to a lit candle or to an animal closed in a sealed container, which would breathe more and more deeply until it died. Also the effects of several "mephitic" airs, like "phlogistic" air, were identical with respect to fire, respiration, and the electric flake, while some differences that could be detected in these phenomena regarded, according to Galvani, their properties and "means," but not their "nature." In the final part of his analysis, Galvani put forward a very important "suspicion":

> Once the ways of action of the causes and of the manifestation of the effects are known, and once the similarity of the electric and fire flames with respiration is established, it is easy to suspect that one identical principle common to the singular phenomena is the electric fluid. (GOS, p. 160)

Such suspicion was strengthened by some "findings of the most recent authors" on the analogies between combustion, respiration, and electricity. Here Galvani explicitly referred to the research carried out in Paris by Volta, Lavoisier, and Laplace a few years before, when Volta had obtained "undoubted signs of electricity [...] from the evaporation of water, from the simple combustions of coal, and from various effervescences, such as those produced by the inflammable air, the fixed air, and the nitrous air" (VEN, III, pp. 285, 296 ff.). Volta had described this research in his memoir on the *elettroscopio condensatore*, published in 1782 in the *Philosophical Transactions* of the Royal Society and 2 years later in the *Opuscoli scelti sulle scienze e sulle arti*.

As had happened for inflammable air, also in this case an influence of Volta on Galvani's research can be detected. In the following chapters we will come repeatedly on the close relationship between these two great Italian scholars and we will underline the continuous exchange and the influence they exerted on each other's work. Contrary to some ways in which the controversy on animal electricity is depicted, often as the story of a contrast between two irreconcilable positions, we shall see that while it was indeed a very heated debate, it was also based on a fruitful interaction between the two antagonists.

The dissertation of April 1786 was a crucial point in Galvani's investigative path. It clearly expressed a unitary view of natural phenomena, which, as we have already pointed out, was an essential element of Galvani's experimental approach, as well as of that of several other contemporary naturalists. In this perspective, the electric fluid regained a central role as the element that linked physical phenomena and the vital functions of the organism. Additionally, the attention that Galvani devoted to the Leyden jar in this research is extremely significant as it certainly led him to concentrate on this instrument again and on its peculiarities, so much so that it became a fundamental point of reference for his physical model of neuromuscular physiology. Finally, the reference to the need to investigate the "forces of the atmospheric electricity" in the final part of his dissertation is a clear indication of the research program he intended to carry out. Soon afterward, in fact, Galvani would resume his work on electrophysiology, to study the effects of atmospheric electricity on muscular motion.

A further contribution to Galvani's new stage of research on electrical phenomena might have come from the publication between 1783 and 1784 of two studies by Lazzaro Spallanzani on torpedoes—taking the form of letters addressed to Girolamo Lucchesini and Charles Bonnet—in the *Opuscoli scelti sulle scienze e sulle arti*, a journal present in Galvani's personal library (see Fig. 5.1). Apart from the relevance of the results they described, Spallanzani's papers are important because they revived the debate on these animals and also because they reported the results of John Walsh's research in great detail. In particular, Spallanzani wrote that in the torpedo Walsh had found "the chest and the back of this animal are in two different electrical conditions, in so far as if the back is in a positive state, the chest is in a negative one, similarly to what is to be found

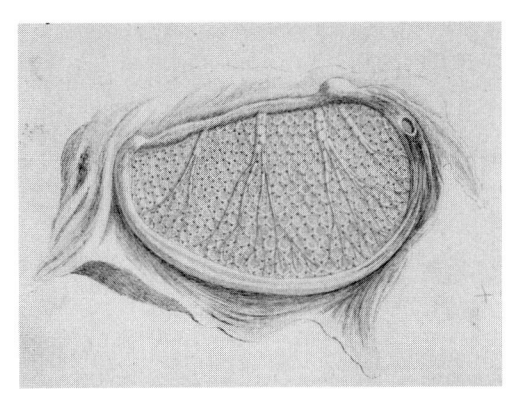

FIGURE 5.1. The surface of the electric organ of a torpedo with its nervous structure in an unpublished sketch made by Lazzaro Spallanzani (Reggio Emilia, Biblioteca Panizzi, Mss. Regg. B 96,°).

in the Leyden jar" (Spallanzani, 1783, p. 98).[1] As evidence of the importance attributed by Galvani to Spallanzani's studies, we can report the content of the tenth note added in the second edition of the *De viribus*, published in Modena in 1792. Here Galvani discussed some experiments where muscular contractions were produced by bringing a metal body close to a frog immersed in water, and referring to the studies on electric fish, he claimed:

> The observed movements of the frog recall the idea of the torpedo, which sometime produces such a great quantity of electricity that the fearful fisherman, hit in his arm with great pain, regrets having caught it; however, this is not a valid reason for diminishing the glory of the discovery of electricity [of this kind of fish], which was obtained by Spallanzani and Walsh after so much effort. (GME, p. 206)

It is then clear that Galvani considered Spallanzani on a par with Walsh as an author of the "discovery of the electricity" of torpedoes. It is worth noting that when he resumed his electrophysiological research, Galvani considered the electric fish, besides the Leyden jar, a fundamental point of reference for his interpretation of the mechanism at the basis of nerve conduction and muscular contraction.

---

[1] Spallanzani's statement that the back of the torpedo is positive during the shock while the underside is negative was hypothetical but turned out just by chance to be right, since the Italian naturalist had not succeeded in measuring the polarity of the fish shock with a physical instrument. After an incorrect result obtained on the island of Malta in 1829 by John Davy (the brother of the more famous Humphry), the question was correctly solved in 1835 by the French scientists Gilbert Breschet and Antoine Caesar Becquerel, who studied torpedoes caught near Venice. Before the two Frenchmen, the Italian physicist and engineer Leopoldo Nobili had correctly established the polarity of the fish shock by working on a torpedo caught in Leghorn. Nobili, however, never published his results (on the subject see Finger & Piccolino, 2011, chapter 24).

5.2 THE EFFECTS OF ATMOSPHERIC ELECTRICITY ON MUSCULAR MOTION AND THE DISCOVERY OF METAL ARCS

On April 26, 1786, only 20 days after presenting his dissertation on the analogy between respiration, the flame, and the "electric flake" at the Academy of Sciences, Galvani recorded in his laboratory journal a series of experiments "on the force of atmospheric electricity in the nerves of cold[-blood] animals." As we have seen, from his chemical-physical research, the hypothesis had emerged that the electricity diffused in the atmosphere could penetrate into the organism and act on the fluids present in it, in order to promote its vital functions, including muscular motion. This hypothesis was in line with the general idea of traditional medicine, according to which organic functions, and human health itself, largely depended on the relationship between the single organism and the environment surrounding it. In its "electrological" version, this hypothesis had already been expounded by authors like Francesco Giuseppe Gardini and Pierre Bertholon. Bertholon, in particular, thought that "this [electric] fluid, so active and penetrating, cannot exist in the atmosphere surrounding us, and in which we are immersed like fish in the water, without acting on us and therefore exerting a specific influence on our bodies" (Bertholon, 1786, vol. I, p. xiv; see Chapter 6.1).

However, there was another important reason that led Galvani to focus his attention on the effects of atmospheric electricity on muscular motion, which was more closely related to his previous electrophysiological research. He intended to verify, in fact, whether the effects and the properties of atmospheric electricity in eliciting muscular contractions were similar or different to those he had found in the first stage of his experimentation with artificial electricity. Once again, this objective was in line with the more advanced research that other scientists were carrying out. As we have noted in Chapter 3, several years before, Veratti had studied the effects of lightning on animal nerves and muscles, by using a discharge produced by the instruments available at that time. In 1782, Volta entitled his memoir on the *condensatore* "*Del modo di render sensibilissima la più debole elettricità sia naturale, sia artificiale*" (*On the Way in Which the Most Feeble Electricity, Both Natural and Artificial, Can Be Detected*) opening it with these words:

> Anyone can understand what advantage to the research on electricity, and especially on the natural electricity of the atmosphere, can come from an apparatus which, by an extraordinary increase of electrical signs, makes observable and detectable that very feeble virtue which otherwise would escape our senses. (VEN, III, p. 271)

Galvani had already identified such an apparatus in the frog, whose contractions, as we have seen, were considered the "least sign" of the presence of electricity and of its action. Now the problem was to check, as Volta had done with the *condensatore*, whether the frog's legs reacted to atmospheric electricity in the same way as they had reacted to the stimuli of artificial electricity. To a modern reader Galvani's research program might seem

A "Fortunate" Discovery

useless or even meaningless, since now we know electricity is always the same whether it is produced artificially, through some sort of instrument, or it occurs "naturally." However, at that time the identity between the different manifestations of electricity was not as obvious as it is nowadays (see Chapter 3.3). Although Benjamin Franklin had demonstrated the electrical nature of the discharge produced by lightning some decades before, there was still an ample debate on the phenomena occurring in the atmosphere. More generally, the various forms of electricity—artificial, atmospheric, animal, or associated with specific bodies like the tourmaline—had not been completely "unified" (a complete "unification" would not take place until the nineteenth century, in particular thanks to Faraday's studies: see Faraday, 1838a, 1838b). The examination of their properties and their effects was still considered as a necessary operation, and also Galvani would compare animal electricity he had discovered with the other forms of electricity known at the time.

On April 26, 1786, at sunset, on the terrace of his house and under a sky laden with "black and white clouds coming from midday," Galvani placed a prepared frog in the glass flask he used in previous experiments, with a long conducting wire that ended in a well (see Fig. 5.2). In his laboratory journal Galvani reported that "at each of four peels of thunder notable contractions of all the muscles of the limbs occurred, and therefore notable tremors and motions of the same [limbs]. [These motions] occurred at the same time as the lightning; thus well before we could hear the thunder" (GME, p. 387). In the following experiments performed that day, as well as in those performed on June 7, Galvani modified the experimental disposition of both the frog and the conductors applied to it, observing that every time lightning manifested itself, some muscular contractions were

FIGURE 5.2. Plate II of *De viribus* (from Galvani, 1791).

produced. In a "warning" concerning "Exp. 11" performed on June 7, he noted the following:

> Warning 1. From all this it follows that the same happens with natural electricity as with artificial electricity, but with more force; contractions excited by lightning have to be compared with the contractions produced by the spark. (GME, p. 389)

In the case of contractions at a distance, every electrostatic spark corresponded to a contraction in the frog's muscles, likewise contractions occurred whenever there was electric lightning in the sky. The only difference lay in the intensity of the electrical stimulus provided by the discharge and, consequently, in the entity of the contractions which, in the case of lightning, were particularly lively even without the devices needed to boost the effects of artificial electricity—as Galvani would note in the second part of *De viribus*. Besides, the action of lightning could be so violent as to leave "all [the frog's] limbs rigid, numb, and stretched out, even 24 hours after the experiment" (GME, p. 387).

Regarding the power of electricity produced by lightning to elicit muscular contractions in the animal, which was demonstrated experimentally by Galvani, it should be noted that this idea (which would be later taken up by other authors) would be the basis of a vast literature. This was so especially in the nineteenth century, when possibility of using electricity or other external agents to bring dead people back to life or to give life to monsters was entertained. In particular, the myth of Frankenstein—created by Mary Shelley following the suggestion of the macabre attempts by Galvani's nephew, Giovanni Aldini, to stimulate the corpses of executed men electrically—still remains a part of our imagination, mainly thanks to its many film adaptations. Some of the famous scenes where Frankenstein tried to bring the monstrous creature to life by using the electricity of lightning recall the fine plate in *De viribus* illustrating the experiments on "stormy" electricity (see Fig. 5.2).

In the second part of *De viribus*, entitled *Le forze dell'elettricità atmosferica nel moto muscolare* (*On the Forces of Atmospheric Electricity in Muscular Motion*), Galvani reported some experiments on the effects of lightning that were analogous to those described in his laboratory journal: "We discovered"—Galvani wrote at the end of this part—"after examining the laws of the observed phenomena as well as investigating the muscular contractions, that the former were maintained in an similar way in experiments involving both atmospheric and artificial electricity" (GDV, p. 376; translated with revision from Galvani, 1953a, p. 58). This conclusion was consistent with the idea, formulated by Franklin and adopted by several contemporary scholars in the field of electricity, that a single electric fluid acted both in the atmosphere and in all bodies. Such a conclusion surely represented a relevant step forward for Galvani's investigation, which also shed new light on the electrophysiological experiments he had previously performed. If the effects of artificial electricity on muscular motion were the same as those produced by the action of atmospheric electricity, then it was reasonable to conclude that the experiments performed in the laboratory reproduced what happened in nature. This was a very important conclusion, which could, among other things, justify all the efforts devoted to this investigation and which urged Galvani to carry on his research work despite the difficulties he had encountered on several occasions.

As it happened, Galvani continued his experimental activity in August and September 1786, studying the effects of atmospheric electricity both with a stormy sky and lightning, and with a relatively clear sky. For example, on August 15 he carried out several experiments under a sky covered by "whitish clouds moving from west to east, without thunder and lightning," while on September 16 there were "extended blackish and whitish clouds moving from west to east, with thunder and lightning" (GME, pp. 393–394). Also the experiment described by Galvani at the beginning of the third part of his *De viribus*—the experiment that involved the discovery of the effects of metal arcs (and was thus crucial for the elaboration of the concept of animal electricity) —was performed in not very clear atmospheric conditions. However, in *De viribus* Galvani tended to picture his experimental course as linear and kept separate the studies conducted on the effects of lightning electricity and on the effects of atmospheric electricity present in a clear sky, indicating that they were two different and succeeding moments along a linear development. Indeed, he wrote: "After we had assessed the effects of atmospheric electricity during a storm, we were inflamed by the desire to investigate the strength of diurnal quiescent electricity" (GDV, p. 377). The experiments concerning atmospheric electricity in stormy weather were described in the second part of the treatise, as a first example of the study of natural electricity of the atmosphere after his research on artificial electricity. Also in this case, Galvani presented these studies in a logical, as well as temporal, continuity with the previously performed experiments on artificial electricity. He thus wrote:

> Having discovered what already we have set forth on the forces of artificial electricity on muscular contractions, we wanted nothing better than to investigate whether so-called atmospheric electricity produces the same phenomena or not. (GDV, p. 374; translated with revision in Galvani, 1953a, p. 111)

It has been noted that Galvani, far from proceeding in an immediate and consequential way from the experiments with artificial electricity to those with atmospheric electricity, interrupted his electrophysiological research for some years. During this period, he worked on other topics that would indeed be fundamental for his taking up again and furthering that particular research path. These examples show how scientific research—by Galvani as well by any other scientists—is very often more articulated and complex than it might seem from the published accounts of scientific works. And while this can make the task of a historian of science more difficult, it certainly makes it more fascinating.

But let us go back to the experiments on atmospheric electricity carried out by Galvani, with his nephew Camillo, in September 1786. They placed some frogs "on the railing" which made up the parapet of the terrace or the light well in Galvani's house and noticed that when they touched the frog's nerves with a pair of scissors, they obtained "several contractions, both when lightning with thunder occurred, and when neither lightning nor thunder manifested themselves" (GME, p. 394). This and similar observations, where contractions occurred in the absence of lightning, attracted Galvani's attention. One of them was described in a manuscript dated October 30, 1786 (see Fig. 5.3):

> In early September, at twilight, we placed therefore the frogs prepared in the usual manner horizontally over the railing. Their spinal cords were pierced by iron hooks, which were used to suspend them. The hooks touched the iron bar. And, lo and behold, the frogs began to display spontaneous, irregular, and frequent movements. If the hook was pressed against the iron surface with a finger, the frogs, if at rest, became excited—as often as the hook was pressed in the manner described. (GOS, p. 163; translated with revision from Pera, 1986/1992, p. 81)

Galvani was impressed by two facts. First, he noticed that muscular contractions occurred even in the case of "mild" electricity, namely when the electricity was diffused in the atmosphere without any apparent manifestation. At this stage of his research, Galvani was convinced that, in the absence of a connection between the frog and an electric source, contractions were provoked by disequilibrium of electric fluid that, by producing a nerve "impulse" or "hit," activated the fluid contained in nerves and, consequently, muscles. Such disequilibrium was evident in the case of a spark or of lightning, but it seemed to be missing in the experiments he was dealing with. The second fact that impressed Galvani consisted of the circumstances in which contractions occurred, that is to say, by connecting the metal hook of the frog with the iron railing on which it was placed.

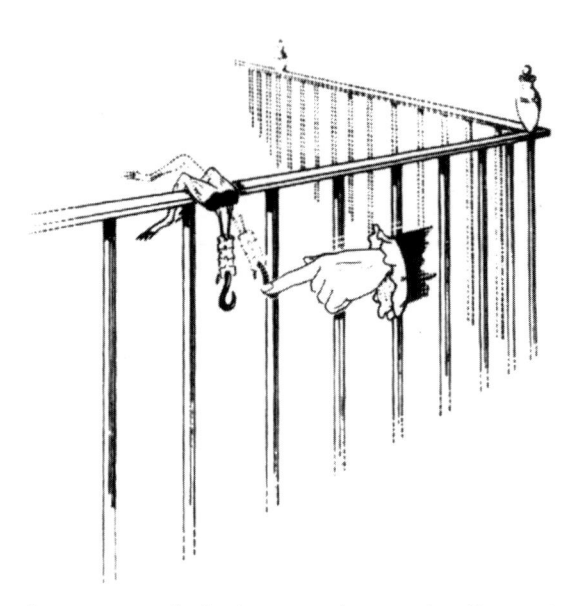

FIGURE 5.3. Galvani's experiment on the frog hanging on the iron railing (from Sirol, 1939).

To explain these observations, initially Galvani supposed that a certain amount of atmospheric electricity had concentrated in the frog and that the contact between the hook and the railing, or between the scissors and the nerves, had elicited the necessary electric discharge to activate the nervous fluid. He interpreted an effect of this kind on the basis of a material notion of electrical atmospheres and, to test his interpretation, he designed an experiment that we may classify as a negative control experiment. More precisely, he decided to repeat the very same experiment in a closed room, in order to exclude any possible influence of atmospheric electricity: If the contractions did not occur, one could conclude that his explanatory hypothesis was correct. However, "against all expectations," the contractions also occurred in the closed laboratory and using different metals (GOS, pp. 163–164).

Once again, as had happened in the case of the contractions at a distance observed in January 1781, an unexpected result opened a new phase of Galvani's electrophysiological research. Contrary to what had happened then, this time the result was much more satisfactory. It is worth noting, incidentally, that even in this case, as occurred with the spark experiment, if one might speak of the intervention of chance in the course of the investigation, it was certainly one of those circumstances when chance favors—as Pasteur noted—a prepared mind. For a far more literal and banal meaning of the word "chance," we may refer instead to the strange accounts reported in two recent *Dictionary of Scientists*—as we saw in the first chapter of this book—according to which the frogs used by Galvani in these experiments were "dead frogs being dried by fixing by brass skewers to an iron fence" (Millar et al., 1996; Muir, 1994; see Chapter 1).

Galvani was immediately ready to see the novelty of the experiment with metal contacts. On September 20, only 4 days after the experiments on the terrace, he was once again in his laboratory in order to study the circumstances in which contractions had occurred. As he had done in the experiments that followed the observation of the contractions at a distance, he modified the variables at play in the phenomenon in order to detect the ones that were necessary for it to take place. In particular, by adopting a protocol very similar to that followed by Walsh in his experiments on electric fish (see Chapter 3), Galvani noticed that when the frog, prepared with a metal hook inserted in the spinal cord, was placed "over various metal plates," the legs of the animal contracted at each contact between the hook and the plate. This did not take place if the hook or the plate was made of glass, resin, or any other insulating material (GME, pp. 397–398).

Under an electrological perspective, which, as we have already seen, was an essential aspect of Galvani's approach to electrophysiological research, this difference was fundamental. Indeed, the fact that contractions occurred with conducting materials and not with insulating substances meant that electricity was at play in the phenomenon of muscular motion: "It seemed to us not unreasonable"—Galvani wrote on October 30—"to admit that the principle from which those movements derived was electrical" (GOS, p. 164). But where could this "electric principle" come from, as in the experiments there had been no sign of electricity, either artificial or atmospheric? Galvani very soon reached the conclusion that this was the experimental evidence, the first after years of research work, of the existence of electricity intrinsic to the frog. In *De viribus*, he explicitly concluded that: "This outcome surprised us greatly and at the same time led us to the suspicion that there was a kind of electricity inherent in the animal itself" (GDV, p. 379).

It was the first time that in his memoir Galvani explicitly referred to animal electricity, as if up to that point he had never thought of this possibility. Actually, as we have seen previously, the hypothesis of the identity of nervous fluid and electric fluid had been advanced publicly by Galvani well before starting his electrophysiological experimentation, and it had repeatedly resurfaced in his investigations. The fact that this hypothesis appeared only in this part of his work indicates that Galvani, on deciding to make his research public, built a different narrative from the one he had adopted before, and that he gave a different meaning to his experiments with artificial electricity. On the other hand, the published report may give a more realistic indication, at a deep level, of what these experiments meant for Galvani. As we shall describe in greater detail later on, the experiments with metals were indeed a turning point in Galvani's path toward the theory of animal electricity, mostly because they brought the Bolognese scholar well beyond a generic idea of a neurolectric fluid with elusive properties; they also allowed him to create a coherent physical model that corresponded to the experimental results he had obtained (the model of the "animal Leyden jar"). The model was able to account for the mechanisms of nerve conduction and muscular contraction, thus solving the experimental problems that had been on his mind for over 5 years.

Galvani's "suspicion" that electricity was a property of the animal and responsible for muscular contractions was strengthened by a series of experiments recorded in his laboratory journal on September 20, 1786. After preparing a frog in the usual manner, Galvani suspended it by its spinal cord (or by a hook inserted in it) so that the legs rested on "the external and upper surface of a hollow metal plate, as on a silver box." Every time he touched the sides of the box with "another metal" he obtained some contractions. The phenomenon did not occur, however, if one person held the frog and another touched the plate, as well as if the two people were connected by means of an "insulating body like, for example, a glass rod." By contrast, contractions reappeared when the connection was established through "a conducting body," like a metal rod, thus forming a "chain." From these experiments, Galvani derived the following "corollary": "The phenomenon [of contractions] seems to depend on a discharge of the muscles made through the conducting arc, similarly to what happens in the Leyden jar" (GME, p. 401).

As a charged Leyden jar could be discharged by connecting the internal metal "armature" with the external one, thus establishing a circuit and an electric flow between them, likewise contractions could be elicited in muscles by establishing a circuit between the muscle and spinal cord through the metal plate and the person who was holding the frog. In both cases it was possible to obtain the phenomenon—the discharge in the case of the jar and the contraction in the case of the frog—by forming a chain of several persons. The same experiment is reported in the memoir of October 30, 1786, where Galvani describes the rhythmic movement of the frog's limbs saying that the frog seemed to be "jumping" when the circuit was rapidly closed by "repeated percussions" of the hand over the surface of the box (GOS, p. 165). In *De viribus*, the experiment is described right after Galvani has first formulated the hypothesis that between nerve and muscle there occurred "the circuit of a most delicate nerve fluid [...] somewhat similar to the electric circuit which is effected in a Leyden jar" (GDV, p. 378).

The analogies between the prepared animal and the Leyden jar, as well as the existence of an electric circuit between the spinal cord and the muscle (suggested by the experiment where the frog was placed on a metal plate), were confirmed by another experiment in which the animal was no longer held by the spinal cord, but by a leg, so that the hook inserted in the medulla touched the surface of the metal plate (see Fig. 5.4). Every time the other leg touched the plate "so that a very short conducting arc is created between the spinal cord and the free limb through the plate, the movements of the free limb become most vehement" (GME, p. 402). The importance of this experiment is demonstrated by the fact that it was reported both in the 1786 memoir and in another memoir written in 1787 and, finally, in *De viribus*, where Galvani described the scene as a sort of "pleasant" spectacle:

> If a frog is held by one leg with the fingers so that the hook fastened in the spinal cord touches a silver plate and if the other leg falls down freely on the same plate, the muscles are immediately contracted at the instant that this leg makes contact with the plate. Thereupon the leg is raised and lifted up, but soon, however, it spontaneously relaxes and again falls down on the plate. As soon as contact is made, the leg is again lifted for the same reason and thus it continues alternately to be raised

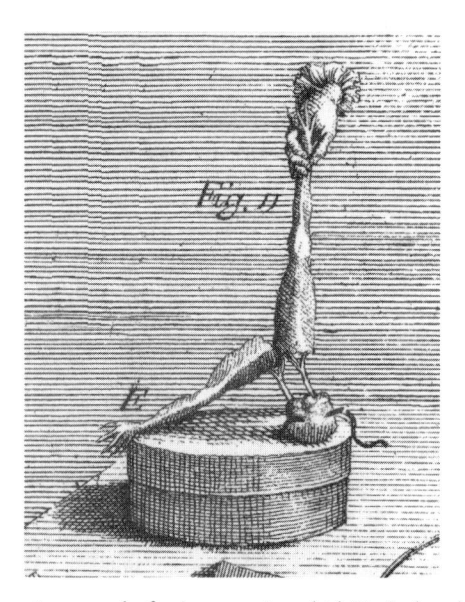

FIGURE 5.4. Galvani's experiment on the frog's preparation which "jigs" when the spinal cord touches the surface of the metal box, the so-called second experiment (detail from plate III of Galvani, 1791).

and lowered so that to the great astonishment and pleasure of the observer, the leg seems to function somewhat like an electric pendulum. (GDV, pp. 379–380; translated with revision from Galvani, 1953a, p. 61)

Because of their importance, the experiments carried out with the frog placed on the metal plate have been renamed by historians as Galvani's "second experiment." In fact, both in this case and in his "first experiment" Galvani was convinced that he was faced with the manifestation of electricity intrinsic to the frog and, consequently, that he had revealed the nature of the nervous fluid responsible for muscular motion. On the other hand, in the second experiment, not only was the evidence supporting this idea much stronger (contractions had been obtained without a spark or any other form of extrinsic electricity), but it was possible to establish an explanatory relationship between the phenomena being observed and what happened in the Leyden jar. The analogy with the Leyden jar suggested a possible physical model and, in particular, indicated that in the animal there must be an electric disequilibrium analogous to that existing between the internal and the external coatings of the jar.

A similar analogy had been advanced in the middle of the eighteenth century concerning the shock produced by electric fish; it was central to the interpretation that John Walsh and Henry Cavendish proposed for this phenomenon just a few years before Galvani started his electrophysiological research. In his letter to Girolamo Lucchesini on torpedoes, Spallanzani reported that Walsh had demonstrated the presence of two different states of electricity between the back and the breast of the animal "in a way analogous to what can be found in the Leyden jar" (Spallanzani, 1783, p. 98). Walsh and then Spallanzani had underlined the functional similarity that existed between the shock of the torpedo and the discharge of the Leyden jar. As was the case with the jar, the shock of the torpedo was transmitted by means of conducting materials, like "a key, a nail or any other similar metal body," or (but less effectively) by using a towel or some other kind of cloth wrapped around the body of the torpedo, which "transmit the shock to the hand, but only if wet; while they do not if they are dry" (Spallanzani, 1784, p. 652). And similarly to dry textiles, insulating bodies like "sealing wax cylinder" or glass were totally unfit for transmitting the shock, as Walsh had noted. Accordingly, he concluded that:

> The Torpedo resembles the charged Phial in that characteristic point of a reciprocation between its two surfaces. Their effects are transmitted by the same mediums; than which there is not perhaps a surer criterion to determine the identity of subtile matter. (Walsh, 1773, p. 474)

Thus, the analogy between animal parts and the Leyden jar was evident in the scientific literature of the time, and interest in it had been revived by the research on torpedoes published by Spallanzani some years before Galvani's observations of the contractions caused by metal contacts. It should also be noted here that these reports by Spallanzani were published in *Opuscoli scelti sulle scienze e sulle arti*, which was among the journals in Galvani's private library. In addition, Galvani had repeatedly related the frog's nerves and muscles to other electrical instruments, like the Franklin square, the electric machine, or the electrophorus. For example, on February 14, 1781, he had noticed that if the air in the flask where the frog was placed (with a long conductor applied to its spinal cord) was extracted, then the contractions were the same as when there was air in the container. One could therefore conclude that:

> The [frog's] conductor does receive nothing or very little from the air; and the contractions are in proportion to the conductor alone, as the electricity of the [electric] machine is in proportion of the conductor which is applied to the back of the machine and which carries the electric fluid to the machine. (GME, p. 273)

The analogy between the frog and the electrostatic machine reappeared in an experiment performed on April 24, 1781, while in some unpublished manuscripts of 1782, Galvani compared the nerves with the "[insulating] substance of the electrophorus." The recourse to the analogy between the experimental preparation and electrical instruments was a typical of Galvani's attitude, both when interpreting phenomena and when designing experiments. It was also an essential aspect of his approach—at the same time electrical and physiological—to the investigation of muscular motion. For a researcher used to reasoning in this way, it was therefore natural to think of the Leyden jar in order to explain the contractions observed in the frog placed on a metal plate, all the more so because this analogy was already extant in the similar phenomenon of the shock produced by electric fish.

If the analogy between the frog and the electrical instruments was a recurring theme in Galvani's thought, the image of the Leyden jar contained an important element of novelty. In his 1786 memoir, Galvani interpreted the contractions observed when the spinal cord and a leg came into contact through the metal plate as follows:

> When the free leg falls down over the plate, a communication between the electricity corresponding to the nerves and the spinal cord and that corresponding to the muscles is established; hence the electric fluid, carried from one part to the other due to the law of equilibrium, flows with strength into the nerves and thus the muscles rapidly contract and the leg rises. Immediately after, when the circuit is completed, [the leg] falls down again and touches the silver plate, so that the circuit is re-established and the leg rises again, and so forth until either the nervous force extinguishes itself, or the free circuit of electricity [*circuitus electricitatis*] between the nerves, or between these and the muscles, is interrupted. (GOS, p. 167)

The contractions were thus due to a motion of the electric fluid that, by being present in different quantities in nerves and in muscles, circulated from the former to the latter and thus re-established the equilibrium. The same happened between the opposite surfaces of a Leyden jar and, according to Walsh, between the dorsal and the ventral surfaces of the torpedo. It is worth noting that the idea of a "circuit" emerged not only in relation to the analogy with the "animal Leyden jar" represented by the torpedo in Walsh's opinion but also to Galvani's chemical-physical investigation, and especially to the notion of a flow of blood and of the inflammable principle in the body (see earlier, Section 1). In particular, as the idea of a circulation of the inflammable principle through the body was capable of explaining both its diffused presence in the body and its effects on vital processes, so the circulation of the electric fluid was able to explain the contractions produced when muscles and nerves were connected through a metal arc. Moreover, since the idea of a circuit was central to the accepted explanation of the phenomena related to the Leyden jar, such an idea implicitly corroborated the analogy between the discharge of the jar and muscular contractions. In the studies on electric fishes the circuit was visualized in a very vivid way by the chains of people— from 2 up to 7 or 8 in the experiments performed by Walsh on torpedoes and up to 27 in the experiments on electric eels (Chapter 3.4; see Piccolino, 2003a and Finger & Piccolino, 2011)—who all felt the shock at the very same moment when contact with each other and with the animal was made. Similar chains were formed by Galvani and his assistants, among whom was Raymond Rialp, the Spanish Jesuit who had played a fundamental role in the experiments with frogs placed on a metal plate (see Fig. 5.5; see Chapter 4.1).

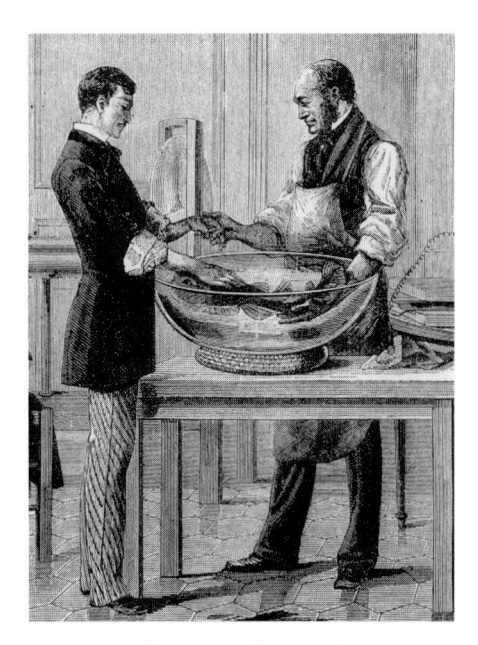

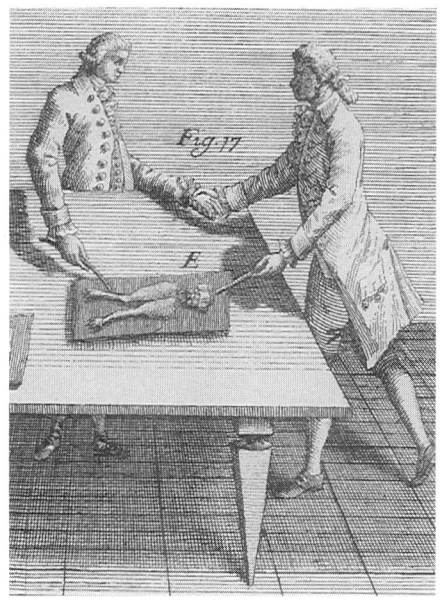

FIGURE 5.5. Comparison between the experiments with "human chains" made by Walsh on the torpedo (left) and by Galvani on the frog (right) (details from a plate of late eighteenth century and from plate IV of Galvani, 1791).

The idea of an electric circuit that emerged at this stage of Galvani's researches seems quite obvious today, as we associate it closely with the notion of electric currents. In Galvani's time, however, the situation was totally different, since there were no instruments capable of producing a constant electrical flow and electricity was usually delivered by directing the "electric flake" or the spark to the target object, often without considering the possibility of an electrical flow toward the ground or other conductors (see Chapter 3.2).

5.3 THE MODEL OF THE MUSCLE AS AN ANIMAL LEYDEN JAR

Once established, the analogy between the Leyden jar and the nerve-muscle system, as well as the idea of an electric circuit, started to influence Galvani's investigation by suggesting the problems for him to focus on and—at least partially—their solutions. In the 1786 memoir Galvani first described the experiment of the frog placed on the metal plate, and then continued as follows: "A circuit of this sort implies a double and contrary electricity, that is to say positive and negative; this requires that we diligently investigate what is the seat of both [these kinds of electricity]" (GOS, p. 167).

If electricity was responsible for muscular contractions, there should be an imbalance of electric fluid, which was in greater quantity in one part (thus positively charged) than in the other (negatively charged). This was exactly what happened in the case of the Leyden jar, where the internal surface was usually positively charged while the external surface was electrically negative. Indeed, the electric machines available at the time usually produced a positive charge on their "primary" conductor that in turn was connected to the conductor of the bottle and, thus, to the internal coating. Hence, it was necessary to establish where the electricity was positive and negative in the experimental arrangement involving the frog and the metal plate. This issue also became crucial in Galvani's electrophysiological investigation because it dealt with one of the most important criticisms of Hallerians to the neuro-electric theory, namely the impossibility of an electrical disequilibrium in the organism. With respect to this problem, however, Galvani was now in a better position than in the early stages of his research. Indeed, the studies on electric fish had shown that an electrical disequilibrium can exist in a living organism, though largely made up of electrically conductive tissues, and also in animals—like the torpedo and the eels from Surinam—that lived in a conductive medium. The fundamental evidence for the attribution of an electrical nature to the fish's shock was based on some of Walsh's experiments where different kinds of bodies conducted—or interrupted—the shock both "in the torpedo and in the phial of Leyden" in the same way. Similarly, in his experiments, Galvani showed that contractions were produced when nerves and muscles were connected through electrically conductive materials, but not when the connecting materials were insulators. Despite the Hallerian objection concerning the conductive nature of animal tissues, these experiments showed that an electric disequilibrium similar to the one existing in the torpedo can exist in the frog.

The way Galvani explored the issue of the electrical disequilibrium in the experimental animal (as it is described in the 1786 memoir) shared the same elements of his general investigative approach, which was based on the application of the principles—as well as of the instruments—characteristic of the electrical research to the field of physiology. First of all, Galvani was interested in excluding the argument that the disequilibrium of the electric fluid—and consequently the "seat" of opposite electricity—was either in the metal plate where the frog was placed, or in the conducting arc used to connect the nerve and the muscle. In this case one would have to conclude that the electricity implied by the phenomenon of muscular contractions was not intrinsic to the animal. However, to claim that "in the same plate, or in the same metallic piece there would be a sort of double pole, one negative and the other positive" did not fit with the known laws of electricity nor with "the experiments performed by many physicists." Indeed, they showed the impossibility of any electric disequilibrium existing between two points of a conducting body "at rest," that is, not influenced by external sources of electricity. Galvani noted that such double electricity in the same conductor was admitted by physicists in the case of tourmaline, but he considered this an exception which "does not seem to be true for any metals" (GOS, pp. 167–168). Besides, he observed that if a frog was placed on a glass plate—an insulating material—instead of metal and the spinal cord and the muscles of the frog were connected by a metal arc, contractions were produced "not unlike the discharge of the Leyden jar or the magical square [i.e. Franklin square] by means of the conducting arc" (GOS, p. 168; see Fig. 5.6).

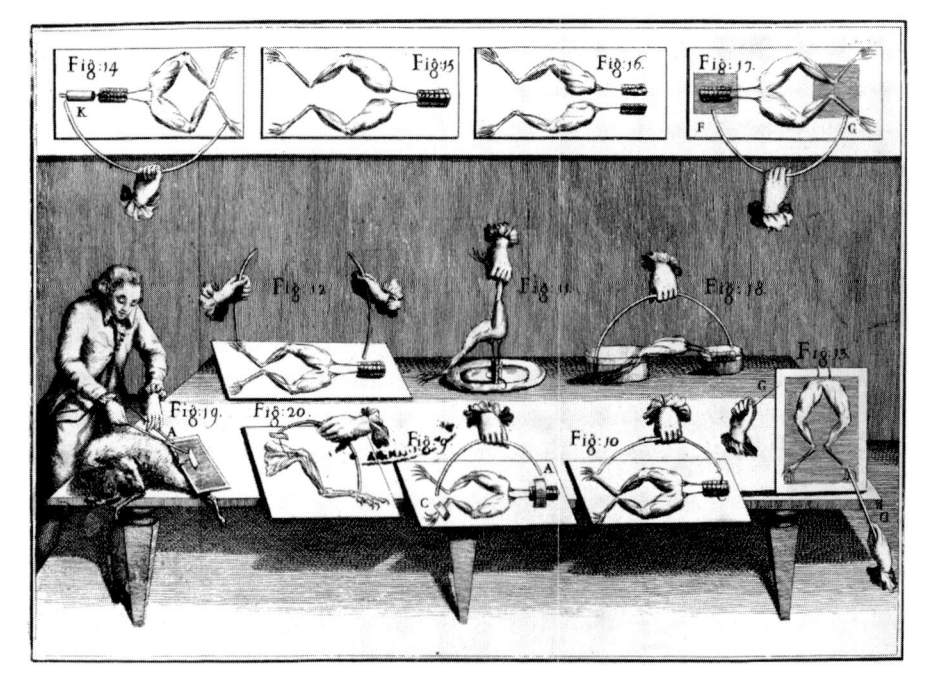

FIGURE 5.6. Various arrangements of the frog, the plate, and the conducting arc designed by Galvani in his experiments with metals (plate II of Galvani, 1792).

As to the possibility that the disequilibrium was in the conducting arc, Galvani noted that the arc he used was "very thin and short"; thus, it was very difficult to imagine how it could contain opposite charges strong enough to produce contractions of constant, if not increasing, intensity with the renewal of the contacts. To support such a possibility, one would need to suppose that the contrary electricity came either from the experimenter who held the arc or from the atmosphere. However, by insulating the arc by means of a "glass cylinder" and by immersing the frog in "olive oil," that is, in an insulating liquid that excluded any possible electric transmission from the external environment, Galvani obtained muscular motions that "were very little different from those we had obtained in open air by keeping the conductor in our hands." In Galvani's view, this experiment provided strong evidence against the supposition of "metallic" electricity, unless one wanted to admit that it was the oil which provided the frog with electricity, thus introducing "something quite new in physics." The very same analogy with the Leyden jar, where the electricity was accumulated on the two glass surfaces and the metal of the discharging arc played a purely conductive role, suggested that the opposite electricity "have their seat in the trunk of the animal," not in the metallic arc (GOS, pp. 170–171).

Galvani provided further evidence against a metallic nature of the electricity responsible for muscular motions with positive arguments for the existence of electricity intrinsic to the animal body (*electricitas animalis* is the original Latin term). First of all, he reported the case of the torpedo and of other electric fishes "in which contrary electricity does develop and accumulate almost spontaneously, something that has not been obtained in metals so far." Moreover, the muscles and the nerves were composed of heterogeneous substances; "thus it does not seem unreasonable—he remarked—to admit that electricity is different in those substances." Here Galvani was clearly referring to his physical-chemical research on animal parts and to the discovery of the different principles that composed them. For him, the fact that the arc was completed in a single metal while there were heterogeneous substances in the animal organism—which were partly conductors and partly insulators—suggested that "the electricity we investigate and which we have discussed so far, is pertinent to the animal and not to the metal." And then he added:

> In the muscles there is a great quantity of matter which, being rubbery, can be apt to develop and keep the electricity, even though there are also conducting materials which are interposed in it; most probably the same takes place in the electrophorus, which is composed by similar substances. (GOS, p. 169)

In this passage we can note again Galvani's use of an analogy with a physical apparatus. An analogy between Volta's electrophorus and an animal preparation would return 10 years later, in 1797, when William Nicholson offered an explanation of the discharge mechanism of electric fish using a similar argument (see Chapter 8.4). Anyway, Galvani considered the possibility that the muscular tissue was the place where the electrical disequilibrium was "developed" and "kept" as a mere conjecture. In fact, this conjecture did not seem useful to uncover the physiological aspects of the mechanism by which the disequilibrium produced the phenomenon of muscular contractions.

The existence of animal electricity was confirmed—in Galvani's opinion—by an experiment that, apparently, suggested a role of metals as the "seat" of opposing electricity. He noted that when he used the same metal, like copper, for both the hook inserted in a frog's spinal cord and the conducting arc, no contractions were produced. This result brought "the suspicion that [...] the difference of metals in the hook and the arc was particularly appropriate in obtaining those movements," a suspicion that was strengthened by a series of experiments planned to this aim. Galvani recognized that he was facing with a "marvellous" phenomenon, which "gave great importance to metals, so that they came to contain a different electricity in relation to their particular nature." However, neither was this observation quite new nor did it constitute a decisive objection against the existence of animal electricity. Indeed:

> We reflected that observers discovered that the difference of metals had but little influence in promoting or preventing the electric discharge of the torpedo; therefore, when examining the question more accurately, it was animal electricity, rather than that of metals, that turned out to be confirmed and recommended. (GOS, p. 172)

The observations on electric fish—animals generally considered as endowed with proper electricity—was only one of the elements that drove Galvani to account for the particularly efficiency of nonhomogeneous metals without referring to the idea of a metallic electricity, as Volta did with his idea of the electromotive power of metals. Paradoxically, while the efficiency of different metals in eliciting contractions was considered by Volta as a fundamental argument against the idea of animal electricity, it was considered by Galvani as an argument in favor of the existence of that kind of electricity and its peculiarity compared to other forms of electricity.

Although Galvani denied the possibility of an active role of different metals in the phenomenon of muscular contractions, the discovery of their efficiency—as he noted in his 1786 memoir—"led us to think that by covering the muscles, or the nerves, or the spinal cord, or the brain, with metal foils of simple or double substance, [...] the contractions would be much stronger and produced by lighter causes" (GOS, p. 180). Such prevision was corroborated by a series of experiments in which the application of some metal laminas—foils made of silver, gold, brass, and most of all tin—to various parts of the frog, was followed by contractions that were stronger than those produced without the use of these armatures (see Fig. 5.7). Moreover, these experiments further supported the analogy with the Leyden jar, whose capacity of storing electricity was markedly increased by the presence of metal armatures.

It is likely that the power of metallic laminas in making the contractions of the frog muscles more effective in the experiments with metallic arcs also played an important

FIGURE 5.7. Plate IV of *De viribus* (from Galvani, 1791).

role in directing Galvani's attention to the Leyden jar as a model of the frog neuromuscular system. On that regard, a significant clue comes from comparing the terminology used by him in describing these experiments in the manuscript memoirs of 1786 and 1787. In the first memoir the metallic sheets used for the electricity involved in the contraction are described exclusively as laminas or foils (*lamine* or *fogli*), whereas in the second memoir they are referred to as "armatures (*armature*)." In the electrical terminology of the epoch, "armature" was commonly used to designate the thin laminas coating the internal and external glass surface of the Leyden jar (Piccolino, 2008).

Not only did the experiments with metal armatures seem to confirm the existence of a circuit of electric fluid between nerves and muscles but also by amplifying the effects of animal electricity they enabled extension of the research from the "prepared frog" to "living animals with intact nerves." This was an important move. While Galvani firmly asserted the validity of comparative anatomy and physiology, he was also aware of the necessity of verifying the experimental results in normal physiological conditions and in warm-blooded animals that were closer to humans. Besides, and here Galvani reasoned like a physician, such a generalization provided the potential to "apply our experiments and efforts [...] to the healing art." He thus repeated his experiments with metal arcs and armatures on several species of birds and quadrupeds, and especially lambs (see Figs. 5.6 and 5.7), also obtaining muscular contractions in these cases and concluding that there was "no doubt left about the existence of a common and universal animal electricity [*de communi atque universali animali electricitati*]" (GOS, pp. 188–189).

5.4 THE FINAL ELABORATION OF THE THEORY OF ANIMAL ELECTRICITY

Galvani's 1786 memoir had the same content as the third part of *De viribus*, thereby constituting a kind of preliminary version of the published work. The major difference between the two concerned the "seat" of contrary or opposite electricity and the meaning of the analogy with the Leyden jar. In the memoir, Galvani concluded that contrary electricity was placed in the nerves and in the muscles (see Fig. 5.8a): "There is no doubt—he wrote—that one sort of electricity is located in the muscle, the other in the nerve; in order to put each of them into action it is sufficient to apply an appropriate conducting body to the nerve and the muscle so as to establish a good communication between them" (GOS, p. 176). To corroborate this conclusion, Galvani reported the observation that the contractions were produced when the nerve and the muscle were connected through a metallic arc. Although this was the experimental arrangement usually adopted by Galvani in his experiments, it should be noted that he realized that contractions could also be produced by connecting either two different parts of the same muscle, or two different points of a nerve or of the spinal cord. However, the nerve-muscle connection seemed to be the most efficient means of producing the phenomenon. Moreover, the idea of electricity circulating between nerves and muscles was in line with the physiological framework of the time, which ascribed a central role to the nerves in the mechanism of muscular motion. Finally, in the experiments with the spark, Galvani had observed that the contractions produced when electricity flowed from nerves to muscles were

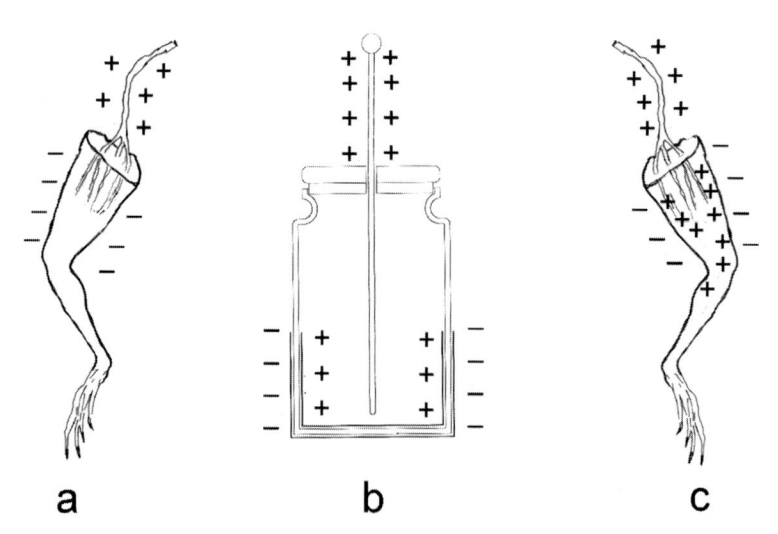

FIGURE 5.8. Galvani's macroscopic models of the muscle as an "animal Leyden jar": (*a*) the electric imbalance is between the nerve tissue (seat of positive electricity) and the muscle (seat of negative electricity); (*b*) the usual distribution of electric charges in a Leyden jar; (*c*) animal electricity is accumulated between the internal and external surfaces of the muscle, while the nerve, by penetrating into the muscle, acts as a conductor.

more evident than the contractions produced in other experimental conditions. Indeed, in these experiments he had noted that the efficacy of the stimulus applied to the nerves increased markedly when a metal wire—the so-called muscle conductor—was connected to the muscles, especially if the wire was also connected to the ground.

With regard to the possibility of obtaining contractions by adopting experimental dispositions different from the nerve-muscle arrangement, it should be noted that the contractions produced by connecting two different points of a nerve constituted an important objection raised by Volta to Galvani's theory. We shall return later in this book to describe in detail the way in which Galvani responded to Volta's criticism; at present we need only remark that in this case—as in the case of the use of heterogeneous arcs—Galvani had preceded Volta's observations and, more important, that he was able to "accommodate" the observed phenomena in the framework of his theory of animal electricity.

While Galvani restated the idea of contrary electricity in nerves and muscles in his 1787 memoir, in *De viribus* he adopted an explanatory model based on a structural and functional analogy between the neuromuscular apparatus and the Leyden jar (see Fig. 5.8c). In this final model, the electric disequilibrium is located between the internal and the external surfaces of the muscle fibers; the nervous fiber hooks in the muscle fiber, exactly like the metal bar that enters the Leyden jar. When the metal arc is applied, a flow of electricity runs from the interior to the exterior of both the physical and the "animal" jar, so that a circuit is established. Note that in this definitive model, the real accumulator of electricity—both positive and negative—in a state of disequilibrium is the muscle, and the electric fluid present in the nerve is an expression of such muscle electricity. Therefore, in *De viribus*, Galvani introduced the following conjecture: "If, however, in such a great obscurity of the matter, one is allowed to formulate a conjecture, my feeling is inclined to place the source of both kinds of electricity in the muscle" (GDV, p. 395).

The definitive formulation of the model of the neuromuscular apparatus as an animal Leyden jar contains some notable differences from the previous model in which contrary electricity was located in the nerve and in the muscle. In the first place, there is a precise indication of the place where the disequilibrium is accumulated and the way it is produced. Furthermore, the definitive model is "microscopic" in nature, in the sense that every single muscle fiber, together with the nerve fiber that hooks into it, constitutes a little Leyden jar. As we shall see, such a model not only allowed the elaboration of a coherent explicative mechanism of the neuromuscular physiology, but it also overcame all the objections raised by the Hallerians against a neuro-electric view. Perhaps it was the conviction of developing a model capable of providing a global neuro-electric explanation of muscular motion that led Galvani—who was reluctant to make his results public—to publish his experiments in 1791.

An indication of the intellectual path that led Galvani from the generic hypothesis of an electric circuit between nerve and muscle to the definitive model of the animal Leyden jar developed in *De viribus* comes from a comparison between the two unpublished memoirs of 1786 and 1787 that we have already analyzed with regard to the different terminology used to designate the metallic laminas applied to nerve and muscle in the experiments with the metallic arcs. The passages in which Galvani refers to the tourmaline are particularly interesting. In the 1786 memoir, the reference to this stone is very brief and is made to underline that the property of accumulating "double and contrary electricity" in the same conductor is a property restricted to this extraordinary stone which cannot be attributed to common metals. By contrast, in the 1787 paper Galvani mentions tourmaline in more detail, describing its structure and using its characteristics as positive arguments for animal electricity. As we have already remarked, the peculiar property of the tourmaline—that once heated, it becomes positively charged at one extremity and negatively charged at the other—was noted at the beginning of the eighteenth century when jewellers, who heated the stone in order to evaluate its worth, observed that it attracted ash (for this characteristic, in Holland tourmaline was referred to as *aschentrikker*, "attracter of ashes"). Following the growing interest in electrical phenomena that characterized the first half of the eighteenth century, this stone became the object of several studies and, in 1756, it captured the attention of Franz Ulrich Aepinus, one of the most important "electricians" of the time (Aepinus, 1756, 1762). The German scholar emphasized that the stone's property to produce a double electricity—positive and negative—when heated "is not to be found in its external aspect, nor in the way it is cut, but it depends on the internal structure and the essential constitution of the stone, as is the case for the magnet" (Aepinus, 1756, p. 106).

It is exactly the dependence of the peculiar electrical properties of tourmaline on its internal structure that was used by Galvani to account for the electrical phenomena of the neuromuscular apparatus in his 1787 memoir. The analogy with the tourmaline is one of those metaphors of a physical kind that are frequent in Galvani's work and that, once formulated, led him to the elaboration of new, more coherent and exhaustive explanatory models (see Fig. 5.9). In his 1787 paper Galvani refers to this analogy in the following terms:

> Our [animal] electricity has much in common with that of tourmaline with regard to its seat, distribution and properties of the parts. Indeed, in this stone there is a double substance, one transparent and rather reddish, the other opaque and colourless; this latter is arranged in strips, the former is situated laterally to these strips. No one ignores that the nerves run in between the strata of the muscular fibres; and that these latter, devoid of blood, are transparent, while the nerves are opaque. In the tourmaline the poles of the double electricity appears along the same opaque line; similarly, in the muscles they [are] in the same direction of the nerves. The double electricity of the tourmaline is not only located in the whole stone, but also in any single fragment of it. Similarly, in the muscles the supposed double electricity is not only located in their wholeness, but in any part of them. (GOS, p. 194)

Galvani considers the strips in tourmaline's morphology as a key element of similarity with muscles, whose structure appears heterogeneous due to the presence of nerve

| Tourmaline crystal | Tourmaline model | Microscopic Leyden jars model |

FIGURE 5.9. Tourmaline (on the left) and Galvani's microscopic models of the muscle as an "animal Leyden jar." At the center the disposition of the nervous fibers located in between the muscle fibers; on the right the finale model in which the nervous fibers penetrate into the muscle fibers.

terminations, and then extends this structural analogy to a functional analogy between the stone and the neuromuscular system. The double electricity may accumulate in the microscopic units formed by a muscular fiber coupled with a nervous fiber, which correspond to the units of a transparent and an opaque strip in tourmaline. The importance of this analogy is that it drives Galvani's attention to the microscopic level of the fibers of both nerves and muscles rather than to the muscle or the nerve conceived as macroscopic entities (see Fig. 5.9). It also allows him to account for the fact that muscular contractions can be produced by stimulating the nerve even if the muscle is lacerated, at least partially, as Galvani observed in several experiments.

From this time onward, it is at the microscopic level that Galvani aimed to determine the site of the electrical disequilibrium, namely the double electricity that he thought he had demonstrated in animal tissues by the experiments on metals. In the period from 1787 to the time of the definitive draft of *De viribus*, he elaborated on this issue, finally arriving at the model of the microscopic "Leyden jar" formed by a muscle fiber and a nerve fiber, which hooks in the former instead of simply being in contact with it. In *De viribus*, the crucial argument for the definitive model of the "animal Leyden jar" is based on both structural and functional aspects of the instrument. As in the physical jar, the hook had the unique role of a conductor and "both kinds of electricity, although of contrary natures, are contained in the same jar," so in the nerve-muscle system it could well be that "the muscle is the proper seat of the electricity we investigated, whereas the nerve plays the role of the conductor" (GDV, pp. 395–396). The idea that the nerve was a simple conductor was corroborated, in Galvani's opinion, by two arguments derived from his experiments. The first dealt with the "the tiny section of nerve remaining in the prepared muscles after dissection," so that it was difficult to imagine that a quantity of electricity could be accumulated in such a small portion of nerve with the power to produce "the large number of contractions" observed in the experiments. The second argument considered the instantaneous muscular response to stimuli as evidence of the fact that the neuro-electric fluid flows "freely and with great speed through the nerves," which had thus to be considered very efficient conductors rather than accumulators of animal electricity (GDV, p. 396; translated in Galvani, 1953a, p. 74).

After establishing that the muscles were the site of contrary electricity and that the nerves were the conductors of the neuro-electric fluid, Galvani explicitly formulated the model of the muscle as an animal Leyden jar:

> If one admits [these conclusions], then perhaps the hypothesis is not inconvenient nor wholly discordant from truth which compares a muscle fibre to something like a minute Leyden jar or to some other electrical body charged with the opposite kinds of electricity, and compares, moreover, a nerve in some measure to the conductor of the jar; in this way one likens the whole muscle, as it were, to a congeries of Leyden jars. (GDV, p. 396; translated with revision from Galvani, 1953a, p. 74)

Compared to the conjecture about an unbalanced electricity between nerve and muscle, this hypothesis was much closer to the Leyden jar, not only conceptually but also visually (see Fig. 5.10). It may well be that the image of the Leyden jar, once it emerged in

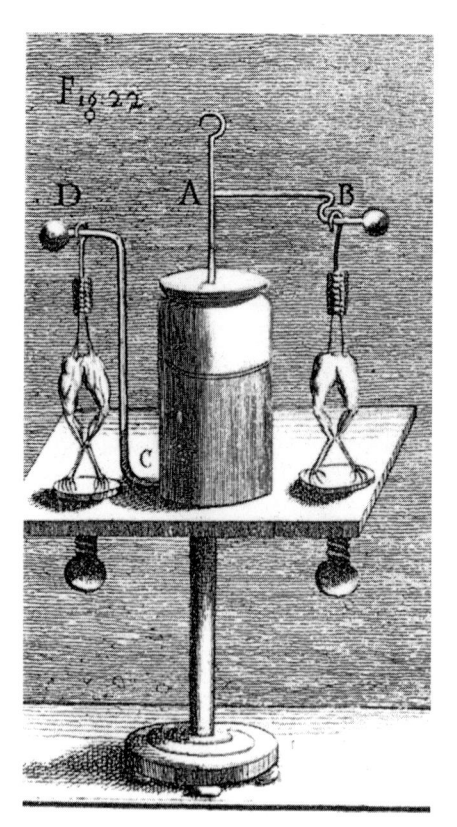

FIGURE 5.10. Visual analogy between the frog's neuromuscular apparatus and the Leyden jar (detail of plate III from Galvani, 1792).

Galvani's mind, induced him to reconsider the idea proposed in the 1786 memoir and to elaborate an explanation of the circuit of animal electricity, which was closer to the image itself. Besides, the new model incorporated the idea of a diffused and microscopic structure that was suggested by the reflection on tourmaline formulated in the 1787 memoir. Indeed, in *De viribus* Galvani added:

> Finally, if one, even briefly, considers the tourmaline stone, which embodies a two-fold and opposite electricity, as it seems from the investigations of contemporary experimenters, one will envision a new argument based on analogy which is not inconsistent with a hypothesis of this kind [i.e., that of the muscle as a Leyden jar]. (GDV, p. 396; translated with revision from Galvani, 1953a, p. 74)

In agreement with the importance assumed by the electrical perspective in Galvani's approach to the study of vital functions, images taken from the field of physics—like the Leyden jar, the electric machine, and the electrophorus—had thus the upper hand over the traditional idea of the nerve as the site of animal spirits (Piccolino, 2008). The analogy between the muscle fiber and the Leyden jar implied a totally new conception of muscular motion. Not only was the electrical nature of the nervous fluid assessed but also the site of the neuro-electric fluid and the way it produced muscular contractions were empirically identified. The fundamental novelty introduced by Galvani was the possibility that under normal conditions the organism contained an electrical disequilibrium and that such an imbalance existed in a singular tissue, without breaking the known laws of electrical phenomena, similarly to what happened in electrical instruments.

By reasoning in such a way, Galvani overcame the Hallerian objection based on the impossibility of disequilibrium between the conducting tissues of the organism. As we shall see in Chapters 9 and 10, Galvani's idea of an electrical imbalance between the internal and external sides of the muscular fiber is apparently very close to the modern explanation of the phenomenon, according to which the electrical disequilibrium responsible for nerve conduction and muscle excitability is produced at the opposite sides of the cellular membrane. As it is well known, the notion of "cell" was not available in Galvani's time, since cell theory was only elaborated in the nineteenth century. Nevertheless, by referring to the concept of "fiber"—developed between the seventeenth and eighteenth centuries by several authors such as Giorgio Baglivi, Boerhaave, and Haller—the Bolognese scholar was able to conceive microscopic elements of the organism whose structure could fit within his theory of animal electricity. As he reported in *De viribus*:

> It is even more difficult to deny the existence of a duplex electricity in every muscular fibre itself if one does not think it difficult, nor far from truth, to admit that the fibre itself has two surfaces, opposite one to the other; and this derives from a consideration of the cavities that many admit are in it, or because of the diversity

of substances, which we said the fibre is composed of; a diversity which necessarily implies the presence of various small cavities, and thus of surfaces. (GDV, p. 396)

After describing the experiments and the arguments for his theory of animal electricity, in *De viribus* Galvani compared this kind of electricity to the other kinds of electricity known at the time. Although animal electricity was characteristic of the animals and their "economy," it shared some properties with both the "artificial and common" electricity, and that of torpedoes and other electric fish. Like the latter, animal electricity was "duplex and contrary, namely positive and negative" in nature, and it was accumulated inside the organism in an unbalanced condition. Galvani explained its action in the production of voluntary motions in the following terms:

In the case of voluntary motion, it is possible that the mind [*animus*] is able, with its extraordinary power, residing either outside of the brain, or as is easier to believe, within it, to stimulate whatever nerve it chooses. This results in the neuro-electric fluid's flowing rapidly from the corresponding muscle to that part of the nerve to which it has been stimulated to return. Once it has arrived and after the non-conducting part of the nerve substance has been overcome by its accumulated strength, it leaves there and it is taken up either by the external moisture of the nerve, by the membranes, or by other contiguous conducting parts. It is then restored through these, just as through an arc, to the muscle from which it departed. This finally results in its flowing copiously to the negative electrical part of the muscle fibres, in accordance with the law of equilibrium, inasmuch as it had probably flowed earlier from their positive electrical part through an impulse in the nerve, as we are pleased to believe. (GDV, pp. 406–707; translated in Galvani, 1953a, p. 82)

Animal electricity was thus unbalanced in the organism—more precisely, in the muscle—exactly as it happened in the case of a Leyden jar. The contraction was produced by the neuro-electric fluid that, in agreement with the laws of electricity, flowed through the nerve from the muscular surface in which it was in greater quantity to the other surface, until the equilibrium was restored. Then contraction ceased, a new electric imbalance was created, and the muscle was ready to contract again. The nerve was, in Galvani's view, a tubular structure made of a conducting material surrounded by an insulating sheath. In such a way it was possible to account for the fact that electricity flowed along determined pathways without being dispersed into surrounding tissues.

With regard to the source of electricity that constantly flowed to the muscles, Galvani referred both to the traditional notion of animal spirits and to his physical-chemical research, claiming that "I should think this [source] would be identical with that indicated by physiologists as being the source of animal spirits, namely the cerebrum" (GDV, p. 402; translated in Galvani, 1953a, p. 78). Although in Galvani's theory the mechanism of muscle contraction in the animal Leyden jar was independent of the brain, the latter still maintained its traditional role as the organ responsible for the main vital functions. In particular, the brain extracted the electric fluid from the blood and put it into the nerves, which in turn functioned as vehicles for its diffusion to all bodily parts, including the muscles. The notion of the blood as the matter from which the electric fluid originated depended on Galvani's research on its composition and on its function as a means of transmission of the various principles in the organism, including the "inflammable air" and the neuro-electric fluid.

However, Galvani admitted that "the manner whereby contractions are induced through the flow of electricity of this kind is very difficult to understand and is veiled in obscurity." He suggested two possible explanations: One came from Haller's theory of irritability and considered the action of electricity as a "mechanical irritation [...] provoking the irritability" of muscle fibers; the other rested on electrical theory and referred to a "very rapid discharge" of the neuro-electric fluid through the muscle fibers that produced "a specific and powerful attraction between the particles composing the muscle fibre after the manner and nature of common electric vapour [...], resulting in the fibre's becoming shorter when these particles are drawn closer to one another" (GDV, p. 403; translated in Galvani, 1953a, p. 79). These alternative explanations could not only satisfy different types of readers, physiologists, and physicians, as well as those who studied electricity, but also faithfully reflect the approach Galvani had adopted in the study of muscular motion, which rested on the integration of a physiological and an electrical perspective.

Although the mechanism of muscle contractions had not been clarified by the experiments at a fundamental level, the electrical nature of "animal spirits" was—from Galvani's perspective—definitively established. By referring again to the contribution of "fortune" in his investigations, Galvani claimed with the pride of the discoverer:

> If it will be so, then the electric nature of animal spirits, until now unknown and for a long time uselessly investigated, perhaps will appear in a clear way. Thus being these things, after our experiments, certainly nobody would, in my opinion, cast doubt on the electric nature of such spirits. For although by reason of our deliberations and numerous experiments we were perhaps the first to present this in public before our Anatomical Theatre, and although many distinguished scholars published the same theory long ago, still we could never suppose that fortune were to be so friendly to us, such as to allow us to be perhaps the first in handling, as it were, the electricity concealed in nerves, in extracting it from nerves, and, in some way, in putting it under everyone's eyes. (GDV, p. 402; translated with revision from Galvani, 1953a, p. 79)

After more than 10 years of research activity, Galvani finally felt confident that he had proved the existence of animal electricity and experimentally confirmed the ideas about the neuro-electric fluid which he had presented during the public function of anatomy of 1780. His theory represented a unification, made on an original basis, of those ideas with the Hallerian notion of the muscle as the site of irritability and, consequently, of the force responsible for muscular contraction. At the same time, moreover, Galvani's theory was able to resolve the objections formulated by Haller and his supporters that had long hampered the development of a neuro-electric notion of animal motion.

6

The Controversy Between Galvani and Volta Over Animal Electricity

THE FIRST STAGE

THE PUBLICATION OF Galvani's *De viribus electricitatis in motu musculari* started a lively debate among a large number of contributors. In Italy and all over Europe, scores of naturalists, physicians, physicists, and simple amateurs engaged in the replication and variation of Galvani's experiments with the aim to confirm, refute, or modify his discovery. In the last decade of the eighteenth century, animal electricity became the most important scientific topic being debated by the medical and scientific community, along with the new chemistry proposed by Lavoisier. The story of this debate is a fascinating one, not least for what it can tell us about the nature of scientific disputes. However, we are still far from having a complete picture of it despite the excellent work carried out, for instance, by Walter Bernardi in Italy and Maria Trumpler in Germany (Bernardi, 1992; Trumpler, 1992; see also the contributions in Bresadola & Pancaldi, 1999).

In the following pages we shall not attempt to contribute to the completion of the picture; instead, we shall reconstruct the fundamental steps in the controversy between Galvani and Volta. Even if we are aware of the limitations of such a perspective, as it does not take into account the other actors in the debate on animal electricity, we believe that it can be useful to get a better understanding of the scientific issues implied in the arguments of the two Italian scientists, and also to correct a few historiographical errors that have been made in the interpretation of both Galvani's and Volta's scientific work. In particular, we shall focus on the works Galvani wrote after *De viribus* in order to examine how he modified his model of the "animal Leyden jar" as a result of Volta's arguments, and of the new experiments he carried out in the 1790s. Likewise, we shall reconstruct Volta's research that relates more closely to the controversy engaged in with Galvani, while avoiding, for the time being, a detailed treatment of aspects of Volta's research that led to the discovery of the battery and the new concept of an electrical current.

6.1 GALVANI'S WORK IN THE SCIENTIFIC CULTURE OF THE LATE EIGHTEENTH CENTURY

We do not know whether Galvani and Volta ever had any direct contact before or during the controversy on animal electricity. In 1780 Volta stayed for a few days in Bologna, where he went to the Institute of Sciences. Sebastiano Canterzani, the secretary of the Academy of Sciences, remembered Volta's visit in a letter of September 20, 1780, addressed to Carlo Bianconi, the secretary of the Academy of Milan:

> I am most obliged to You, illustrious Sir, for giving me the possibility of personally meeting Mr. Alessandro Volta, whose celebrated name I knew before for his great discoveries in physics! The special esteem provided for similar people renders their company most satisfactory, in particular when they possess such a likeable character and exquisite manners as Don Alessandro. But he stayed in Bologna very briefly, only two days. I went with him to the Institute, where he spent several hours, and I introduced him to Mr. Monti, Veratti, Matteucci, and some others of our professors.[1]

We would like to think that on this occasion, besides Canterzani, Monti, Veratti, and Matteucci, Volta also met Galvani, who at the time was professor of anatomy at the Institute. It is worth noting that in just a short time, Galvani would start his electrophysiological experiments, while in the previous month of February in the anatomical theatre of the Archiginnasio he had presented his ideas on the role of electricity in vital functions. It is therefore quite plausible that Galvani would want to be present to meet a man of science like Volta, who had already published some fundamental contributions in the field of electricity. And yet nowhere in their writings is there any reference to the fact that they had met or knew each other personally. At any rate, Galvani and Volta always held each other in high esteem, even when their opinions began to diverge more and more as the controversy on animal electricity progressed. In a letter to Volta, which he enclosed to the second edition of *De viribus,* Giovanni Aldini—Galvani's nephew—sent his uncle's "kindest regards" to the Como physicist, adding that Galvani was "an admirer of your industry and successful experiments" (VEP, III, pp. 181–182). Galvani himself, in his works, referred to Volta as a "most learned physicist and experimenter" and "one of the most famous physicists and experimenters of our century." In 1797, when the differences in their interpretations of phenomena seemed irreconcilable, Galvani stressed Volta's

---

[1] S. Canterzani to C. Bianconi, Bologna, September 20, 1780, in BUB, ms. 2096, b. I.

"erudition, and depth of wit," referring to him as the author of "many beautiful and finest experiments" (GEA, p. 55). On his part, Volta always considered Galvani's experiments "very fine" and his discoveries "rather admirable," while he defined their author a wise experimenter, in opposition to those "jugglers in physics" who had dealt with animal electricity before him. In the spring of 1792 he hailed Galvani's discovery of animal electricity as "one of those great and brilliant discoveries, which deserves to be considered defining an era in the field of physical and medical sciences," comparing it with Franklin's discovery of the electrical nature of lightning (VEN, I, pp. 15, 24). And there was no lack of expressions of esteem and appreciation for Galvani and "his very fine experiments" or for "the many nice experiments he has made," even when Volta had quite different opinions on the nature of that electricity that the Bolognese doctor "obstinately" called "animal." It is also evident, as we shall see, that each of the two contenders paid great attention to the opponent's results and to their interpretations, not only to find possible weak points but also to develop their own hypotheses and make them more suitable to the experimental data as they were emerging.

A clear demonstration of the high esteem in which Galvani and Volta held each other's writings can be found in the fact that they always felt it necessary to answer and refute every single argument raised by their opponent and, what is more important, they both felt the need to go back to the experiments to get the necessary elements to rebut their opponent's criticisms. In the first of the five *Memorie sull'elettricità animale* (*Memoirs on Animal Electricity*) of 1797, after arguing for his notion of animal electricity through an analogy with electric fish, Galvani wrote, referring to Volta: "However I do not want to base my claims only on analogy arguments, and, as Mr. Volta uses experiments to prove the truth of his theory and the falsity of mine, so it is fair that I pursue a similar path" (GEA, p. 5).

Both Galvani and Volta were endowed with great argumentative skills; this sometimes resulted in them presenting their opponents with their own objections, and raising criticisms that apparently had no chance of being countered. However, the real importance of the controversy lay in the fact that it induced the two scientists to develop new facts and new experimental evidence in support of their own theses. It was indeed a great scientific controversy that deserves close analysis not so much for its rhetorical aspects (even if they marked some important phases) but especially for its scientific and epistemological features.

De viribus was probably completed between the end of 1789 and the beginning of 1790, but it only appeared in the seventh volume of *De Bononiensi Scientiarum et Artium Instituto atque Academia Commentarii*, dated 1791. In actual fact it was published at the beginning of the following year (Bernardi, 1992, p. 59). The delay in publication was a typical feature of the *Commentarii*, which was the official publication of the Bologna Institute of Sciences that collected the research works of its members. For this reason and also for its relatively small circulation, the *Commentarii* was not a very effective means of scientific communication. Galvani thus decided to print a few copies of the *De viribus* separately, probably in 1791, and he sent these to various scientists; one of them reached

Volta in Pavia, in March 1792. A second edition of the treatise, edited and annotated by Galvani's nephew, Giovanni Aldini, was published in Modena in the autumn of 1792; it comprised 750 copies, which was a remarkable number at the time.

Nonetheless, there were other channels through which Galvani's research was being circulated. These included direct testimony of people who had participated in the experiments on animal electricity, correspondence emanating from Bologna, and the reviews in scientific periodicals of the time. In Germany, for example, the first news of Galvani's research was received in August 1792 via a student who had spent some time in Pavia; there he had witnessed the first discussions on the *De viribus*. In Paris news was learned from Eusebio Valli, a physician from Casciana (in the countryside near Pisa) who had studied in Pavia, and who replicated Galvani's experiments at the Académie des Sciences in July of the same year. In England Galvani's research was made public by Tiberio Cavallo, who in 1793 read two letters at the Royal Society sent to him by Volta. This "Account of some discoveries made by Mr. Galvani, of Bologna; with experiments and observations on them" (VEN, I, p. 169) was published in the society's *Philosophical Transactions*.

Besides the difference in the times and the ways in which it became known, *De viribus* caused a variety of reactions among scientists. As the Italian historian Walter Bernardi has shown, the controversy with Volta was neither the first nor the only dispute in which Galvani was involved, apparently against his will (Bernardi, 1992, 2000). Both from Bologna and from various other centers of the Italian peninsula there were immediate criticisms from the supporters of Hallerian irritability. They charged Galvani with reviving Tommaso Laghi's old neuro-electric hypothesis, without having first overcome the objections raised against it some decades before by authors like Felice Fontana and Leopoldo Caldani (see Chapter 3.3). The importance of these attacks led Galvani and his collaborators (particularly Aldini) to prepare an introductory dissertation and a series of notes for the second edition of *De viribus*. One of the notes dealt with the alleged impossibility of creating an electrical imbalance between animal tissues due to their electrically conductive nature (one of the Hallerians' major objections). It argued that this objection was contradicted by the examples of tourmaline and of electric fish, which produced electricity even when completely immersed in water, a conducting liquid. These examples and Galvani's and Volta's experiments proved that "animal humidity does not prevent electrical excitation; on the contrary, it makes an easy way for electricity to obey the soul and, on its command, to move toward various places in order to put muscles in motion" (GME, p. 207). In another note there was a direct attack on the Hallerian theory of irritability, based on the principle of the simplicity of nature. As both animal electricity and irritability were considered to be properties of muscle and had analogous effects, they either needed to be reduced to the same principle or one of the two had to be excluded. Since irritability was a "not yet well known" principle, as Haller himself admitted, Aldini (together with Galvani) challenged its supporters to prove its existence and its "established laws," as Galvani had done for animal electricity; otherwise they would have to abandon their theory in favor of animal electricity (GME, pp. 212–213).

Moreover, it was remarked that, if the theory of animal electricity developed in *De viribus* represented an alternative to the doctrine of irritability, it was an altogether different matter from the neuro-electric conceptions of muscular motion proposed decades or even just a few years earlier. According to the introductory dissertation to the second edition of *De viribus*, the novelty of Galvani's theory did not lie so much in confirming the concept of animal electricity or the neuro-electric hypothesis of muscular motion, as in having pointed out new characteristics of this electricity that "governs the animal machine." Among these characteristics was the capacity of animal electricity to manifest itself "without the intervention of any artificial electricity, without friction or strokes, in both cold- and warm-blooded living animals." In addition, it was stressed "that [animal electricity] is present in the forces of life until the end and flows and circulates easily and regularly between muscles and nerves." It was experiments, wrote Aldini, that "indicated to Galvani the celebrated path to follow in order to discover the electricity intrinsic to the animal," and it was the appeal to experiments that definitely distinguished Galvani's treatise from all the works that had been previously published on the topic (GOS, pp. 222–223).

Let us consider, for example, the work of the Frenchman Pierre Bertholon; in 1780 he won (together with the Piedmontese physician Francesco Giuseppe Gardini) a prize for "medical electricity" awarded by the Lyon Académie des Sciences, Belle-Lettres et Arts. In 1786 Bertholon published an enlarged second edition of his work, promoted by his publishers as "the first complete body of doctrine on electro-animal economy." The examination of the relationship between electricity and living bodies was divided into two parts: The first described the influence of "artificial" and "atmospheric" electricity on the human body and claimed the existence of an "animal" electricity being produced by the organism, while the second one examined "the electricity of the human body in states of disease"—that is to say, the application of electrical knowledge and instruments to diseases and their treatment (Bertholon, 1786).

The organization of Bertholon's book was thus similar to that of *De viribus*, so much so as to raise the suspicion of a direct influence of the first one on the second. This suspicion was strengthened by Galvani himself, as in his treatise he repeatedly cited Bertholon, giving him the merit, among other things, of introducing the phrase "animal electricity" (which had indeed appeared previously in the writings on the torpedo by Walsh and Hunter; see Piccolino, 2003a). Moreover, the two naturalists shared some general ideas about the relationships between electricity and the organism. Bertholon, for example, maintained that "artificial electricity has the greatest influence on muscular movements" and that atmospheric electricity produced the same effects on the animal body as artificial electricity. Above all, he declared that "beyond the electricity that the atmosphere communicates to human bodies, there is another that is peculiar to them," which he called spontaneous or animal electricity. The latter was produced within the body by heat or by friction between the solid parts or between these and the fluid parts, and joined the fluid flowing in the nerves resulting in a "nervous-electric" fluid that caused various effects, including muscular motion (Bertholon, 1786, vol. 1, preface and pp. 117–118).

It is indeed the apparent similarity between some of Bertholon's and Galvani's physiological ideas that makes the great difference in their approaches to the study of the living world even more striking, to our eyes as well as to the majority of their contemporaries, and it makes Galvani's research even more innovative. In Bertholon's book there were no experiments to demonstrate the effects of artificial or atmospheric electricity on the human body, even less on muscle contractions. The conjectural, rhetorical, and anecdotal character of Bertholon's arguments is in clear contrast with his statements on the importance of facts and experiments in the study of "animal economy." Thus, dealing with spontaneous electricity in the human body, Bertholon stated that "the truth one wants to establish must be proved by facts and observations," but immediately afterwards he provided a list of observations that were often secondhand and of a dubious validity. To the list belonged "natural electrical commotions felt or produced by some persons, spontaneous combustions, sparking lights which come out from the eyes of people affected by hydrophobia or by some violent passion, [...] electric lights which shine spontaneously on the body of some persons during night without any previous friction" and, at long last, an experiment designed by Bertholon himself, according to which an insulated person, put in contact with a conductor, exhibited signs of electricity. Unfortunately, he noted, not all "the members of the human species are equally apt to these sorts of experiences," but this only meant that the lack of the signs of spontaneous electricity was due to "some obstacles," and not to the absence of such electricity. Bertholon thus concluded:

> Here is the sum of the facts, observations and experiences which are related to the spontaneous electricity of the human body, i.e., to that sort of electricity which is not the effect of the communication of atmospheric electricity nor of the electricity produced by an electric machine. (Bertholon, 1786, vol. 1, p. 140)

In Bertholon's work, as well as in that of other authors of the time like Gardini, there was no clear distinction between a merely physical electricity which can be produced in an animal (or any other) body by means of friction or other trivial causes, and electricity produced by specific physiological processes, in order to carry out specific functions within the animal economy. In the perspective of these authors, electricity becomes an essential element of the human (and also animal and vegetable) microcosm as well as of the physical macrocosm, its variations causing changes in mood and temper—exactly as was the case with humors in the Galenic system—and its disorders being the cause of any sort of disease and pathological state. Immersed in an atmosphere rich in electricity, wrote Bertholon, man "is like a fish or rather a sponge immersed in water"; therefore, he cannot avoid being affected by variations in the electricity surrounding him and entering the body through the pores of the skin or through the lungs. As a consequence, diseases should be reclassified according to the excess or deficiency of the "electric fire" in the body. And to give strength to these "new systems," which, in fact, revived ancient science in a new form, one resorted to a series of extraordinary tales. One of these involved a lady from Cesenatico (a town on the Adriatic coast) who burned due to an excess of electric fire, or the story of a religious man from Florence who survived for a few days after a spontaneous fire suddenly enveloped him—a fire due, again, to an excess of internal electricity. There were also stories about noble ladies with a high degree of "sensibility" to the variations of atmospheric electricity, who would suffer atrocious pains or would become extremely irritable on the days of abundant electricity, and reports about people with particularly abundant electricity who emitted a powerful light during night. Last but not least, as electricity is life and sex is an act essentially related to the transmission of life, here was the gossip about men so powerful in vital electricity as to give electric shocks when they reached sexual orgasm, as if they were torpedoes or other electric fish.

While Bertholon supported his theory of animal electricity with anecdotal, unique, and extraordinary evidence, Galvani devoted a large part of his *De viribus* to a detailed description of animal preparations, experimental arrangements, and results, with the aim of rendering his experiments replicable. Moreover, he adopted a series of strategies, partly material, partly literary and rhetorical, which had been developed in modern science to serve both as a guide in scientific research and as a common ground on which to assess its results and implications and validate its conclusions (Eamon, 1990; Shapin & Schaffer, 1985). Among these strategies, an important role was played by testimony, which implied the possibility of extending a particular experience to other people so as to transform it into a "matter of fact" not depending on the account and authority of a single individual. Displaying an experiment to witnesses, as happened in public demonstrations organized in literary or scientific societies or in the *salons* of the erudite nobility, reinforced the credibility of the experimenter and the truthfulness of the experimental results even in the eyes of a public that did not assist in the performance. The same aim was pursued by quoting in the published works the names of those who assisted in the experiments, especially when these were well known and authoritative.

In Galvani's case, the reproducibility of results did not represent a particular problem, since the experiments with frogs and metal arcs were successfully replicated by most scientists of the time. Despite this, the Bolognese physician resorted to the strategy of testimony in order to convince his readers about the truthfulness of his reports. For instance, in *De viribus* on the occasion of the observation of muscular contractions at a distance (his so-called first experiment) he referred to the presence of other people. In the third part of his memoir, he explicitly cited the "most erudite Spanish Rialp, once belonging to the Society of Jesus" as one of his assistants in the laboratory, who played an active and very important role in the experiment on the frog in a silver cup, Galvani's so-called second experiment (GDV, p. 378). This shows how Galvani used testimony, especially in the case of experiments that were new and original, and thus more exposed to doubts and perplexities on the part of the reader. This strategy was adopted by Galvani not only in *De viribus* but also in his following works. For example, in *Trattato dell'arco conduttore* he wrote that the experiment on the direct contact between nerve and muscle (his "third experiment") had been "successfully repeated last summer in the presence of learned physicists" (GAC, p. 84). In his *Memorie sulla elettricità animale*, directed to Lazzaro Spallanzani, Galvani used testimony to answer Volta's criticism that the crucial experiment on muscular contractions without the use of metals gave doubtful and inconstant results:

> I can assure that I obtained the motions [of the frog's legs] not a few times, as Volta claims, but in many, many experiments, so that in a hundred times the effect had not happened just once; and these experiments have been recently replicated in our Institute of Sciences in the presence of many learned scholars, and other people well versed in this sort of things, and they never failed.

To reinforce the truthfulness of the experiment, and therefore the factuality of the phenomenon, Galvani then added:

> To confirm the truth, I succeeded in obtaining the kind attention and assistance of several illustrious friends to my experiments; among them I think it suffices to cite the most learned secretary of our academy, Sebastiano Canterzani, to whom I never omitted to show any of my experiments, and to communicate any of my conjectures before publishing; but more than any other I could quote, You [i.e., Spallanzani] who are such an excellent and absolute master in every scientific field, as in the difficult art of experiment, and who, for my good fortune, have been not only a witness to this experiment, but an approving judge. (GEA, pp. 5–6)

Both Canterzani (a mathematician well known in the Republic of Letters both for his research and his institutional role in the Bologna Institute of Sciences) and Spallanzani (a celebrated man of science with international connections) were certainly authoritative and trustworthy individuals in matters of science. Galvani was well aware that his reference to these two scholars reinforced the veracity of his scientific report and his credibility as an experimenter.

In eighteenth-century experimental philosophy there was another kind of testimony that aimed at persuading the readers of the validity of the experimental reports, and one that was perhaps more important than visual testimony. This technique, defined by historians as "virtual witnessing," consisted in transmitting to readers an image of the experimental scene so realistic as to convince them that an experiment had actually been performed in that way and yielded that specific result (Shapin & Schaffer, 1985, esp. Chapter 2). Virtual testimony could be pursued both through writing techniques and through the use of detailed and naturalistic illustrations. For instance, Robert Boyle—who was one of the first to develop this strategy—included very realistic representations of instruments such as the air pump in his works and of other elements of the experimental scene (like the people involved) that were not essential to the understanding of the text but reinforced the credibility of the report. As to the language used by Boyle, and which he himself defined as rather "prolix," here is what he wrote in the warning *To the Reader* of his 1660 treatise on experiments concerning the elasticity of air:

> On my being somewhat prolix in many of my Experiments, I have these Reasons to render, That some of them being altogether new, seem to need the being circumstantially related, to keep the Reader from distrusting them. [...] the most ordinary reason of my prolixity was, That foreseeing that such a trouble as I met in making those trials, and the expence of time that they necessarily require (not to mention the charges of making the Engine, and imploying a Man to manage it) will probably keep most Men from trying again these Experiments: I thought I might do the generality of my Readers no unacceptable piece of service, by so punctually relating what I carefully observ'd, that they might look upon these Narratives as standing Records in our new Pneumaticks, and need not reiterate themselves an Experiment to have a distinct idea of it, as may suffice them to ground their Reflections and Speculations upon. (Boyle, 1660, unnumbered)

Galvani made ample recourse to the method of virtual witnessing in *De viribus*. We may say that the structure of this work was designed not so much with the aim of offering readers a new conception of the role of electricity in neuro-muscular physiology but of presenting the intense and rich development of the author's experimental research in a visual way. In the brief introduction Galvani explicitly wrote: "I thought my efforts would be rewarded [...] if I presented a brief and accurate account of the discoveries in the same order of circumstance that chance and fortune in part brought to me, and diligence and attentiveness in part revealed" (GDV, p. 363; translated in Galvani, 1953a, p. 45).

The story begins quite abruptly with a very vivid description that presents to readers' eyes the "room for experiments" at the moment of the fortuitous production of the spark which makes the frog legs' contract even if the animal was not connected to an electric machine. The reader is put in front of a doctor who, after preparing the frog, was reflecting to himself while several assistants were working in the laboratory, one touching the animal preparation with a scalpel, a second rotating the disk of the electrical machine, and a third observing a strange phenomenon and calling Galvani's attention to it. *De viribus* then continues with a report, organized in chronological sequence, of the experiments that followed the first "chance" observation, the results obtained, and the circumstances in which the experiments were performed; these were often accompanied by references to witnesses or details underlining the "reality" of the places (such as "the house of the most illustrious Jacopo Zambeccari" where an important experiment with metal arcs had been carried out).

Some of Galvani's experiments are illustrated in four plates included in *De viribus* that were meant to help the reader focus on the fundamental circumstances of the experiments described in the text but were also designed to give an impression of "reality" and "factuality" surrounding Galvani's research. In the second plate, for instance, the terrace where Galvani performed the experiments on "atmospheric electricity" is represented with very realistic details, such as roof tiles, a dormer on the left side, the pulley to raise water on the top of the well on the right side, the clouds in the sky, and the sun shade on the floor (see Fig. 5.2 in Chapter 5). In the third plate several experimental arrangements with metal arcs are illustrated, while in the fourth plate other experimental arrangements are shown together with the representation of some gentleman in elegant dress who participated in the experimental work (and who were to be considered trustworthy due to their social status) (see Figs. 6.1 and 5.7 in Chapter 5). Even more realistic than the published plates are the watercolors that served as a model for the illustrations (and which are now kept in the archives of the Bologna Academy of Sciences). While in the first plate of *De viribus*, for instance, Galvani's room for experiments has no ceiling, this is represented in the original watercolor, in which the realistic dimension is further reinforced by the use of shading, the decoration of the furniture, and a marked use of perspective (compare Figs. 1.3 in Chapter 1 and 4.1 in Chapter 4).

The illustrations in *De viribus* contrast strongly with those published in Galvani's *Memorie sulla elettricità animale*, which are much more schematic and didactic. One reason is that *De viribus*, being the first of Galvani's works on animal electricity, deployed every means available, including visual representations, to convince the reader of the veracity of the experimental results and of the credibility of Galvani as an experimenter. Once accepted by his colleagues as reliable—and this happened soon after the publication of *De viribus*—Galvani could abandon some of the rhetorical strategies adopted in

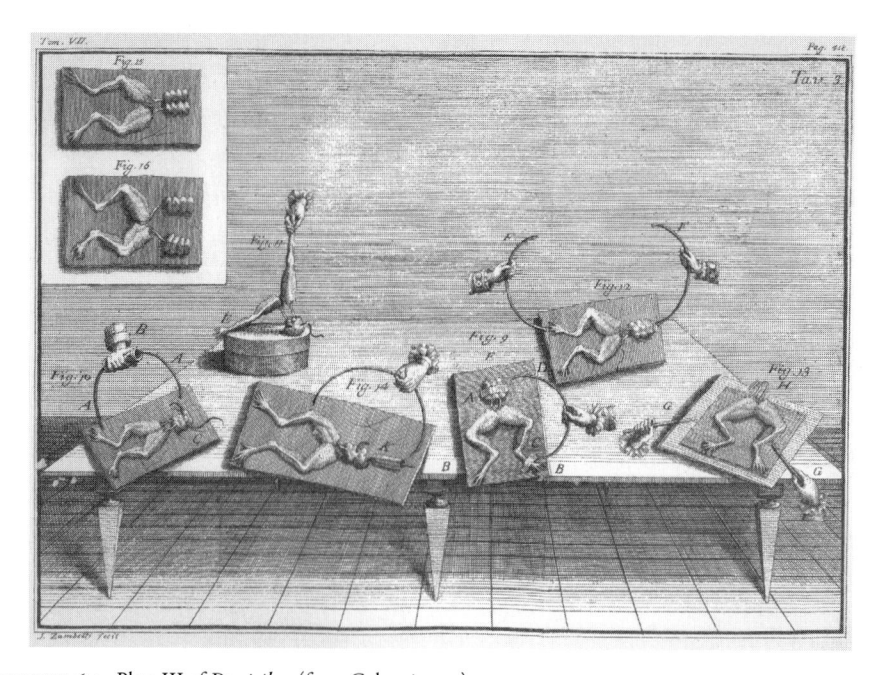

FIGURE 6.1. Plate III of *De viribus* (from Galvani, 1791).

his first memoir, either eliminating the illustrations as in his *Trattato dell'arco conduttore* or attributing different functions to them as in his *Memorie*.

Another strategy adopted by Galvani to persuade his readers that his descriptions concerned real facts and experiments and not inventions consisted of his frequent reference to chance, to surprise for something that happened unexpectedly (or for something expected but did not happen), to expressions of marvel, satisfaction, or amusement. In some cases—as we already noted—Galvani represented in a particularly vivid way the experimental scene, as when he connected the frog to the electrical machine through "an iron wire more than a hundred arm long" and observed "with great wonder" that, at the moment the conductor of the machine gave a spark, "the severed frog moved and almost jumped, even at so great a distance." Or take the experiment of the "electrical pendulum" that we considered earlier (see Chapter 5.2) and that we like to repropose as a good example of the style in which *De viribus* is written:

> If a frog is held by one leg with the fingers so that the hook fastened in the spinal cord touches a silver plate and if the other leg falls down freely on the same plate, the muscles are immediately contracted at the instant that this leg makes contact with the plate. Thereupon the leg is raised and lifted up, but soon, however, it spontaneously relaxes and again falls down on the plate. As soon as contact is made, the leg is again lifted for the same reason and thus it continues alternately to be raised and lowered so that to the great astonishment and pleasure of the observer, the leg seems to function somewhat like an electric pendulum. (GDV, pp. 379–380; translated with revision in Galvani, 1953a, p. 61)

Galvani's literary style differed significantly from that of those who preceded him in the study of animal electricity, such as Bertholon and Gardini. This difference was remarked upon by many readers of the time. In the second issue of the *Opuscoli scelti sulle scienze e sulle arti* for 1792—an important scientific journal published in Milan—a review of *De viribus* appeared that then served as a model for a number of other reviews published both in Italy and beyond the Alps. This review opened with the following words:

> That the electric fluid should play a role in muscular motion was suspected by many [authors]; the phenomena of the torpedo and of the trembling eel, in which electricity so clearly manifested itself, greatly increased this suspect. However, no one could find a clear demonstration [of electricity's role] in the other animals until now, as it is offered by Mr. Dr. Galvani in his nice dissertation De viribus electricitatis in motu musculari. (Soave, 1792, p. 113)

After repeating some of Galvani's experiments in the spring of 1792, Volta formulated a similar opinion. In his *Memoria prima sull'elettricità animale* (*First Memoir on Animal Electricity*), dated May 5, 1792, he stated that "the existence of a real and true animal electricity [...] is evidently proved in the third part of this work [i.e., *De viribus*] with many well combined and accurately described experiments" (VEN, I, p. 15). The novelty of Galvani's research was thus the experiments on the conducting arcs and metal foils or "armatures," which distinguished it from "those badly understood or at least equivocal experiences and observations" that had characterized previous works in this field. Here Volta explicitly referred to some authors, including Bertholon, "who wrongly take for animal electricity that excited by the friction of animals' fur, men's hairs and dresses, and confuse the artificial extrinsic electricity with a natural electricity intrinsic to the living bodies." The difference between these authors and Galvani was analogous to that between those who only supposed the electrical nature of lightning, like Jean-Antoine Nollet and other electricians, and Benjamin Franklin, who instead rendered this atmospheric electricity "manifest and touchable." As the merit of this discovery was Franklin's, so it was Galvani who had "all the merit and paternity of this great and stupendous discovery" of animal electricity (VEN, I, pp. 18–24).

Volta initially accepted Galvani's research with enthusiasm, so much so that he wrote of having "changed [my mind] from incredulity to fanaticism" with regard to the discovery of animal electricity. Incredulity depended on the fact that, before *De viribus*, it was very difficult even to suppose that the electricity of the torpedo and other electric fish could be common to all animals, which in normal conditions did not manifest any electric sign (shocks, sparks, attractions, and repulsions, etc.). In fact, thanks to his experimental arrangement of the prepared frog and the metal arc, Galvani was the first to be able to make visible an "electricity so weak, that we can not feel any shock from it nor make it sensible to the most delicate electrometer" (VEN, I, p. 25). For Volta, the proof that Galvani had found a new electrical phenomenon came from the fact the muscular contractions were produced by connecting nerve and muscle through a conducting arc, but it did not happen if the connection was established through an insulating material: "it is well known to anybody even with little knowledge of electric science"—Volta wrote—"that the proper and only function [of the conducting arc] is to re-establish the equilibrium of the electric fluid, previously imbalanced" (VEN, I, p. 16). As we shall see, shortly afterward Volta radically changed his opinion on this point, contesting Galvani's view of animal electricity and introducing a new concept in physics, namely the electromotive power of metals.

6.2 VOLTA'S EARLY RESEARCH ON ANIMAL ELECTRICITY: QUANTIFICATION, MUSCULAR PHYSIOLOGY, AND THE "SPECIAL THEORY OF CONTACT ELECTRICITY"

Volta was 47 years old in the spring of 1792; he held the position of professor of experimental physics at the University of Pavia, which was the most prestigious institution in the Habsburg Empire, and had recently become a member of the Royal Society of

FIGURE 6.2. A portrait of Volta as a young man (from Volta, 1918–1929, vol. 3; original engraving by Raffaello Morghen, based on a drawing made by Luigi Sabatelli and Pietro Ermini).

London (see Fig. 6.2). His fame within the Republic of Letters came mainly from his studies on electricity, which earned him the title of the "Newton of electricity," as some naturalists like Jean-André Deluc and Georg Christoph Lichtenberg called him. In 1775 he had invented the electrophorus—an instrument that, when charged, provided a seemingly inexhaustible supply of electricity—and afterward had formulated some fundamental new concepts, such as those of capacity and tension, which are still at the heart of electrical science (see Fig. 6.3). His research led him to focus his attention on weak electricity, that is, on those phenomena that did not manifest any visible sign of electricity (such as attractions, repulsions, electric sparks, and sounds) but were considered genuinely electrical, such as those involving the electricity present in the atmosphere and that developed by water during evaporation. To make weak electricity—both "atmospheric" and "natural"—detectable, Volta invented the "*condensatore*," an instrument that could reveal very small quantities of electricity, thanks to the play of variable capacity. As we shall see, this instrument would play an important role in the controversy over animal electricity.

Volta's interest in weak electricity is very important in understanding his reaction to Galvani's *De viribus* (see Pancaldi, 2003). The contractions of Galvani's frog manifested an electricity so weak that it could not be measured by any of the ordinary electrometers; an evaluation of the quantity of electricity needed to induce the contraction could be obtained only with the *condensatore*. This meant that Galvani's prepared frog could be considered as "an animal electrometer, incomparably more sensible than any other most sensible electrometer" and thus could represent a new and powerful experimental tool to study weak electricity. After successfully repeating Galvani's experiments both on cold- and warm-blood animals, Volta focused on the question of the "quantity, quality, and modality" of animal electricity. His approach in this investigation was the same as the one

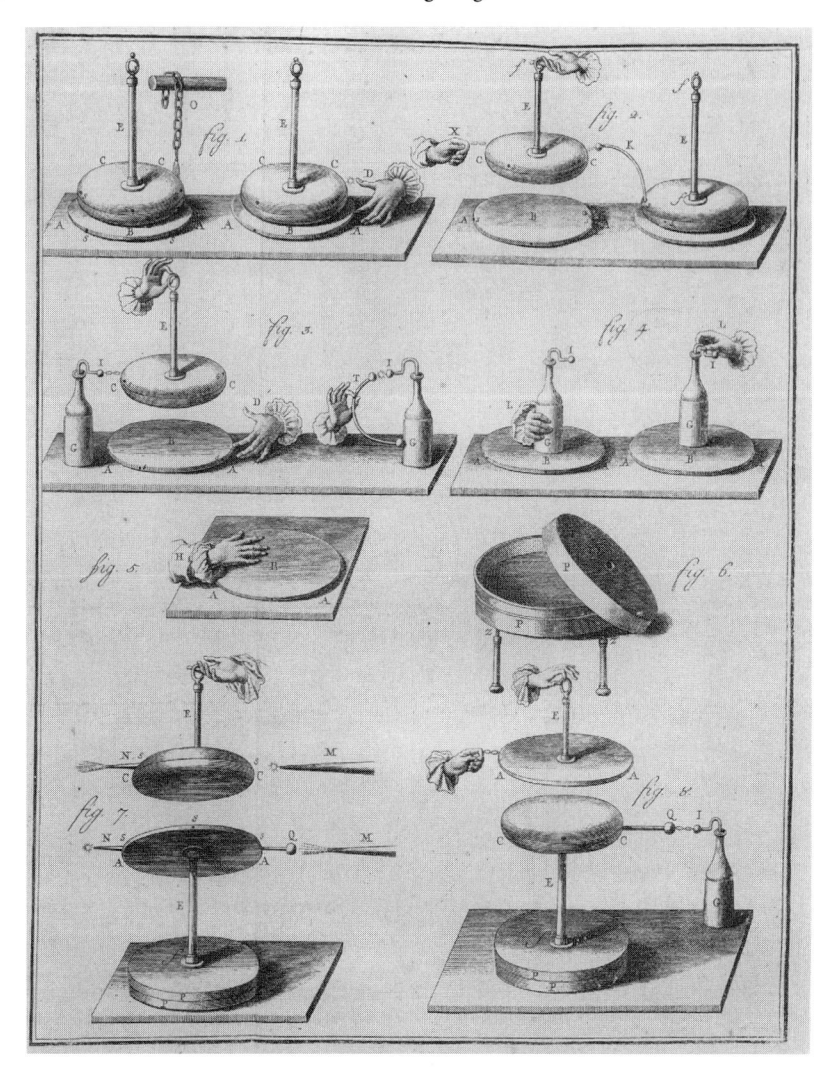

FIGURE 6.3. Volta's electrophorus in a plate of his *Seguito della lettera al signor dottore Giuseppe Priestley*, 1775 (from Volta, 1918–1929, vol. 3).

he had adopted in previous research: "What good can one expect, especially in physics, if things are not reduced to degrees and measures? How can one evaluate the causes, if not only quality, but also the quantity and intensity of effects are not determined?" (VEN, I, p. 27). Volta's statement suitably summarized the "quantitative spirit"—the values of order, systematization, measurement, and calculation—that characterized scientific investigation in the latter part of the eighteenth century and distinguished it from traditional natural philosophy (Fox, 1974; Frängsmyr, Heilbron, & Rider, 1990; Heilbron, 1993).

Volta's first step toward the quantification of animal electricity was the measurement of the quantity of electricity sufficient to produce muscular contractions in the frog's legs. This was also one of the early questions examined by Galvani, who had concluded that contractions were indeed the least sign of an unbalanced electricity (see Chapter 4.1). Volta, however, made a step forward by designing an experiment that was not described in *De viribus* or in Galvani's laboratory notebooks. He took a minimally charged Leyden jar and connected its positive pole (probably the hook that stuck out from its interior) to the nerve and the negative pole (probably the external coating or "armature") to the muscle; then he inverted the poles. He found that a much smaller quantity of electricity was needed to produce the contractions in the first case than in the second, thus concluding that the nerve—or the internal surface of the muscle—was the seat of negative electricity, while the external surface of the muscle was positively charged. Volta's conclusion, based on the assumption that the flow of electricity between nerve and muscle was favored by the contact of the animal parts with those oppositely charged parts of the jar, was contrary to what Galvani had claimed in *De viribus*, even though the Bolognese savant had presented his idea not as a fact but only as an "extremely convincing hypothesis" (GDV, p. 400; translated in Galvani, 1953a, p. 77).

Galvani's hypothesis concerning the seat of the electrical disequilibrium depended on the analogy between the neuro-muscular system and the Leyden jar, which normally was positively charged on the internal surface and negatively charged on the external. Incidentally we may note that, as Volta claimed, inside the muscle (and nerve) fiber there is actually an excess of negative charge compared to the exterior. However, the "truth" of Volta's result might be just the issue of chance. As a matter of fact, contrary to what he claimed, modern experiments show that it is easier to stimulate a nerve if we apply a negative stimulus to its external surface than a positive one (as Volta did). From that point of view one could conclude that Volta has derived a "true" result from an incorrect experiment. In this regard, we must, on the other hand, be aware that it is very difficult to reconstruct all the circumstances of a past experiment, even when its description is accurate, as is usually the case with Volta and Galvani. Indeed, it is often hard to identify the instruments and materials used, as well as to understand the part of "tacit knowledge" about experimental procedures and methods that is fundamental in the laboratory activity but is not rendered explicit in scientific writings (Polanyi, 1958).

In May 1792, Galvani responded to Volta's alternative view about the seat of animal electricity in a letter addressed to Bassiano Carminati, a physician and colleague of Volta in Pavia. The letter was published both in the periodical "Giornale fisico-medico" and in the second edition of *De viribus*. For Galvani, the fact that the nerve reacted more readily when stimulated by positive electricity—as Volta had found—could depend on the disequilibrium created by the extra charge brought in the animal preparation by the artificial electricity of the Leyden jar, and not on the "natural" imbalance existing between muscles and nerves. In other words—according to Galvani—the contractions observed in Volta's experiment could be due to a secondary discharge consequent to the "surcharge" created within the animal tissue by external electricity. If this was true, then—Galvani wrote—"it would seem that, notwithstanding the beautiful experiments carried out by the most illustrious Mr. Volta, the hypothesis of an excess [of electricity] in the internal part and surface of muscular fibre, and of a defect in the external one, could be maintained" (Galvani, 1792, p. 74).

Galvani continued by saying that Volta's research could suggest a new explanation of a very important "physiological point," that is, "the physical cause of voluntary motions." As in Volta's experiments artificial electricity in a condition of surcharge, when directed to the nerve, produced a sort of impulse that set the electricity contained in the muscle fiber into action, so one could suppose that a similar action was exerted by the will on the electricity contained in the brain or in the muscle (Galvani, 1792, p. 74). Although Galvani considered this explanation more dubious than the one he had suggested in *De viribus*, the concept of an extra electrical charge would later return in his work.

It is important to note that Galvani's speculation about the physiological mechanism of muscular motion was influenced by Volta's ideas. Far from being attracted only by those aspects of Galvani's research that related to the study of inorganic electricity, Volta was very interested in animal economy and the explanation of physiological functions such as sensation and animal motion (see Chapter 8). In particular, in his *Memoria prima* Volta proposed an explanation of voluntary motions that was mainly based on experiments carried out on "whole and intact animals" instead of on Galvani's prepared frog. From the observation that under these conditions it was only possible to produce muscle contractions by applying coatings ("armatures") made of different metals to the animal, Volta had derived some important conclusions. He had elaborated particularly on the existence inside the animal tissue of a continuous, maintained flow of electricity present during rest. In his views, in the absence of physiological or experimental excitation, this maintained circulation of electricity was so weak that it did not result in the dissipation of the internal electric disequilibrium produced by "the organization of the forces of life" (VEN, I, p. 33). This occurred because animal tissues and fluids, being poorer conductors than metals, slowed down the continuous flow of electrical fluid along the nerves and muscles. In this way Volta could provide a solution to one of the main Hallerian objections against the neuro-electric hypothesis (see Chapter 3).

Moreover, with the conception of a continuous flow of electricity, he was able to suggest a way of explaining the mechanism of muscle contractions. Indeed, muscles did not move "as long as the natural and harmonic motion of the electric fluid is not disturbed," but when for some reason "a turmoil or disturbance [happened] in the harmonic circulation, fluctuation or motion of the electric fluid inside the animal organs," then the contraction occurred. Among these causes Volta listed both the artificial application of metal coatings and arcs, and the natural action of the will, which can "increase, or decrease, or stop, or reverse the flow of the fluid towards those parts, i.e. the muscles, that intend to excite the motion" (VEN, I, p. 33).

Volta's reflections on the experiments with intact and whole animals were not restricted to physiological considerations, however. The physicist from Como deemed them to be "more instructive" than Galvani's "as they lead us to penetrate somehow the natural flow and condition of animal electricity in the whole, healthy living body" (VEN, I, p. 33). In a manuscript of the same period Volta focused on "the need for *dissimilar armatures*" in order to produce contractions, by remarking that this was not easily explained by Galvani's theory and that he had not been able to find a reason for this fact at that time (VEN, I, p. 40). In fact, the problem of the efficacy of different metals in exciting muscular contractions became the central issue of *Memoria seconda sull'elettricità animale* (*Second Memoir on Animal Electricity*), also published in the "Giornale fisico-medico." This work was dated May 14, 1792, that is, 9 days after the first memoir, a clear sign of the intense pace of Volta's experimental research in that period.

In the *Memoria seconda* Volta drew on the examination of the minimum quantity of electricity needed to excite muscle motions and discussed some of Galvani's experimental arrangements described in the first two parts of *De viribus*. In particular, Volta concluded that the observation of muscular contractions at a distance, Galvani's so-called first experiment, could be explained as the result of the action of the electricity produced by the electrical machine rather than as evidence supporting the existence of an electricity intrinsic to the animal body (see Chapter 4.2). Galvani thought that in this case the contractions were due to the flow of positive electricity from muscles though the nerves toward the "aerial strata" around the conductor of the electrical machine made electrically negative because of the discharge of the spark (which would subtract positive electricity from the conductor). Volta, on the other hand, considered that the phenomenon was due to the "action of the electrical atmospheres": the frog and its conductors were electrified because they were immersed in the "sphere of action" of the machine's conductor; when a spark was elicited at the discharge of the machine, the electrical atmosphere was destroyed and electricity accumulated in the frog flowed to the conductor to re-establish the equilibrium and thus contractions occurred. In this case the frog behaved like any charged body and there was "nothing surprising" in Galvani's observation or in his experiments on artificial and atmospheric electricity (VEN, I, pp. 46–47).

Volta's criticism went well beyond Galvani's first experiment, however. It concerned the foundation of Galvani's theory of animal electricity, namely the analogy between the neuromuscular system and the Leyden jar. Though Volta had accepted this analogy as a "plausible and fascinating explanation" of muscular motion, now he was of quite a different opinion:

> The most plausible and fascinating explanations, which seem in agreement with the fist general appearances, are rarely confirmed by a more rigorous examination of particular phenomena; and when we try to extend a nice discovery, which has presented us, to great and magnificent consequences, we are often obliged to make a step backward and to renounce to great part of our conceived plans. (VEN, I, pp. 57–58)

What Galvani had proved was the action of electricity on the nerves, but his idea of an electric circuit between the internal and external surfaces of the muscle through the nerve was—according to Volta—completely conjectural and not supported by convincing evidence. This was a serious methodological objection to Galvani and addressed a real weakness of his theory. As we have seen, Galvani had also found that nerves were more excitable than muscles and had supposed that the electrical disequilibrium was to be found between these two different tissues (see Chapter 5.3). The analogy with the Leyden jar, however, had made him change his mind and led him to consider the efficacy of nerves' stimulation as a consequence of their conducting property. We may note here that methodological questions were repeatedly put forward during Galvani-Volta controversy, which is thus a very instructive episode for understanding the epistemological aspects in the emergence of modern experimental science.

Volta based his criticism of Galvani's theory on two sorts of experiment. In the first instance he isolated the crural nerve of a frog (or the sciatic nerve of a lamb) and applied to it two metal coatings at a distance of "one or two lines" from each other (without any direct contact with muscles); then he connected the two coatings to a very weakly charged Leyden jar. Contractions developed as occurred when one armature was on the nerve and the other on the muscle. The contractions with the two armatures on the nerve appeared even in the absence of any communication with the Leyden jar, by simply connecting them through a metallic arc. Or even by putting them in contact without the intermediary of the arc (VEN, I, pp. 58–59).

From these experiments Volta concluded that "the stimulating action of electricity acts *primarily* on the nerves […], while the motion of the dependent muscles is a *secondary* effect of nerves' excitement." This conclusion was "different from that of Mr. Galvani" and was not invalidated by the observation of contractions excited by an arc (or artificial electricity) applied only to the muscle because—in Volta's opinion—in this case electricity acted on the nervous filaments contained in it. For Volta, moreover, when the stimulus was applied to the nerve, it was not possible that contractions were due to a flow of electricity from nerves to muscles because in this case one would need a "strong electricity, a live and sharp spark" rather than "a most feeble electricity, as it is animal electricity, which is undetectable by the most exquisite electrometers" as it occurred in the experiment with the application to the nerve (VEN, I, pp. 61–62). Incidentally, we note here that modern electrophysiology has proved that the electrical stimulus is more effective when applied to the nerves than to the muscles. Because of this higher nerve excitability, electricity acts primarily through the excitation of the nerve fibers present inside the muscle body, even when the stimulus is applied to the surface of a muscle.

The second type of experiment that Volta used to distance himself from Galvani's theory was even more original. The physicist from Como applied two metal coatings to his tongue and established a contact between them: He then felt an acid taste, similar to that produced by stimulation of the tongue by an electrified conductor. This fact, and the absence of any contraction or motion of the tongue, "is sufficient to prove," he wrote, "that the nervous papillae of the tongue, and not its muscular fibres, are those which are immediately affected in both cases by the electric fluid, the penetration of which excites and lightly stimulates them" (VEN, I, p. 62). For Volta this experiment confirmed the idea that the primary action of the electric stimulus acted on the nerves, as well as showing that external electricity could produce its effects without a close anatomical and physiological connection of nerve and muscle. It also suggested that the physiological response to an external stimulus depended in an essential way on the nature of the stimulated nerve and not on the nature of the stimulating agent. As a counterproof, Volta applied the metal coatings to the motor nerves of a lamb's tongue and observed some convulsions. He thus concluded that muscle contractions occurred when the "nerves of motion" were stimulated, while sensations were produced by stimulating "the nerves of sense." This idea was very original and pointed to an important physiological mechanism, which would later be explained by Johannes Müller with his theory of "specific nerve energies" (see Chapter 8.2).

After arguing that the seat of animal electricity was the nerve, and not the muscle, in his *Memoria seconda* Volta focused on the conditions of his experiments, particularly those concerned with the arc applied to the nerve as well as those on whole and intact animals. As in all cases it was necessary to use "dissimilar armatures," that is, coatings of different metals, to produce muscle contractions, Volta studied the relative efficacy of various metal couples in stimulating the contractions, thus arriving at a sort of "scale of metals" divided into classes. In this study he adopted an approach based on the systematic variation of experimental conditions, similar to what Galvani had done in his electrophysiological research. Unlike Galvani, however, Volta was now quite convinced that, whatever experimental arrangement one could adopt, the use of different metals was the only means that remained valid for obtaining muscle contractions, independently of the point of stimulus application and the specific metals used. In Volta's words, "it is the diversity of metals that counts" (VEN, I, p. 71).

With the publication of *Memoria seconda* Volta definitely discarded Galvani's theory of animal electricity. As he wrote in a letter sent some months later to Galvani's nephew, Giovanni Aldini, he now believed that:

> It is on the nerves, and solely on them that electricity acts, both in the case of a mild artificial electricity and of animal electricity; it is not at all necessary that the electric fluid flows through them [i.e., the nerves] to the muscles; even less so that there follows any discharge between nerve and muscle, or between the internal and external surface of the latter [...]: in sum, the electric fluid is not the immediate cause, even if as a stimulus, of muscular motions, but only a mediate, occasional and remote cause, as his proper action consists in stimulating and exciting the nerves. (VEN, I, pp. 151–152)

Galvani's explanation of muscle motion, based on the analogy between the muscle and the Leyden jar, was thus rejected by Volta, who however still acknowledged Galvani with the great merit of having open a new and important path: "There remain the materials [of Galvani's edifice], namely the beautiful outcomes of his original experiments and the discoveries that followed; oh yes, there remain such precious materials for another fabric, more solid if not finer, which may be raised" (VEN, I, p. 152).

When Volta wrote these words in his *Memoria terza sull'elettricità animale* (*Third Memoir on Animal Electricity*), dated November 24, 1792 and published in the "Giornale fisico-medico," he had already began to build such a new "fabric." The issue of the *Opuscoli scelti sulle scienze e sulle arti* that followed the one with the review of Galvani's *De viribus* reported the news that "the most illustrious Prof. Volta, by continuing the experiments on animal electricity—a subject which can now be considered his own—makes new discoveries everyday." In particular, from his experiments on the tongue and metal coatings Volta had reached the very important conclusion that metals "are to be considered no more as simple conductors, but as real motors of electricity" (Opuscoli, XV, 1792, pp. 213–215). In a letter sent to a correspondent in the summer of 1792, Volta was already convinced that there was no animal electricity in the experiments in which it was necessary to use different metals to produce the contractions of muscles, while still attributing to "a real and true animal electricity" the phenomena observed under the conditions of Galvani's experiments on the prepared frog (VEN, I, p. 117).

At an early stage of his research on animal electricity Volta seemed to admit the coexistence of the two possibilities—that is, that involving the action of dissimilar metals and that implying an intrinsic animal electricity—that could explain different experimental results. As we shall see, this sort of double possibility would tend to disappear from Volta's writings as the controversy with Galvani developed. However, though the physicist from Como arrived at resolutely denying the existence of a natural disequilibrium between nerves and muscle, he continued to claim a physiological role of electricity in muscle motion. In particular, the electric fluid was moved by the action of the will and flowed from nerves to muscles to produce the contraction. This mechanism did not work for involuntary muscles such as "the ventricle [i.e., the stomach], intestines, the heart," which he found could not be stimulated by contact with metal arcs (VEN, I, p. 125). Incidentally, Volta's claim that electricity did not act on such muscles would be soon contradicted by the experiments of other naturalists like Felice Fontana (see Chapter 8.1).

Volta was well aware that the property of metals for "producing" and "promoting" an electrical disequilibrium was "a new virtue of metals, never supposed yet, which my experiments have led me to discover" (VEN, I, p. 117). Indeed, it was not possible to explain the contractions excited by the application of the conducting arc to two dissimilar coatings by referring to the known conducting property of metals, as contractions did not happen if the coatings were made of a single, strongly conducting metal. Physics had thus acquired a new concept, which was quite different from the classification of materials into insulators and conductors and which led to the consideration of electrical phenomena in a new light. To the concept of discharge, which had dominated electrical research in the previous decades, one had now to add that of "transflow" (*trascorrimento*), "torrent," and "current"—to cite some of the terms used by Volta to describe the circuit of electricity between two metals. If one may say that, from the point of view of the history of physics, Volta's theory of the electromotive power of metals was a step toward the emergence of electrodynamics, in light of the controversy on animal electricity it had a fundamental role in the development of Galvani's view of muscular motion, as we shall see later.

Volta's "special theory of contact electricity"—so called by historians to distinguish it from the "general theory of contact electricity" that Volta developed at a later stage—represented a real alternative to Galvani's theory of animal electricity. Though it had been formulated to account for the contractions produced by the application of the conducting arc to the nerve or the muscle of whole and intact animals, as well as for the effects observed in the experiment on the tongue, it could also explain the experiments carried out on Galvani's prepared frog. Galvani himself had noted that after some time it was no longer possible to produce muscle contractions by the contact between nerve and muscle, but it was then possible to get them by using different metals. For Volta this was further evidence that what really counted for muscular motion was not the presence of an electrical fluid in a condition of imbalance inside animal tissues but the diversity of metals. For Volta, moreover, it was impossible to be sure that when Galvani used the same metal for the conducting arc and the coatings, that metal was perfectly homogeneous. If this was true—Volta asked quite rhetorically in his *Memoria terza sull'elettricità anmale*—"what remains of that animal electricity which Galvani claimed and seemed to have proved with his very beautiful experiments?" (VEN, I, p. 147).

At this point a question may be raised: What was the path that led Volta in a few months to change his opinion from an enthusiastic acceptance of Galvani's theory of animal electricity to the rejection of the analogy between the muscle and the Leyden jar, and to the denial of the existence of animal electricity? A first answer lies in Volta's interest in weak electricity and its quantification. This led him to focus on those experiments in which electricity was so weak that it was not possible to obtain muscular contractions but with the use of different metals. None of these experiments—and this is a second aspect—implied the obligatory use of Galvani's "prepared frog" and this fact probably made Volta insensible to the model of the "animal Leyden jar," which presupposed a precise spatial relation between the artificial instrumental apparatus (the metal arc) and the animal structures involved (muscle and nerve fibers).

Volta had a different attitude toward the animal preparation and the experimental situation from that adopted by Galvani. In particular, Volta did not give any special relevance to the experimental arrangement that Galvani considered fundamental for producing the contractions, namely, the connection of nerve and muscle through the conducting arc. For Volta all that mattered was that electricity was introduced in some way to the animal body, as when he applied a weaker and weaker artificial electricity to the frog that produced "more and more tenuous sparks." Then he observed that to excite the contractions:

> it is not necessary to directly hit any part of the animal with such weak sparks, as it suffices that they strike between the electrified conductor and another metal one, which communicates with the frog's body either directly, or by the interposition of

a third, fourth [conductor], etc.; in sum, [it suffices] that the frog somehow functions as a communicating link between these conductors, so that the electric fluid is constrained to pass also through it. (VEN, I, p. 44)

Moreover, since his early repetition of Galvani's experiments, Volta objected that animal electricity, being associated with life and will, had to be studied in living animals instead of dead and "prepared" frogs (Pancaldi, 2003). Hence, the special role he gave to experiments on whole and intact animals and to those on living tissues such as the tongue. He considered his experimental procedure not only more instructive than Galvani's but also more original and extensive, accusing his opponent of having carried out his experiments in a somewhat restricted way, without the needed variation of experimental arrangements. In his first letter to Tiberio Cavallo, which was published in the *Philosophical Transaction of the Royal Society* for 1793, Volta wrote:

I do not know whether he [Galvani] has carried out other [experiments], but those which he has made public in his work are restricted in an excessively narrow circle; every time they consist in uncovering and isolating the nerves, and in establishing a communication between these nerves and the depending muscles through electrical conducting bodies [...]. However, by varying this sort of experiences in multiple ways, I showed that both these conditions, to uncover and isolate the nerves and to simultaneously touch these and the muscles in order to produce the pretended discharge, are not at all necessary. (VEN, I, p. 181)

6.3 GALVANI'S *TRATTATO DELL'ARCO CONDUTTORE*: THE CRITICISM AGAINST VOLTA AND THE NOTION OF A CIRCUIT OF ANIMAL ELECTRICITY

In his response to Volta, also Galvani started from methodological considerations, underlining how in the experiments with metal arcs the physicist from Como had not given due consideration to some aspects of the conditions the animal was in at the time of the experiment. According to Galvani, in order to understand the role of a conducting arc in the phenomenon of contractions, it was necessary, first of all, to assess the "natural muscular force" of the animal. If, indeed, the same type of arc produced contractions in one animal but not in another, that could depend not on the characteristics of the arc, but on the "varying strength of the animals." Likewise, and that was the case pointed out by Volta, the variable efficacy of different arcs applied to the same animal could be due to the variation in the strength of the animal, and not to the difference in the arcs. It was therefore necessary, argued Galvani, to use animals always prepared in the same way and endowed with the same strength, but, above all, it was paramount to "replicate the same experiments more and more times." Disregarding these cautions had been, according to Galvani, "the source of the different outcome of the experiments, and of the disagreements and various opinions of the experimenters" and might explain, in principle, the different degree of efficacy of the various types of arcs used by Volta in his experiments (GAC, p. 10).

Galvani distinguished three degrees of "animal force"—maximum, medium, and minimum—and, to ensure a reliable reproducibility of experiments, recommended avoiding the minimum-force degree, cautioning that:

> It is not convenient to perform the experiments, especially those implying comparisons, in the third degree, that is the minimum degree of the said force, as the re-occurring of contractions through a new type of arc in this case may be not the effect of this [arc]; so that the conclusions of a more powerful activity of the given arc could be completely erroneous. (GAC, p. 9)

Here Galvani was clearly referring to Volta's experiments in which the special efficacy of bimetal arcs was particularly evident. On this account, we would like to point out that the choice of the type of animal, on which Galvani insisted, was one of the fundamental criteria of the experimental method developed by Lazzaro Spallanzani and other great naturalists of the time. Spallanzani, for example, in his research on blood circulation carried out in the 1770s had used "preferably the water newt, as it seemed very apt to manifest and clear up those phenomena thanks to the easy preparation of [its] vases, their great transparency, and the most vivid purple-red colour of [its] blood." The choice of this animal, like that of the frog by Galvani, had been particularly felicitous, so much so that Spallanzani commented: "In truth, from the examination [of the newt] I acquired so much physiological knowledge, that I doubt whether a different animal had conceded the same to another observer after the discovery of circulation" (Spallanzani, 1932–1936, vol. 1, p. 43).

Galvani's methodological discourse on the animal to be used in experiments can be found at the beginning of a book entitled *Dell'uso e dell'attività dell'arco conduttore nelle contrazioni dei muscoli* (*Of the Use and Activity of the Conducting Arc in the Contractions of the Muscles*), organized in the form of a *Treatise*, and published in Bologna in April 1794. It is a work full of statements concerning the correct experimental method and the "right laws of philosophising" of Newtonian origin (see Fig. 6.4). Although it was published anonymously, its author was certainly Galvani, as several readers had supposed at the time it came out. The choice of anonymity might partially reflect the contribution of other scholars—Galvani's colleagues or collaborators—in writing the book. This would explain both the presence, among Galvani's manuscripts, of preliminary versions of the *Trattato* in different handwriting, and some significant stylistic differences to previous works. The recourse to anonymity, which has been considered by most historians as a mark of Galvani's reluctance to be directly involved in the polemic generated by *De viribus*, was also probably dictated by the desire to look like an impartial observer, outside the controversy, so as to increase the credibility of his arguments and to be able to

FIGURE 6.4. A copy of the anonymous work *Dell'uso e dell'attività dell'arco conduttore* given by Galvani to a colleague, on which it is written: "the author of the work is the illustrious Dr. Luigi Galvani" (AASB, Fondo Galvani, cart. II,©).

criticize Volta from a sort of "impersonal" point of view. This interpretation would well suit Galvani's resolute and not at all submissive character (not to mention that he was quite ambitious), which we have outlined in the previous chapters and which makes him more similar to Volta than the traditional image of him as a reserved man, unwilling to establish contacts outside the Bolognese milieu might suggest.

We may note that the methodological criticisms expounded by Galvani in the first part of the *Trattato*, and also in other parts of this work, all quite evidently directed at Volta, were only partially justified. Despite his tendency to adopt preparation methods and experimental arrangements that differed from those privileged by Galvani, Volta did in fact attribute great importance to experimental conditions. He considered, for example, the variability of the action exerted by external stimuli on animal preparations: "as this [action] depends on the varying strength and disposition of the same animal, on it being more or less well prepared, recently or less recently, on ambient temperature, etc." (VEN, I, p. 380). Moreover, he distinguished, on the basis of his observations on the animal's "electric vitality," "four degrees or stages of death." These were characterized by the vivacity of muscular motions and by the ease with which they could be excited by electric stimuli of various types, and he did care about how such vitality depended on the manner the animal was killed (VEN, I, pp. 31, 81 ff.). Furthermore, the fact that Volta preferred experimental arrangements different from Galvani's, giving particular importance to experiments with whole animals, is rather understandable. This was in part due to his tendency to design relatively easy-to-perform experiments, which did not require special skills in their preparation, also because in this way the reader would be in the position to replicate them and verify the results. About the contractions obtained in a "living and intact" animal, Volta wrote: "As to the [animal] preparation, these experiments are easier to perform than [those] carried out in Mr. Galvani's way, as no dissection of the animal is needed; and they prove much finer and more pleasant" (VEN, I, p. 64).

Surely Volta was surprised when he noticed that contractions could be excited in animals which were alive and not prepared, or "lacerated" in various ways (a problem which, in fact, Galvani does not examine closely in *De viribus*), and his surprise was followed by the feeling that this was indeed an original observation, a discovery of his own. It was similar to what had happened to Galvani when he placed the conducting arc between the nerve and muscle of the prepared frog "in the usual manner"; Volta placed the bimetal arc on the skin of the "alive and intact" animal, often armed with thin metal "shirts" or "dresses," and this became the ideal point of reference for interpreting the electrical phenomena involved in neuromuscular physiology.

Returning to the *Trattato*, we can notice that its expository structure is more systematic than that in *De viribus* and that it is divided into numerous parts (12 "chapters") according to the themes being considered rather than to the logic and the chronology of the experimental research referred to. As its title suggested, the central theme of the work was the "conducting arc," an expression used to refer both to the experimental device (the arc, usually made of metal, used to discharge the electricity accumulated in a state of imbalance) and to the circuit through which the discharge took place. In turn, this circuit was meant to refer both to the use of electrical apparatus (electric machine, Leyden jar, and Franklin square) and to the role of animal preparation. The fact that in the title there is no reference whatsoever to the themes of neuromuscular physiology, which in fact constitutes the essential topic of the work, is not, we believe, at all fortuitous. The basic assumption on which the whole logical-experimental structure of the *Trattato* is based is, in fact, that electricity normally acts on a body when this is part of a circuit (or "arc") of conducting materials through which the electric fluid flows under the form of "current" or of "torrent" (this being a term quite frequently met with in the *Trattato* and may betray a different hand from Galvani's in the writing of the work; see Atzori, 2004, 2009). This assumption, which we have seen emerging in Volta's writings during his early studies on animal electricity, represented a significant step forward, from electrostatics (the form of electric science that quite naturally corresponded to the instruments which had dominated the eighteenth century, in particular the electrical machine and the Leyden jar) toward electrodynamics, the new science of electricity that will have its most emblematic instrument in Volta's battery. In the case of neuromuscular physiology, Galvani pointed out at the beginning of his *Trattato*, the arc (meant as a circuit through which the electric fluid flows) is a necessary condition for the contractions to be produced, even when it may not be evident to an inattentive observer (GAC, p. 3).

The *Trattato* is important both for the logical arguments being developed in it, and for the experimental results it refers to. As far as the first point is concerned, it is necessary to underline how in this work Galvani changed his own hypothesis about the "animal Leyden jar" in some significant respects. He did this by developing a conception that looked more suitable to account for the characteristics of the phenomena being studied and that, at the same time, allowed him to counter more effectively the objections raised by Volta on the basis of the efficacy of bimetal arcs. Paradoxically, as we shall see, it was, at least partly, by elaborating on ideas coming from Volta, that Galvani developed his new conceptions. The experimental results referred to in the *Trattato* are also particularly relevant because they showed that it was possible to obtain muscular contractions by circulating animal electricity without using arcs made of metal or other materials extrinsic to the animal. In particular, there is an experiment in which contractions were produced by the direct contact between nerve and muscle, without any conducting body in between. As we shall see, at one stage, it looked as if these experiments regarding "contractions without metals" could bring the controversy to an end in favor of Galvani, thereby removing the objections that Volta had raised on the basis of his observation on bimetal arcs.

To respond to Volta's objections in a thorough and precise manner, Galvani realized that it was necessary to take up "the issue from its beginnings," and he developed a significantly new conception of his interpretative model of muscular motion. Despite the polemical nature of his arguments, and the frequent "lessons" on the scientific method clearly addressed to the physicist from Como, Galvani decided to raise the level of his exposition and go beyond the simple rebuttal of his adversary's objections. Indeed, the explanation of the special efficacy of heterogeneous arcs fitted into a more general question regarding the problem of the "conducting arc" and neuromuscular physiology, accounting for observations that Galvani had considered only marginally in his *De viribus*.

Though Galvani confirmed Volta's conclusion that an arc made of different metals was particularly efficient in exciting muscle contractions, he underlined that this characteristic was not an indispensible element for their production. Not only could arcs made of a single metal (or also of nonmetal conductors) produce contractions, but even more important, contractions could be produced simply by establishing a direct contact between nerve and muscle, without using any external element. Here Galvani referred to previous observations that, in his view, did away with any possible doubt about the existence of "an intrinsic electricity, naturally unbalanced and proper to the animal"; but, most of all, he intended to present a new experiment that, he said, he had already replicated "last summer in the presence of learned physicists" (GAC, p. 84). In its most convincing version, the experiment consisted of holding a "prepared frog" by hand or with an insulating body, and then bringing the nerve and muscle of a thigh into contact (by using another insulating body in order to manipulate them): "at the moment of the contact contractions of both legs will occur" (see Fig. 6.5). This experiment, argued Galvani, was "decisive," since it excluded any external coating or arc, as well as any metal or other substance, apart from animal parts. It thus proved that muscular contractions were produced by an electrical imbalance existing in the animal. More precisely, according to Galvani in the animal there was a special "machine" endowed with these three properties: (1) "to contain two opposite [kinds] of electricity"; (2) "to keep constantly and essentially separated and isolated these two sorts of electricity"; (3) "to keep hold of its electricity so tenaciously, that it does not allow the exit of [its electricity]

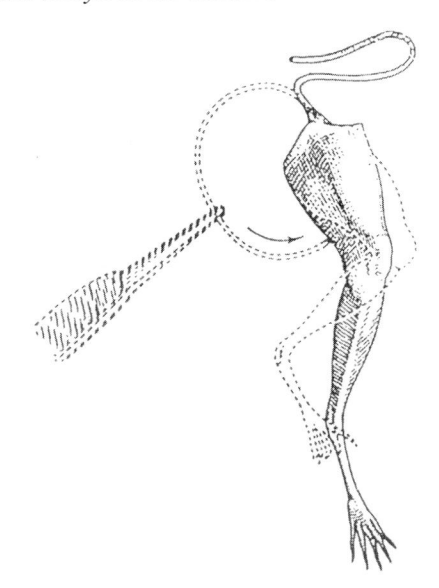

FIGURE 6.5. Galvani's experiment on the contractions produced by the direct contact between nerve and muscle, without the use of any metal, the so-called third experiment (from Sirol, 1939).

(so to manifest it with contractions) if the latter is not brought back to the same point by means of the arc." As these properties corresponded to those that characterized the Leyden jar, it was thus possible to conclude—as Galvani had already stated in *De viribus*—that the muscle was an "animal Leyden jar" (GAC, pp. 77–78).

The experiment on the direct contact between muscle and nerve—Galvani's so-called third experiment—represented a really formidable move in the difficult game which was being played by Galvani and Volta, a game whose strong points seemed to lie, from the very beginning, more in new experimental results than in mere arguments of logical and rhetorical nature. However, Galvani was well aware that if he wanted to definitely refute his opponent's theory of the electromotive power of metals, he could not simply show the possibility of producing muscular contractions without using an artificial arc. This was also because his direct contact experiment succeeded only under particular circumstances (for instance, it was necessary to use—as Galvani noted—"a lively and freshly prepared frog") and thus could appear not completely valid in Volta's eyes. He could argue that Galvani's explanation referred to a particular kind of contraction, obtained under very specific experimental conditions. What Galvani needed was a general interpretation that could account for all the experiments and phenomena observed up to that point.

The new theoretical conception proposed by Galvani in his *Trattato* did not just offer a coherent explanation of the special efficacy of bimetal arcs in the production of muscular contractions but accounted for a number of observations that were difficult to explain on the basis of the initial model of the animal Leyden jar developed in *De viribus*. Among these observations was the fact that contractions could be elicited not only when a circuit between the excitable tissues of the animal was established through the conducting arc but also when the circuit was suddenly opened or interrupted. The latter condition had been observed by Galvani in his previous investigation, but it was Volta who had stressed its importance and used it to obtain a prolonged and particularly effective stimulation of animal tissues (adopting a method that would later be named "Volta's alternatives"; see Chapter 8). If contractions were the direct (or indirect) result of the discharge of the "animal Leyden jar," as Galvani had supposed in *De viribus*, it was difficult to find out why they could be produced when the discharge circuit was interrupted. Another observation that had already emerged in Galvani's early experiments with metal arcs suggested the possibility that contractions occurred also when a variation or "mutation" of the contacts between the elements of the metal arc, or between this latter and animal tissues, was produced, as, for instance, when pressure was exerted or "the places or the instruments of the contacts or the same contacts were changed, even in a barely sensible manner"—as Galvani wrote in an unpublished memoir of 1786.

The central aspect of the new explanatory model elaborated by Galvani aimed at accounting for the experimental results obtained in the laboratory as well as for the physiological processes involved in muscular contractions. It was the idea that a continuous circulation of the electric fluid existed in the animal, "a continuous torrent," which flowed from the interior of the muscle through the nerve fibers—defined by Galvani as the "internal arc"—and then exited from nerves to reach the external part of the muscle fiber through the wet tissues and the "involucres" or "membranes" that surrounded the nerve and the muscle (the so-called natural external arc). According to this idea—which Galvani had first outlined in his 1792 letter to Carminati—the electrical circulation did not produce any evident muscle contractions under normal conditions, for "when this torrent flows placidly and evenly, then the muscle seems to be at rest." As to the electric flow through the nerve tissue, Galvani conceived the nerve as a hollow conductor surrounded by an insulating membrane, in a way similar to what he had written in *De viribus*. However, he now thought of the internal conducting matter of the nerves not as a "hypothetical lymph," but rather as a solid substance composed of "some oily parts mixed with many conducting parts." The presence of oily parts was needed to account for the characteristic of nonperfect conductor attributed to the nerve matter (a characteristic observed since Galvani's early experiments) and could contribute to "moderate the impetus of the electric torrent, without stopping its flow." In other words, this characteristic of the nerves explained why in normal conditions the electric flow was "placid and even" and did not produce any contractions (GAC, pp. 126–128). In this regard we may note that in 1781, Felice Fontana had claimed that the interior of nerve fibers was solid, supposing it was made of a glutinous and elastic substance (Bentivoglio, 1996).

To explain how a moderate flow of electricity through the insulating nerve sheath was possible, Galvani referred to a physical model, which is very interesting from two points of view. First, to a modern electrophysiologist it seems to anticipate the notion of ionic channels, that is, the molecular structures that are one of the most important discoveries of modern life sciences (see Chapter 10). Secondly, and more important, this physical model shows at its best one of the most interesting aspects of Galvani's scientific attitude, namely his reasoning through analogies taken from different fields of natural investigation and his creativity in designing artificial models to explain natural phenomena. A similar attitude, which characterized Galvani's research from the very start, was shared by the most important naturalists of the eighteenth century, such as Volta and Henry Cavendish, who had designed an artificial model of the torpedo by which the mechanism of producing an electric shock could be visualized (see Chapter 8). In his *Trattato* Galvani proposed the following "small machine" (*macchinetta*):

> Let's take a glass jar in the form of a flask, which is more convenient to build, and armed [i.e., covered with a metal coating] both internally and externally as usual; a conductor connected to the internal surface must exit from the flask's neck as long as one likes, then it must be all plastered with some insulating matter like wax and be put into contact with the external armature. Let's now make little holes in different points of the plastering which relates to the conductor and then pour water or another conducting fluid on all the plastering so that the fluid enters the holes and gets into direct contact with the conductor. In this way it is sure that there is communication between the internal and external surfaces of the jar through the fluid. (GAC, p. 136)

This "small machine," a perfected version of the animal Leyden jar developed by Galvani in *De viribus*, could artificially reproduce the normal conditions of the electric flow within the nerve-muscle system. The flask represented the muscle fiber, while the conductor communicating with its internal and external surfaces represented the nerve, which was covered by an insulating but perforated sheath that allowed a moderate passage of (electric) fluid from the interior to the exterior of the muscle. This mechanism was designed to keep an electrical disequilibrium in the system, at the same time avoiding an excessive accumulation of electricity, which could produce "a serious danger of lesion and alteration" of both nerves and muscles (GAC, pp. 138–139).

The usefulness of obstacles to the electric flow in the new electrophysiological conception of Galvani was probably related to his reflections on some phenomena encountered by physicists of the time in their electrical experiments. To bring about powerful and evident electrical effects (like sparks, "stars," cracklings, and the other typical manifestations of the electric force), it was necessary to produce a sudden discharge of electricity accumulated at high voltages in the electrical machine, in its conductor or in the Leyden jar. To make these phenomena quite evident, appropriate precautions had to be taken, by putting obstacles to the dispersion of electricity in the charging stage, by using adequate insulators, and by contrasting the influence of humidity. We may say that the obstacles existing in the continuous flow of the electric fluid through the "natural arc" represented, in Galvani's view, something analogous to a leaky dam and the stream of a river. When a continuous and placid base flow is present, a great "force" (nowadays we would call it potential energy) is accumulated and maintained in the fluid element, but at the same time the dispersion produced by the leak can avoid the excess of the mass of water that otherwise could become a source of danger.

In Galvani's model, contractions occur when the "placid electric torrent" flowing through the natural arc formed by the muscle and nerve tissues is altered by increasing its "speed and force" (GAC, p. 143). In experimental conditions this could be obtained by applying a conducting arc between the nerve and the muscle, which indeed modified the internal component of the natural arc by acting on the external one. In Galvani's words, the artificial conducting arc would "deviate the said torrent from the nerve and force it to make a sort of jump or leap both from one part of the nerve to one other, and from one part of the nerve to the muscle; and apparently this cannot be done without increasing its impetus and force, and consequently also the [impetus and force] of the internal torrent." A similar mechanism acted not only when the artificial arc was applied to the circuit of nerve-muscle tissues but also when it was disconnected from the circuit or when the contacts between these elements were in some way manipulated and altered. In these cases there was also an "alteration of the flow of the external torrent, and therefore of the internal [one]," which was the condition under which contractions were elicited (GAC, pp. 143–145).

We may note here that though the explanatory model developed by Galvani in the *Trattato* was more extensive than that of *De viribus* (as it accounted for a larger number of observations) it was also less deterministic in the fact that it could be interpreted both in terms of a specific electricity intrinsic to the animal body as well as of a common electricity which flowed through the animal tissues as through any other body. Moreover, it did not have a great impact on the debate over Galvani's theory of animal electricity, due to the fact that many contemporaries could not read Italian, the language in which the *Trattato* was written (while *De viribus* was in Latin, the international language of eighteenth-century culture). For instance, German scholars such as Christian Heinrich Pfaff, Alexander von Humboldt, and Johann Wilhelm Ritter, who made important contributions to Galvanism, knew Galvani's theory only in the first version developed in *De viribus*, which they could read either in the original Latin or in a German translation published in 1793 (Trumpler, 1992).

In the *Trattato* Galvani gave special attention to the experiments with the conducting arc made of different metals because Volta had based his main objections to animal electricity on those as well as his theory of the electromotive power of metals. Galvani's explanation of the great efficacy of this sort of arc in producing muscle contractions referred to the new model of the animal electric flow. On the one hand, a composite arc increased this flow; on the other hand, the contact points between its heterogeneous parts created some "obstacle" that contributed to increasing the strength of the electricity and that—being removed (or in some way altered) by a "mutation of the contacts"—could bring that sudden alteration of the electric flow in the internal arc that was needed for the genesis of the contractions. In this view, the special power of the heterogeneous and composed arc was thus related to the greater number and variety of contacts, as well as to a greater possibility of contact change, which in turn depended on the heterogeneous nature of the substances involved, and was not related to an electromotive power of metals, as Volta had supposed (GAC, pp. 146–147).

Another experimental arrangement considered by Galvani in the *Trattato* was the application of a bimetal arc to two points of the nerve, an arrangement which Volta had used to refute the model of the animal Leyden jar. Galvani argued that it was possible to form a continuous arc between nerve and muscle, even when there was apparently an interruption in the discharge circuit, through, for instance, some "light humidity" proceeding from animal tissues and depositing itself on the insulating parts of the experimental apparatus. He thus proposed the notion of "occult arc," that is, a complete and real circuit that was not immediately evident to the experimenter "unless he is very attentive and seriously reflects on the finest circumstances of the experiment" (GAC, p. 44). For Galvani this idea became an appropriate explanation to account for various experimental outcomes in which contractions did not seem to depend on the circuit of electricity between the nerve and the muscle, and allowed him to respond to Volta's objections also from a methodological point of view:

> In truth, as it is firm and proved by many facts that the arc applied from the nerve to the muscle elicits the contractions, before admitting a new and equally efficient one which is applied to other parts or to the same part, namely to the same nerve, it seems necessary—in view of the prudent laws of philosophising—that this new arc cannot refer to the first one, i.e., that it cannot ever communicate with the muscle through any conducting body: but in this case it is certain that this communication is established through the humidity which extends itself from the nerve to the muscle; therefore to suppose that the above mentioned contractions are excited by the mere arc applied from one nerve to another [nerve], or from a part of the nerve to another part of the same nerve, seems to be contrary to the right laws of philosophising. (GAC, pp. 53–54)

These "right laws of philosophising," which implied that one should not invoke a new cause to explain new phenomena that could be explained by known causes, were often referred to by both Galvani and Volta during their controversy on animal electricity. It may be noted here that in the case of the arc applied to the nerve (as in similar cases) Galvani's concept of an "occult arc" was in fact an ad hoc hypothesis, as it was very difficult to eliminate from the experimental scene such elements as "the proper and intimate humidity of the animal" that constituted the "occult" part of the arc. As we shall see, ad hoc hypotheses were used also by Volta, and some of them would play an important role in the development of the investigation of the two Italian scholars and of their debate (Pera, 1986/1992).

A similar methodological criticism was moved by Galvani to those physicists (again including Volta) who had not fully investigated the properties of the discharge of the

Leyden jar and had criticized his model of the animal Leyden jar on the basis of some properties found in the latter, but not in the former, as for instance the greater efficacy of heterogeneous "armatures" or arcs than homogeneous ones, and of some specific metals (GAC, pp. 77–78). On the one hand, Galvani invited the physicists to extend their experimental investigation on the Leyden jar in order to find out if some new conditions, such as the composite nature of the conducting arc or the choice of a specific metal for the coatings, increased the power of its discharge as it happened in the physiological mechanism. On the other hand, he stressed that his model of the Leyden jar was a metaphorical image, and not a truthful representation, of the nerve-muscle system. Here we may note that the value of images and metaphors used in scientific investigation and communication is often to point to a research path or to a theoretical conception in a suggestive way, without an exact identification with what they are meant to represent. It is thus misleading to say that Galvani's analogy was false because it did not perfectly correspond to the experimental outcomes. Rather, it had the same meaning of Bohr's image of the atom as a miniaturized solar system, which did not imply Bohr's claim of a perfect correspondence between the laws of the physical macrocosm and those of the atomic microcosm. As we shall see, Volta also used this kind of metaphorical reasoning at a crucial stage of his research path, which led him to the invention of the battery (see Chapter 8).

The new model developed by Galvani in his *Trattato* allowed him to give a coherent explanation of the contractions of voluntary muscles, which had been his central question since the beginning of his electrophysiological investigation. In Galvani's view, which developed the suggestions proposed by Volta in his writings (see earlier), these contractions were produced by "the increased force of the said electric torrent," that is, of the basal flow of electricity through the natural arc, because of the activity of the soul. In Galvani's words, the "soul by acting on the brain in its marvellous and incomprehensible manner may determine and send a greater quantity of electricity through the nerve to that muscle it wants to set in motion, so as to increase the force of the internal torrent." The soul could also act on the biological electric current by changing the "contacts" in the nerves. As in *De viribus*, in the *Trattato* Galvani did not speculate further on the action of the soul, which was "an immortal entity that will ever remain well beyond human understanding," but focused on the "material agent" of muscle motion, which "seems undoubtedly to be, as we have tried to prove, an electricity specific and proper to the animal" (GAC, p. 162; see Chapter 4).

In autumn of 1794, shortly after the *Trattato*, Galvani published (again anonymously) a short *Supplemento* (*Supplement*), in which he developed some experimental and methodological arguments. In particular, he focused on the phenomenon of the contractions obtained without using any metal, and especially by directly connecting nerve and muscle. He insisted on some precautions to be taken in the experimental arrangement, such as leaving the animal tissues almost intact, making a very small section of the thigh muscles and the crural nerve, so that "the underlying muscles could receive on themselves the same and correspondent crural nerves" in the way of a conducting arc (GSA, p. 3). Galvani's insistence on this experiment and its experimental circumstances shows the importance he assigned to it in the controversy with Volta but also his worries about the possible failure of its replication. In this regard, it is worth noting Galvani's attention to the details of the experimental dispositions he designed in his laboratory. In fact, the model of the animal Leyden jar developed in *De viribus* was somehow the result of a particular experimental setup, which in turn referred to a particular anatomical organization, namely a specific set of nerve fibers that penetrated into the muscle fibers of the corresponding muscle. Galvani insisted on this point in the *Supplemento*, thus marking his difference to Volta's attitude, which from the very beginning of his research on animal electricity had not conformed to the Bolognese physician's experimental indications.

For Galvani one of the conditions that allowed observation of the strongest contractions was to establish a direct, or even indirect, contact between a nerve and the specific muscle which it innervated. This was due to "a pre-existing, natural and determined organisation and connection [of the nerve and its corresponding muscle], so that animal electricity is naturally kept unbalanced" (GSA, p. 8). In the *Trattato* he speculated on this "particular organisation" of the nerve-muscle system by referring to the peculiarity of animal electricity and to the existence of a specific "machine" in the animal body:

Such a disequilibrium exists in the animal either naturally or induced by artifices. If naturally, then one should confess that in the animal there is a particular machine which is apt to produce such a disequilibrium, and this electricity should be called animal so to denote not an electricity whatever, but a specific one applied to a particular machine. (GAC, pp. 70–71)

The concept of machine, and the related view of a complex organization, were among the fundamental aspects of a new notion of living beings that had emerged in the seventeenth century. This was due to the path opened by Cartesian philosophy and to the work of naturalists such as Marcello Malpighi. If the animal organism was formed by machines that worked on the basis of mechanical laws analogous to those of the machines built by humans, then it was paramount to understand the structure and organization of animal tissues and of the organic machines that composed them. This conception had led to a renewed interest in anatomy, and especially in microscopic anatomy, a field in which Malpighi had given extraordinary advances. As we have seen in previous chapters, Malpighi's ideas had strongly influenced the scientific milieu that revolved around the Bologna Institute of Sciences and had been developed by Bolognese naturalists such as Jacopo Beccari and Giuseppe Veratti, both mentors of Galvani. The latter had investigated both human and comparative anatomy since his early years and had become a professor of anatomy at the University and at the Institute (see Chapters 3 and 4).

However, there were some notable differences between the simple machines alluded to in seventeenth-century science (based on the use of ropes, levers, channels, sieves, wheels, etc.) and those considered in the eighteenth century, which were much more complex (such as the electric instruments used by Galvani). Moreover, though the structural complexity of the animal machine envisaged by Galvani implied the study of its form and the disposition of its parts (thus an anatomical investigation), this knowledge was not sufficient to explain its functioning, which in fact required a more dynamic and active mode of experimentation. In this regard, Galvani's scientific approach took on and developed the physiological method that Albrecht von Haller had defined as "*anatomia animata*" or living anatomy (Duchesneau, 1982; Steinke, 2005).

The notion of machine represents an important difference between Galvani and Volta's attitude to the study of animal electricity and can explain what was at stake in their controversy much better than an anachronistic opposition in terms of a "biological" against a "physical" point of view (as claimed by Pera, 1986/1992). Galvani seemed to attribute to the disposition and organization of the animal machine a fundamental role (both in structure and function) that is much less evident in Volta, at least in the initial stage of his research. One of the many apparent paradoxes in the Galvani-Volta controversy is that Galvani, an anatomist and physiologist, developed an explanatory model of the role of electricity in muscular motion that was in agreement with the laws of electricity known at the time, while Volta, a physicist, did not hesitate to invoke quite new physical laws when he was faced with phenomena that did not seem to fit into established knowledge. Galvani brought out this point himself when in the *Trattato* he wrote:

> Here it is that this phenomenon, which at first sight seems contrary to Galvani's theory, is in fact very favourable to it, being totally coherent with its principles, and these latter quite consonant with those [principles] known and established by the physicists; I do not know if the same can be said of the principles proposed by the most learned Mr. Volta. (GAC, pp. 106–107)

Going back to the *Supplemento* and to the experimental arguments developed by Galvani in this work, we may note two observations made here for the first time which denote the experimental acuteness and accuracy of the Bolognese physician. The first observation concerned the importance of using the extremity of a severed nerve to make the contact between the animal tissues in order to obtain muscular contractions without the use of any metal (GSA, pp. 13–14). This observation, which Galvani did not explain but would replicate in a work published some years later, in 1797, can be understood well only in the light of modern knowledge of electrophysiology (see Chapters 9 and 10). The severed part of a nerve or a muscle represents a less resistant path toward the intracellular environment, which has a negative potential of a little less of $1/10$ volt in relation to the extracellular environment. The connection between the extremity of the severed nerve and the extracellular environment, which can be realized through the contact of the former with the external surface of a muscle or of another tissue (as Galvani remarked), produces an electrical flow that is able to start the process of electrical excitation of the nerve fiber. This is because of the potential difference that exists between intracellular and extracellular fluids. Instead, if a contact is established between the surface of the muscle and an intact part of the nerve, no electric flow is produced, as normally there is no sensible potential difference between two points of the extracellular environment located near intact tissues; this condition does not allow the excitation of the nerve and thus the production of the contractions. When heterogeneous coatings or "armatures" are used, then the nerve fibers are actually excited, even though the extremities of the arc are applied on two points of the extracellular environment that are at the same potential; however, in this case, the source of the electrical disequilibrium is the electromotive power of metals discovered by Volta, as the potential differences produced by the couples of metals usually used in these experiments (copper-zinc or tin-silver) are greater than biological ones (about 1 volt).

Galvani's experimental ability and acuteness are testified by a second observation contained in the last pages of the *Supplemento* that referred again to the circumstances of the experiment on the direct contact between nerve and muscle. When preparing the animal, the surface of the muscle should not be lacerated, as this "would much disturb and often even prevent the success of the experiment" (GSA, p. 21). As the one examined earlier, this observation can also be accounted for by the modern knowledge of electrophysiological phenomena. The severed extremity of the nerve is at a negative potential in relation to the extracellular ambient because of the influence of the intracellular negativity. If this nerve extremity is brought into contact not to an intact muscle but to a severed one (which is also at a negative potential), no electric flow is produced as no potential difference is established; as a consequence, no excitation would ensue.

Before leaving Galvani's works of 1794, it is worth mentioning the last pages of the *Trattato*, in which Galvani applied his electrophysiological notions to the problem of disease, as he had done in the last section of *De viribus*. Indeed, he had not abandoned the conviction he had learned from his teachers and that related to a "rational" conception of medicine, that scientific investigation had the goal of improving "the medical art" (GAC, p. 167). On the basis of his theory of the natural arc, Galvani referred to "changes in the substance or in the contacts" of the nerve or muscle tissues as possible causes of convulsive motions. In particular, "morbid conditions" could consist of "sharp particles" that affected the nervous matter or the organic fluids so as to produce "preternatural" changes in the contacts of the internal arc or in the electrical properties of animal tissues. Hence, this accounts for the insurgence of convulsions or even of muscular paralysis (GAC, p. 154). In this way Galvani joined together "humoral" concepts deriving from the medical tradition with those introduced by seventeenth-century corpuscular matter theory, under the hat of his neuro-electric view of animal functions.

With the publication of the *Trattato* and its *Supplemento* Galvani believed he had answered Volta's objections against his theory of animal electricity and refuted his opponent's ideas. Reversing the meaning of one of Volta's judgements, he thus claimed:

> If things are as I said, if such electricity is truly and completely proper to the animal, and not common and external [to it], what will it be of Mr. Volta's opinion, who on the basis of the alleged experiments has pretended to completely exclude animal electricity, and to limit Galvani's discovery to the mere invention of the most exquisite electrometer in the animal? (GAC, p. 123)

In the next chapter we shall see how Volta reacted to these words of Galvani and we shall examine the final stage of the controversy that opposed the two Italian scholars in the last decade of the eighteenth century.

7

The Controversy Between Galvani and Volta Over Animal Electricity

THE SECOND STAGE

IN PUBLISHING THE *Trattato dell'arco conduttore*, Galvani hoped that "the experiments and arguments" he was adducing were such as to convince everybody of the existence of "an electricity proper to animal," as he expressed himself in a letter addressed to Lazzaro Spallanzani, dated April 26, 1794. A copy of the *Trattato* was added to the letter with the hope that the famous naturalist could give an opinion on it.[1]

The choice of Spallanzani as an interlocutor was made with intent. First, he was one of the most famous scientific scholars of the period, a protagonist in the studies on animal reproduction, and an author of fundamental works on the physiology of respiration and digestion. He had been studying a great variety of phenomena, from volcanoes to sea animals, from the geological conformation of the territory to the "sense" of bats. Moreover, since 1769 he had been professor of natural history in Pavia, the University where Volta also worked, where the criticism to Galvani and Volta's theory of metallic electricity were well received. Furthermore, Spallanzani was familiar with the subject and was ready to give an attentive and critical evaluation. As a matter of fact, even if he had sided with Galvani, he had nevertheless raised some doubts about the existence of animal electricity. These doubts had, however, been dissipated during a visit to Bologna. On this occasion Galvani had the chance to show him some of his experiments. Moreover, Spallanzani had carried out studies in the field and published, in the period of 1783–1784, two memoirs on torpedoes; these resonated in the scientific circles of the time (Spallanzani, 1783, 1784; see Chapter 5 and Piccolino, 2001).

---

[1] Letter of L. Galvani to L. Spallanzani, Bologna, April 26, 1794, in Spallanzani, 1984–2012, *Carteggi*, V, p. 43.

We can cite a comment Spallanzani made in a letter of October 1797 as evidence of the careful and objective view he held with regard to Galvani's works. Spallanzani remarked that the experiments on spark production in the torpedo referred to by the Bolognese doctor had indeed been carried out exclusively in the "electric gymnotus." This was an appropriate remark because a spark from the torpedo's shock was not produced until about 40 years later by Santi Linari and Carlo Matteucci (Finger & Piccolino, 2011; Piccolino, 2011).

In his response Spallanzani praised the "new pamphlet" and its author (soon identified as Galvani). He remarked that with the *Trattato* Galvani had succeeded in "confuting victoriously the objections, and to highlight the useful consequences." These words would have sounded very sweet to Galvani's ears because in a letter dated August 1, 1794, he answered in this way:

> I give thanks to Your very Illustrious Lordship for the letter you sent me, which could not be more courteous and appreciated. Concerning the controversies and doubts on animal electricity, this letter produces a fulsome calmness in my soul, which was indeed rather restless. I was very much concerned that, with the possibility of many opposing opinions, this new branch of animal physics would remain without some of those advantages that hopefully sometimes it could produce. In hearing now your judgment, a judgement of complete certitude and authoritativeness, as favourable to the afore-said animal electricity, and to the expressed laws, I no longer have any doubt about the truth of the thing and of the usefulness that is desired for it.[2]

In this passage the personal involvement of the scientist in his research and in his conception is expressed in a clear way, even with its emotional dimensions. Seen from the perspective of the researcher, science is a much more human enterprise compared to how it might appear from the outside. Scientists are not isolated individuals, with no other relationship than those they entertain with the object they are studying. Above all, they are individuals who interact with others, and they long for the recognition of their own work by others. This is particularly so in the case of controversies, when they are more exposed to discussion and criticisms on their ideas and results. In these circumstances they look for the support and consent of their colleagues, especially of the more authoritative among them. This consent is necessary, not only to give support to their contributions but also as an element of self-confidence and as drive to pursue further the scientific endeavor.

In this chapter we will deal with the final phase of the controversy between Galvani and Volta. We will do that mainly by analyzing the writings of the two scholars subsequent to the publication of the *Trattato dell'arco conduttore* (among them particularly Galvani's *Memorie sulla elettricità animale*, which appeared in print in 1797). Besides

---

[2] Letter of L. Galvani to L. Spallanzani, Bologna, August 2, 1794, in Spallanzani, 1984–2012, *Carteggi*, V, p. 44.

considering some new "crucial" experiments proposed by Galvani and Volta, we will pay particular attention to the influence exerted on the other by them. We will also analyze some points that, in our opinion, have not been well clarified by historiography, as—for instance—the relation between animal and common electricity, and the importance of the physiological perspective in Volta's research.

7.1 VOLTA'S "GENERAL THEORY OF CONTACT ELECTRICITY"

Galvani (together with Spallanzani and various of their colleagues) was convinced that the *Trattato dell'arco conduttore* would solve all doubts and put an end to any controversy, clarifying the problem of the existence of animal electricity and the mechanism of muscle contraction. This was not, however, a unanimous opinion. Volta was quick to read the *Trattato* and used it to restate his views on the arguments raised by Galvani. Volta did that by inserting his comments in a letter sent to the Piedmontese abbot Anton Maria Vassalli, perpetual secretary of the Turin Academy of Sciences, and published in the August 1794 issue of the *Giornale fisico-medico*. In this letter he also considered the experiments adduced in support of the animal electricity theory by Galvani's nephew, Giovanni Aldini, who had recently published his *De animali electricitate dissertationes duae* (*Two Dissertations on Animal Electricity*). In this work some experiments were described showing that, in contrast to Volta's assertions, frog muscle contractions could be produced even using a perfectly homogeneous arc made by liquid quicksilver (Fig. 7.1).

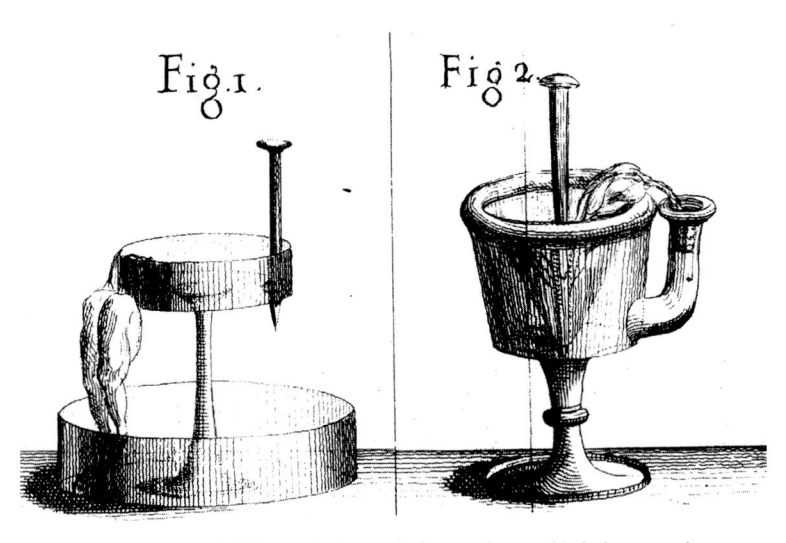

FIGURE 7.1. The experiment of Aldini, with the quicksilver used to establish the circuit between nerve and muscle (from Aldini, 1794).

Volta's letter to Vassalli (the first of five written between February 1794 and the autumn of the following year) contained two partially distinct arguments. The first, making up the body of the letter, developed the theory of the electromotive power of metals; in the second were formulated a series of criticisms and objections to Galvani's new experiments. In support of the "metallic electricity," Volta reiterated the importance of his tongue experiment and of the experiment with the arc applied exclusively to the nerve (described—as mentioned—in his *Memoria seconda sull'elettricità animale*). He did not value, however, the counterarguments raised by Galvani in the *Trattato*. On the contrary, concerning the acid or alkaline sensation produced by the application of different metals on the tongue (a phenomenon considered by Galvani "untrustworthy and inconstant"), Volta replied that the judgement of the author of the *Trattato* was a "miserable refuge," invoked only in order "not to be obliged to draw the consequence, that I draw, and that is self-evident." In Volta's opinion even Aldini's experiments with quicksilver and that of Galvani with the direct nerve-muscle contact were inconclusive: the first because quicksilver was not necessarily a homogeneous metal: the second because the experiments appeared to be "inconstant" and, above all, "doubtful." This was because it could be attributed to a mechanical irritation produced by the contact between animal tissues, instead of by a genuine passage of electric fluid (VEN, I, p. 271 n(*a*) and p. 274 n(*c*)).

The tunes and terminology used by Volta in the letter to Vassalli, and by Galvani in the *Trattato*, were similar, and this did not seem to allow for any possible reconciliation between the two opposing conceptions of animal or metallic electricity. Both proposed ad hoc arguments in order to counteract their adversary's objections. Both, moreover, accused the antagonist of lacking the necessary attention when conducting the experiments. On his side, Galvani made recourse to the idea of an "occult arc" between nerve and muscle, that is, a somewhat invisible arc and thus an indemonstrable one. This was in order to account for Volta's experiments on the arc of the nerve alone (within Galvani's own theory that both nerve and muscle needed to be included within the current circuit). Galvani had, moreover, raised various doubts about some of Volta's results, by saying that they were far from being so constant and unequivocal as pretended by the physicist of Pavia.

As indicated earlier, on his side Volta discounted Galvani's "third experiment" by invoking a mechanical stimulation at the moment of the contact between nerve and muscle, a stimulation that was hard to conceive in the experimental conditions indicated by the Bologna doctor. Furthermore, he accused the Galvanians of proposing some "experiments, beautiful in themselves, but indeed capable of convincing only those who would be lingering exclusively on them, without going much deeper."

Above all, both Galvani and Volta were convinced that their adversary infringed the first and most fundamental law of the scientific method. That is, the prohibition of unnecessarily multiplying the causes for a phenomenon for which either one of them was convinced to have found the sufficient cause. Volta accused Galvani of assuming, without need, the existence of a type of animal electricity different from that ordinarily present in all physical bodies. On his side, Galvani accused Volta of making recourse, again with no need, to a new property of metals, that of putting electricity in motion.

Galvani and Volta shared a similar language, the grammar of which was that of modern experimental science and the lexicon that of the natural philosophy of their era. Nevertheless, their attitudes remained opposed. At least two reasons accounted for that. The first was at the same time epistemological and psychological. It was due to the presence in either of the researchers of a strong idea that directed their own research and influenced the interpretation of both their own results and the results of their adversary.

Totally absorbed in their respective research programs, Galvani and Volta were sometimes incapable of discerning the novelties contained in the results and the interpretation proposed by the antagonist.

On the one hand, in the *Trattato* Galvani sometimes shows that he has not fully understood the meaning of Volta's theory of the electromotive power of metals. For instance, in one passage he negates the possibility of an electric disequilibrium between heterogeneous metals. This is because "being metal conductive bodies, the electricity of one part of the arc should necessarily be in a condition of equilibrium with the electricity of the other part." In this way he is making reference to the classical laws of electricity which, however, Volta had explicitly got through with his new theory of the electromotive power of metals. On the other hand, Volta did not fully appreciate the new model of animal electricity developed by Galvani in the *Trattato*. In the third letter to Vassalli, Volta opposes his concept of "circulation, i.e., of a continuous flow of electric fluid" between different conductors, to Galvani's theory based on the concept of "charge or unbalance, and consequent discharge in animal organisms." As we have remarked, this objection could be applied to Galvani's first model of the animal Leyden jar. It was not appropriate against the model developed by Galvani in the *Trattato*, based on the idea of a continuous circulation of animal electricity and on its modification as possible cause of muscle contractions.

The second reason making reconciliation between Galvani and Volta difficult was of a more scientific nature, and depended on the complexity of the physiological mechanisms involved in their experiments. As already mentioned (and as it will be clarified in Chapters 10 and 11), the mechanisms will be clarified only by the research of Alan Hodgkin and coworkers in the twentieth century. The model of the involvement of electricity in the excitability of nerve and muscle issuing from this research is completely beyond the limit of eighteenth-century science, to which both Galvani and Volta belonged. Neither Galvani's nor Volta's theories could account satisfactorily for the way electricity is involved in nerve and muscle physiology. Therefore, they could not explain in any exhaustive way the phenomenological aspects that were emerging from the experiments they were actively pursuing in the years of the controversy. In this situation both Galvani and Volta pushed their endeavors along in two directions: on one side they tried to bring elements, of both experimental and conceptual natures, to support their respective theories and, on the other side, they tried to point to the phenomena that the antagonist's theory could not explain.

Galvani and Volta were both well aware of the complexity and difficulty of the experiments in which they were absorbed. In the *Supplemento al Trattato dell'arco conduttore*, Galvani devoted particular attention to clarifying the care and precautions to be adopted in carrying out experiments on animal electricity. For Galvani, it was not enough to be a good experimenter, to be skilled in manipulating the animal preparations and instruments. Another important quality was needed: the perseverance in repeating and varying the experiments and in accurately controlling the conditions under which they were conducted. Indeed, Volta also invoked an enduring perseverance as a quality when he said "it is necessary to see the experiments, to make and remake them, by changing form and manner, as I have done myself, in order to be perfectly convinced of the results" (VEN, I, p. 294). To curiosity and utility (two values characterizing the researcher of the Enlightenment) Galvani, Volta and other great experimenters of the epoch (like Albrecht von Haller and Lazzaro Spallanzani) added a new virtue—perseverance—which distinguished the true researcher from the amateur. The need for repeating and varying the experiments, in order to understand all circumstances and determining factors, obviously required the availability of considerable time. It is possible that such a requirement, felt so deeply by the protagonists of our story, has been one of the reasons for the eventual emergence of the professional scientist in the nineteenth century. Although with notable exceptions, this professional scientist would tend to appear more "ordinary" and less "heroic" than the natural philosopher of the previous era.

Returning to the controversy, it was clear to Volta that assessing an important experiment like that of the direct contact between nerve and muscle could not be rebutted by a simple remark, as was the case in his second letter to Vassalli. This experiment performed by Galvani (and other similar ones published by Eusebio Valli also in 1793), did appear as a straightforward refutation of the theory of metallic electricity. As a matter of fact, they excluded the need of any metal, be it homogeneous or heterogeneous, in order to obtain the contractions. Because of their importance, these experiments convinced many scholars, who had taken Volta's side, to change their opinion and accept Galvani's views on animal electricity. As expressed by Volta himself in his third letter to Vassalli, dated October 27, 1795 (but published in the next year): "these experiments impressed many people, and drew them toward Galvani's banners when they had already subscribed, or were going to subscribe, to my totally different conclusion" (VEN, I, p. 289).

Not only Galvani but also Volta cared for the judgements of other scholars concerning his opinions. Moreover, he agreed that he had been too precipitate in his first evaluation of Galvanian experiments, particularly when he said that it could never be possible to produce contraction in the absence of metals (and other bodies having similar properties, like coal).

How could Volta emerge from this situation that appeared like a dead end for him? A possibility was to attempt to reconcile the two opposite doctrines, by admitting that in the experiments with heterogeneous metals, the movement of electric fluid was due to electricity extrinsic to the frog, whereas in the case without metals (or with a homogeneous metal), contractions were produced by animal electricity. Volta himself alluded to such a possibility but rejected it because of a reluctance to multiply the causes.

Far from admitting the need of an animal type of electricity in the circumstances of the direct muscle-nerve contact, what Volta actually did was to modify and extend his conception of the electromotive power of metals. It sufficed to assume that electricity could be moved by the contact of two dissimilar conductive bodies, without them needing to be metals. Volta was thus generalizing his theory of metallic contact, as he recognized.

> In this way the principle is generalized, that any time two different conductors are connected, an action arises, which pushes more or less the electric fluid; in such a way that, as far as the circuit is closed between three of them, whatever they are, with the proviso that they are different from each other, some current is constantly excited, either modest, or weak or very weak. (VEN, I, p. 300)

In other words, a current could be produced not only by connecting together different metals (referred to henceforth as "conductors of the first class"), between them or with the Galvanian frog, but also by the contact of all "humid" conductive bodies of different natures (as were the nerve and muscle in Galvani's "third experiment").

Undoubtedly at the moment it was formulated, this "general theory of contact electricity" (as it was subsequently called by historians) was evidently an ad hoc hypothesis. However, it would turn out to be of paramount importance for the comprehension of the mechanism underlying the generation of electric phenomena in living organisms. We will discuss that in Chapters 9 and 10 where we will deal with Nernst's theory of electrochemical potential and Bernstein's "membrane theory" of bioelectric phenomena. Suffice it now to say that the paths of scientific discoveries are often unpredictable and even ad hoc theories can prove to be of importance for scientific progress.

The new theory of Volta appeared to be at the same time simple and comprehensive: With a single principle, it sufficed to account for all phenomena put on the stage by the various protagonists of the controversy on animal electricity. Moreover, the new theory could account for the difference between the experiments with metals, and those made without using metals. In their researches, both Galvani and Volta had become progressively aware of the importance of coating (or "arming") both nerves and muscles with laminas of different metals. This experimental arrangement proved to be particularly important with weak frog preparations, that is, with preparations that did not show any contraction when tested in the ordinary way. Galvani had confined himself to saying that the "armatures" simply increased the force of contractions. In his opinion this was a distinguishing feature of animal electricity compared to the metallic one. To support his view, he had enlisted the example of electric fishes in which, according to him, different metals were particularly efficacious in evoking the production of shocks (see Chapter 6.3). On his side Volta attributed the diversity to the fact that metals, or first-class conductors, were "exciters," or "motors" of electricity that were much more effective than humid bodies, that is, second-class conductors.

Besides his conviction that the general theory of contact electricity answered all the objections and experiments of Galvani's' supporters, Volta believed that he was opening a new field of experimental investigation. In the last three letters to Vassalli (and also in other works written in the period 1795–1797), he described accurately all the combinations in which the contact between different conductors could produce an electric current. This line of investigation, together with the reflections on electric fishes that we will examine later, would bring him, within the space of few a years, to contrive his electric battery—the instrument that would open the field of electrodynamics.

But what of Galvani? Despite the implications of Volta's new theory, there still remained a path for him to pursue. Volta himself suggested such a possibility in his fourth letter to Vassalli (left unpublished in the period) with, however, the remark that it was an impractical option.

> There could perhaps be a possibility to straighten up this pretended animal electricity, that I declare as nonexistent, and that with many experiments I think to have knock down; I have substituted it with the other principle that I have contrived, of a purely artificial electricity, i.e. a form of electricity moved by an extrinsic cause. It would be necessary that adversaries would show the appearance of convulsions in frogs etc, with the making of conductors, all of the same species, by no way dissimilar the one from the other; this they would never be able to do. (VEN, I, p. 325)

As was the case with the direct nerve-muscle experiment, when he had been obliged to retract his statements, even on this occasion the physicist of Pavia showed himself to be hasty and unwary. After little more than 1 year, Galvani would publish (as we shall see in the next paragraph) an experiment just like the one indicated by Volta. This was indeed a crucial experiment in the history of electrophysiology.

The publication of Galvani's *Trattato dell'arco conduttore* and the new experiments of the nerve-muscle contact did not only result in the elaboration of the general theory of contact electricity on Volta's side. They induced him also to deepen the physiological questions implied in the experiments with Galvanism. Furthermore, they represented a starting point on a path that would bring humid bodies as a possible source of electromotive power to the center of Volta's attention, and not exclusively as conductors and detectors of electricity (as the frog tissues had been in the first phases of his research on animal electricity; see Chapter 8.4).

In 1795, in the same period of the letters to Vassalli, Volta took care to clarify to some correspondents his thoughts about the mechanism of muscular contractions both in dead preparations as well as in live and entire animals. At that time these letters were not published, and thus they did not play any important influence on the course of the controversy. They are, however, of great importance in order to get a more complete picture of Volta's research and to understand better some aspects of the path eventually leading him to the invention of the battery.

In August 1795 Volta wrote a letter to Francesco Mocchetti, a doctor who had been his student at the University. In this letter Volta retraced the main steps of his research. He emphasized the discoveries made "since the springtime of 1792," on the need for dissimilar metals in order to obtain contractions in "entire and intact" frogs, and on the importance of the direction of "electric current," which was much more effective if it flowed from nerve to muscle than vice versa. This observation, also noticed by Galvani, had induced Volta to elaborate a "conjecture":

> This discovery makes it clear that the course of electric fluid in nerves most favourable to the excitation of motion in muscles is that made in the same direction in which the power of will acts and manifests itself on the muscles under its control. That is, by going down from the head or trunk toward the branches. This could induce us to conjecture that the electric fluid is the one used by the soul in order to produce motions in voluntary muscles. (VEN, I, p. 364)

In another letter, written a short time before and addressed to Orazio Delfico (another of his students), Volta went more deeply into analyzing the possible action mechanism of will, by ascribing to it the capacity of "giving a small push to the electric fluid, and of determining a very bland current in the corresponding nervous branches or filaments." This current was evidently very "bland" since no instrument had been able to detect it, and it could flow through nerves "as in any other conductor." As a matter of fact—Volta continued—nerves "are in no way hollow tubes or channels, whereby any subtle fluid could flow, no matter how subtle it could be. The *animal spirits* [...], an animal fluid or agent, never defined or understood, comes now to be known, if we say that it is the common electric fluid" (VEN, I, p. 341).

At this point Volta clarified, perhaps for the first time explicitly, the fundamental difference existing between the frog prepared "in Galvani's manner" and the living animal, when considered from the point of view of the electric mechanisms.

> To put an end, if the electric fluid has an influence on animal movements, it does have it under the domain of will. The will commands to the electric fluid, which is stationary in all bodies and in all the parts of the living animal, and particularly in the nerves of voluntary motion, which are perhaps more conductive than other parts. On the other hand, in dead animals, or in the excised parts, in which the action of will has ceased, electric fluid remains equilibrated and quiet, as in any other Conductor whatsoever. (VEN, I, p. 365)

In dead animals or excised members (i.e., in the condition of the experiments with frog preparation) conducting arcs could, in Volta's opinion, replace the action of the will, which is no longer operative. The contraction could then be brought about by "fitting together" two different conductors, according to the general theory of contact electricity.

The difference between the mechanism of contraction in the live and dead animal induced Volta to propose to Mocchetti the following consideration:

> What I have mentioned here suffices to make it clear that I do not exclude all forms of *Animal electricity*; on the contrary I ascribe to it the most noble part in the Economy, whereas I show to be inexistent the one put on the stage by Galvani, and supported by his adherents: one based on a supposed unbalance of electric fluid between nerves and muscles, or between the interior and the exterior of muscles. (VEN, I, p. 364)

As we shall see, Volta would again propose this thesis as a hypothesis of possible reconciliation between his position and that of Galvani. As a matter of fact it was just an apparent reconciliation, as far as it negated the central point of Galvani's theory. This was the existence within the organism of a particular machine, having the function of keeping electricity in a state of disequilibrium and of producing its discharge in order to put muscle in motion. Moreover, for Galvani there was no sense in distinguishing the physiological behavior of dead and live animals. For him, in the frog preparation, despite the absence of will, the animal machine continued working for a while. This was a clear indication that its operations were not a direct consequence of the permanence of life (life in the sense of expressions as "soul" or "will"). Indeed, in this case Volta's attitude, dictated at least partially by the needs of the controversy, corresponded to the most ancient and conservative conception of the medico-biological thinking of the epoch; that is, the conception which rejected the application of the results and interpretations derived by the study of dead animal preparations to living beings, and particularly to humans. These conceptions had been replaced by the most advanced scholars of the time, and among them particularly Spallanzani and Galvani himself (see Chapter 4.1 and Chapter 8.4). In this context we might remark that in the *Memorie sulla elettricità animale*, Galvani would show that in electric fishes, the ability to give a shock would remain for a certain time even after taking out the heart, that is, "the principle of life" (see section 7.3 below).

In the next paragraph, we will come back to Volta's views on the relation between life and muscular motion. It is important to remark here how freely he invoked immaterial entities (such as souls and will), in order to attribute the capacity of putting in motion electric fluid "in a physical way." This is despite his status as a physicist (in the modern sense) attributed to him by historians. On the other hand, an author like Galvani, considered by many a supporter of "vital spirits," entertained a view that was more "physicalist" and more fruitful for possible developments in the field of life sciences.

7.2 GALVANI'S REPLY TO VOLTA'S CRITICISMS AND THE 1797 *MEMORIE SULLA ELETTRICITÀ ANIMALE*

After the publication of *Trattato dell'arco conduttore* and *Supplemento*, Galvani remained silent for about 3 years. In this period he was, however, not inactive. Besides continuing his experiments on the conductive arcs, he concentrated his interest on the torpedo. To this end, in August 1795, he made a long journey along the Adriatic coast, near Rimini, in order to get the fishes. As he would explain afterward, "for a long time I wished to examine this [animal] electricity in some of those animals, in which there was no question about its presence and about the mentioned circulation." Such desire had long been frustrated by the impossibility of keeping these marine creatures in the laboratory given "the impossibility of having them transported live from those distant places." Eventually, in 1795, he had the chance to go "for leisure, together with some honest friends, to the beaches of the Adriatic sea" first in Senigallia and afterward in Rimini (GEA, p. 64). Despite the short period he could spend there, from May 14 to May 19, Galvani succeeded in making an important series of experiments that he recorded in a notebook.

The results of these experiments were included in the *Memorie sulla elettricità animale* addressed to Spallanzani and published in September 1797 in Bologna. This work represents the last public document of Galvani, who would die about 1 year later, on December 4, 1798. In it we find his answer to Volta's criticisms against the direct contact between nerve and muscle, new objections to Volta's experiments of the arc applied exclusively to the nerve, and some arguments against the general theory of contact electricity. We also find an attempt to describe in a complete and clear way the presumed circulation of the electric fluid in all the crucial experiments of the controversy. This is not, however, all. His idea was that science should not only progress on the basis of logical reasoning but also, and principally, by answering experiments with new experiments. With this in mind, Galvani described a new "crucial" experiment based on the direct connection between nerves: Accordingly, he made an arrangement in which contractions were produced without any heterogeneity of different materials. It seems a fitting retort to Volta's objection based on the general theory of contact electricity, and to the challenge raised by Volta in the fourth letter to Vassalli.

The *Memorie* represents therefore a fundamental text, a sort of scientific legacy of Galvani. In it the scholar of Bologna portrays all the qualities needed for doing experimental science (knowledge, skill and manual ability, curiosity, utility, perseverance) in their highest degree. By reading them, we cannot but share the judgement of Spallanzani:

> For his novelty, for the importance of its doctrines, for the nobility and delicacy of its experiments, for the subtle analysis and the solid criterion accompanying them, for the felicity of the explanations of the most complex and abstruse phenomena, for the clarity and brilliance with which it is written, this work appears to me as one of the most beautiful and valuable of the eighteenth century Physics.[3]

---

[3] Letter of L. Spallanzani to L. Galvani, Scandiano, October 25, 1797, in Spallanzani, 1984-20, *Carteggi*, V, pp. 53–55.

By addressing himself directly to Galvani, Spallanzani wrote, "with it you have erected a building that, because of the firmness of its foundations, will last for the centuries to come."

With reference to Spallanzani's eulogy we need to consider that Galvani's examination of the nerve-to-nerve contact has been considered a fundamental experiment for the foundation of electrophysiology. If we consider, moreover, that electrophysiology is one of most advanced sectors of contemporary biological research, it cannot be said that Spallanzani's praise, expressed soon after the publication of the *Memorie sulla elettricità animale*, was an overstatement.

In the first memoir Galvani considered Volta's objections (expressed in the second letter to Vassalli) to the experiments on the direct nerve-muscle contact. As mentioned, Volta judged the experiments as "untrustworthy and inconstant," since the contractions did not appear in a constant way and were, anyhow, weaker than those produced by metals. Volta had also remarked that contractions could be the consequence of mechanical irritation induced by the contact between nerve and muscle. Galvani's response was that, if experiments were carried out with the necessary care, contractions were almost constantly produced. As to the difference with the experiments involving metals, he undermined its significance by making recourse to a physical analogy pointing to the importance of presence of the effect rather than of its strength. In a charged Leyden jar we obtain a strong discharge by using a metallic arc, whereas, by using an arc made up of moist wood, the discharge is very weak. In Galvani's opinion, we will err by far if, from this difference, we would make inferences about the charge accumulated in the jar. Volta's objection that mechanical irritation was the cause of the contraction in the nerve-muscle experiment was also manifestly wrong according to Galvani. In the *Trattato* he had already remarked that the experiment succeeded when the connection between nerve and muscle was established through "a bland contact." Moreover, in the *Supplemento* he had refuted the mechanical explanation by showing that if the nerve was left falling onto a flat surface of glass, sulphur, or marble, there were no contractions, despite the stronger intensity of the mechanical shock.

Coming back to this argument, in the first memoir to Spallanzani, Galvani remarked that the experiment succeeded even if the manipulation was made with "an extreme slowness," in such a way that the resulting contact between the nerve and muscle was "bland and very light." It was without success if the nerve was put in contact with a small insulating body despite its roughness and hardness. It did not occur either if the nerve was put into contact with a piece of muscle excised from the animal and manipulated in a way that there was no humidity that could be responsible for an electric communication with the animal body. This was clear evidence that, in order to have the contractions, it was necessary to insure an adequate electric circulation between the nerve and muscle of the animal preparation, whereas a simple mechanical contact was ineffective.

In a crescendo of laboratory creativeness aimed at "further dissipating these shadows, and these fears" of mechanical irritation to the nerve or muscle, Galvani proposed other important variations of the experiment. In one of them, a small strip of tissue made of "tendon or flesh" was first applied to both the nerve and muscle; the contact was afterward established by bringing the two strips closer together. Contractions were produced despite the absence of any direct mechanical action on either the nerve or muscle.

As Galvani had previously remarked, in many of these experiments, contractions could appear not only at the closure of the circuit but also when the circuit was suddenly interrupted. In particular, if some precautions were used in the experiments using the tiny tissue strips, it could happen that "contractions occur the one after the other, not only more felicitous and ready, but sometimes, so frequent, that the leg involved in the experiment would exhibit a kind of electric carillon" (an expression that he had already used with a slightly different spelling—*carriglione* instead of *cariglione*—in the *Trattato* to describe a similar phenomenon occurring in the case of the first "contractions without metal") (GEA, pp. 11–14).

Another experimental setup used by Galvani to remove the problem of mechanical stimulation was based on the use of small pieces of muscle tissues excised from the animal. One of those pieces was placed in contact with the nerve and the other with muscle of the leg remaining in situ. When the circuit was completed by manipulating with a small glass rod a third piece of muscle (or even a small piece of frog skin), contractions were produced despite the absence of any movement in the preparation. He concluded by saying:

> Everybody sees therefore that the arc is here made up exclusively of animal substances. He sees, moreover, that any uncertainty or doubt is removed, together with any suspicion of mechanical stimulus, and, finally, that is proved at once that the circulation of electricity depends exclusively on the animal machine that is thus activated. (GEA, p. 14)

With these experiments Galvani seemed to answer Volta's objections purposely and with the force of experiments. In these particular circumstances, he was probably addressing the challenge contained in a note added by Volta to the second letter to Vassalli, where the physicist of Pavia claimed that a mechanical artifact could not be excluded even when Galvani did his nerve-muscle experiments with the greatest care, bringing the two tissues to a delicate contact rather than to a hit:

> Let us conclude then that these experiments do not prove anything, because they leave the suspicion of a mechanical stimulus. In order to exclude a suspicion of this type, one should come to experiments in which the frog nerves and muscles do rest quietly, and are not touched or pressed in any other possible way. (VEN, I, p. 281n)

In the *Trattato* Galvani had already used the phrase "animal machine" to indicate the existence inside the animal body of a contrivance capable of producing and maintaining an electric disequilibrium. However, in the *Memorie sulla elettricità animale* the use of this phrase will represent one of the main points, particularly a differentiating element of animal electricity with respect to the idea of "common" electricity. In the *Trattato* he had also already disposed of many of Volta's objections to the possible presence of mechanical artefacts as sources of the stimuli for muscle contraction with a series of suitable experimental arrangements. In the *Memorie* he continued in this line by refining his experimental methodology and thus making his arguments stronger. For Galvani, therefore, every new experiment represented, on one side, the way to answer a specific question or objection, and, on the other side, a moment of apprenticeship that was useful for learning and improving new manipulations and conditions. These could result in addressing new problems and gathering new knowledge.

Galvani's answer to Volta's initial objection against the experiment of the contact between nerve and muscle can undoubtedly be considered detailed and timely. It did not, however, take into account the new arguments developed by Volta with reference to his new general theory of contact electricity. Galvani considered these new arguments in the second memoir, where he directed his attention to the possibility that contraction could originate, as Volta proposed, by the heterogeneity of the nerve and muscle tissue. It was for him a groundless supposition, because it was possible, as he was going to prove, to produce contractions with no heterogeneity whatsoever. The description of the new experiment was as follows (Fig. 7.2):

> I prepared the animal in the usual way; then I cut the one and the other of the sciatic nerves near their exit from the vertebral canal; afterwards, I divided, and

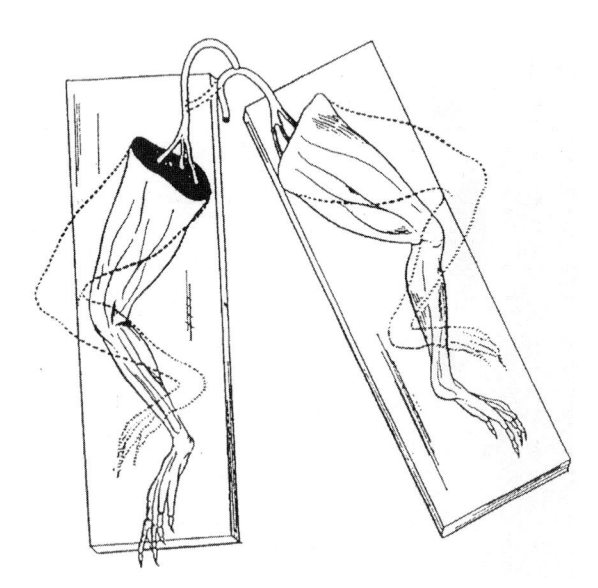

FIGURE 7.2. Galvani's experiment of the contraction brought about by the direct nerve-muscle contact (from Sirol, 1939).

separated a leg from the other, in such a way that any of them would remain with its corresponding nerve; in the following I bent the nerve of one in the shape of a small bow, and after that having lifted with the usual small glass rod the nerve of the other leg, I let it fall on that nervous bow. In doing that one should observe the precaution that, in its fall, the nerve should touch in two points the other nerve bent as a bow; and, moreover, that the small mouth of the first nerve is one of the two points. I saw then the moving of the leg corresponding to the nerve that I let fall onto the nerve of the other; sometimes even both legs. The experiment succeeds while the two preparations are totally isolated one from the other, and have no reciprocal relation, except for the touching of the nerves. (GEA, p. 16)

In this way Galvani had succeeded in obtaining contractions by forming an arc made exclusively of nerve substance. He could then ask in a rhetorical way: "What heterogeneity could then be invoked in support of the occurred contractions, given that only nerves come to contact?" The obviously negative response could only lead to the following conclusion:

It appears thus to me that there is another series of contractions which can be obtained without stimulus, without metal, and without any minimal suspicion of heterogeneity; these are then produced by a circulation of an electricity intrinsic to the animal, and naturally unbalanced within the animal. (GEA, p. 17)

Galvani was thus giving a direct answer to the challenge issued by Volta in the fourth letter to Vassalli. With his experiment he had indeed succeeded in obtaining contractions by using—in Volta's words—"conductors all of the same species, in nothing dissimilar the one from the other."

This experiment is generally considered a "capital" experiment for the birth of electrophysiology. Indeed, neither Volta nor the other adversaries of animal electricity could propose any substantial objection capable of undermining the evidential value of the experiment as support of an electricity intrinsic to the animal. It could not bring the phenomenon toward a simple physical interpretation, not even that of the new Voltian physics of the "general theory of contact electricity." Volta could not even invoke the agency of mechanical artifact at the moment the circuit between the nerves was closed. The counterargument was expressed by Galvani, who anticipated this kind of objection, with these words:

But why then by hitting one of the same nerves on an arc much harder and rougher made of non-conductive materials, as those made up of sulphur or glass contractions are not produced? And still in this case the stimulus originating from the hitting should be much stronger. (GEA, pp. 16–17)

To considerer the matter in detail, there was indeed an aspect in the new experiment that could have left Galvani open to his adversary's criticism and particularly to one of the subtle moves frequently used by Volta in order to undermine Galvani's apparently unquestionable results. For the experiment to succeed it was necessary that one of the points of the nerves included in the arc should necessarily be the "small mouth" of the sectioned tip. In his objection to the experiments with metallic arcs, Volta did not hesitate to invoke the least and most indemonstrable heterogeneities in the matter of the arc or in the way of fitting together the various parts. It was certainly possible, in the new nerve-nerve experiment, to suppose a difference in the composition, and consequently, a heterogeneity between the matter of the cut "small mouth" and the substance of the intact nerve surface. This would have represented an important element if favor of Volta if he intended to undermine the value of the new capital experiment.

Surprisingly, however, at the time Galvani's new experiment did not receive much attention either on Volta's side or on the side of the many scholars who were actively interested in animal electricity. This was partially due to the fact that the *Memorie* were written in Italian, a language rather poorly known among the scientific community of the period. Another, more important, reason was that in the meantime Volta was publishing results of experiments which were providing substantial evidence to support the existence of the metallic electricity he had supposed. In two letters written in August 1796 to the German scientist Carl Gren, and published in various journals, both in Italy and in other European countries, Volta presented results in which he could show the action of the metallic electricity in the total absence of the frog preparation.

Before considering Volta's new and fundamental experiments, we should comment just en passant on the need to have the sectioned nerve mouth within the electric circuit in Galvani's experiment described in the second memoir. As a matter of fact, at least in an a posteriori perspective, the need for this apparently heterogeneous contact was not against, but was decidedly in favor of the existence of animal electricity, in Galvani's sense. By cutting a nerve one exposes the internal part of the fibers, which is at an electric potential different from the external part, as we shall see in detail in Chapter 10. It is this difference in electric potential that plays a stimulating role in the nerve-nerve experiment. This is because of the existence of electricity in a condition of disequilibrium, accumulated between the interior and the exterior of the excitable fibers, a notion that corresponds closely to Galvani's conceptions.

In the first letter to Gren of 1796, Volta restated the three main experimental conditions in which it was possible to obtain the contractions in the frog legs or the effect on taste and vision. These were (a) the contacts between two metals (i.e., two conductors of the first class) and a humid conductor (meaning the animal or human body), (b) the contacts between two humid conductors (i.e., two conductors of the second class) and a metal; and (c) the contacts between three humid conductors. In the second letter he went on to describe an entirely new type of experiment based on the exclusive use of metals. These new experiments had convinced him that, in contrast to what he had first supposed, the main cause of the electric fluid motion was "the mutual contact of those same metals" and not the contact of metals with humid conductors. Moreover, and importantly, the existence of electricity due to the contact of different metals could be demonstrated without any recourse to the frog or to any physiological detector. It could indeed be detected by using an instrument invented a few years earlier by the English scientist William Nicholson, the "multiplier" or "duplicator," or by a method based on the use of an instrument invented by Volta himself, the *condensatore* combined with a sensitive electroscope (an apparatus that he called a "micro-electroscope").

We will come back to these experiments later. Let us now remark, however, that through this new line of investigation, Volta's research started diverging from that of Galvani and became more closely linked to the investigation of the other more typical "electricians" of the epoch, who were particularly engaged in demonstrating weak electricity in its various forms by using purely physical devices. Moreover, with these experiments Volta became convinced that the contacts made exclusively between metals were a more powerful motor of electricity than those also involving humid bodies, a notion that would influence the future progress of his research. Finally, as noted particularly by Giuliano Pancaldi, the use of Nicholson's multiplier, based on the operation of a series of metallic circular discs, contributed to the mutation of the shape of the metal conductors in Volta's laboratory. In his previous experiments he had used mainly arcs or cylinders, as illustrated in the figure added to the first letter to Gren (Fig. 7.3). He now started using

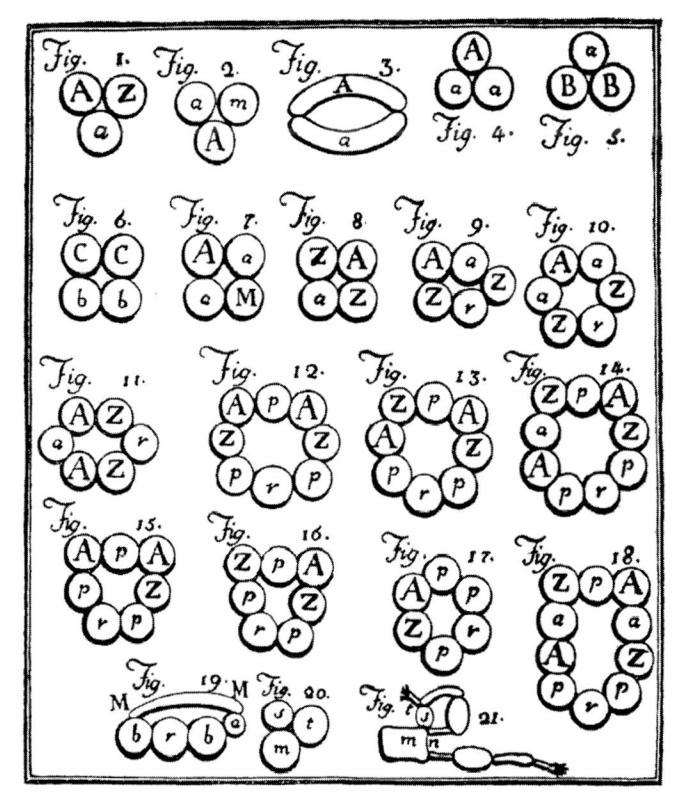

FIGURE 7.3. Various combinations of different electric conductors used by Volta in his experiments and described by Volta in the letter to Gren of 1796. Notice the disk shape of most of the conductors (from *Estratto di lettera al professor Gren di Halla* in Volta, 1918–1929, vol. 1).

mainly discs made of metal or other materials, according to the circular shape needed for the functioning of Nicholson's multiplier (Pancaldi, 2003).

At the time he was writing his *Memorie sulla elettricità animale*, Galvani did not know about these recent developments of Volta's research. After presenting (in the second memoir) the experiment of the contractions obtained with the nerve-nerve contact and discussing its implications as support of the theory of a natural form of electricity present in the animal body, Galvani adumbrated the possibility of a reconciliation between his own views and those of Volta. This move corresponded to what Volta himself had written in his *Memoria terza sull'elettricità animale* (and in some letters to Francesco Mocchetti and Orazio Delfico of 1795). He wrote:

> It could still remain only the doubt that, besides these contractions dependent on this natural electricity, there could be others depending on an extrinsic or common electricity; or on the electricity existing in those same metals used as arcs or armatures. This last type of electricity could be supposed to be different in the arcs or armatures due to natural causes because of the diversity of metals, or it could be assumed that it is [the electricity] spread in animal bodies as in any other conductive body, and put in disequilibrium because of the force of heterogeneity of metals. (GEA, p. 17)

Even though this supposition could have reconciled—as Galvani remarked—his own views with those of Volta, nevertheless he was convinced that the conception of his adversary met so many difficulties that it could not be maintained. This attitude corresponded symmetrically to the one expressed by Volta in his third letter to Vassalli, where, after reasserting the value of his the electric power of dissimilar conductors, the physicist of Pavia concluded:

> Why then should we make recourse to another principle, supposed but not proved, of an electricity proper and active in the animal organs, if it suffices—as I shall show and make tangible—the only principle of the action of dissimilar conductors, a principle demonstrated by so many experimental proofs, clear and self-evident, and incomparably more numerous? (VEN, I, p. 293)

Neither of the two scientists seemed thus inclined to any real compromise, nor were they willing to renounce some minor point of their own interpretation which would incorporate at least partially the adversary's explanation. These were not the only reasons for the difficulties in reconciling two different forms of electricity necessary for exciting nerve and muscle in the frog experiments. Both scholars had an understandable attachment to their own theory as would be expected. For both Galvani and Volta the respective theories had been progressively defined and refined through a long period of experimental research. Few scientists who have been involved in the complexities of a difficult but fascinating research for years are ready to abandon their interpretation or theory for a different view if this is incapable of fully accounting for the totality of the observed phenomena. This was the case in the controversy between Galvani and Volta. Galvani's theory of animal electricity did not account for the experiment of metallic electricity, and particularly for the last experiments in which Volta had succeeded in obtaining a purely physical measure of this electricity. On the other side, Volta's theory could not account for many of the experiments without metal, and particularly for Galvani's experiment of the nerve-nerve contact.

To refute Volta's theory, in the second memoir Galvani returned to some points already examined in the *Trattato dell'arco conduttore* and developed them further. Even though he admitted the possibility of a certain heterogeneity in armatures or arcs composed of the same metal, he did not believe that heterogeneity was sufficient to move the electric fluid with enough power to produce the contraction of frogs in a chain with two or more persons, as shown by the experiments. It was safer to think therefore of "a structure in the muscle similar to the Leyden jars, or to a comparable machine, by means of which—as we know—even a minimal quantity of electricity is capable of making a long journey, even in the presence of so many impediments" (GEA, pp. 18–19).

Faithful to his methodological principles that militate against purely verbal discussions, which are particularly fruitless when dealing with hardly verifiable problems (like the presumed "heterogeneity" of homogeneous arcs), Galvani decided to make recourse in this case to experiment. One of his strategies was to create artificial heterogeneities in otherwise homogeneous matters and study their effects. In none of these experiments were the results such as to support Volta's contention about the absolute need for a difference between the metals or other conductors used in the arc in order to produce contractions.

Galvani was, however, convinced that, despite his new results, an arrow still remained in Volta's bow, that is, the experiment of the arc applied to two points of the same nerve. As he recognized, this experiment made "a great impression in my mind, when I saw first such a phenomenon." Moreover, he was aware that the argument of the occult arc proposed in the *Trattato* did not clarify all doubts. In writing his new memoirs he was now convinced that he had now carried out some really decisive experiments, capable of showing that even in these experiments the electric flow did include the muscle, too. They therefore made untenable Volta's suggestion that the contraction obtained with the arc applied to two nerve points could be due to an external electricity stimulating a presumed "nerve force."

One of the experiments that Galvani presented for the first time in the *Memorie* consisted of separating a sciatic nerve from a frog, and applying it in the form of an arc between the nerve and muscle of another preparation. Afterward he applied two heterogeneous armatures to the sectioned nerve and connected these armatures by means of a metallic arc. According to Volta, under these conditions the electric flow should occur only through the sectioned nerve. This nerve could not influence the muscle with its nerve force because it was not anatomically innervating it, and as a consequence no contraction should be produced. The results, however, contradicted this prediction, making clear, in Galvani's views, that Volta's interpretation of his experiment of the arc applied uniquely to the nerve was definitely wrong (GEA, p. 28).

Furthermore, there was another experiment that was even more convincing. In a prepared frog, dissimilar armatures were applied to a crural nerve, which was then cut in the portion amid the armatures (see original *Fig. 4* in our Fig. 7.4). In Volta's opinion, no contraction would be produced because there was no nerve portion between the armatures. However, even in this situation the application of a metallic arc resulted in contractions in both legs (GEA, p. 38). Besides confirming the inconsistency of Volta's conceptions, in Galvani's view this experiment supported the need for an electric circuit between nerve and muscle in order to produce contractions. In these particular circumstances, this was, according to him, the course of the electric flow: from the thigh, to the lower portion of the sectioned nerve *I*, then through the arc to the spinal cord, and finally through the nerve stump *H* it returned to the thigh. To support this view, Galvani showed that contractions disappeared by cutting the nerve stump *H*. After this manipulation, they reappeared if the two extremities of the sectioned nerve *H* were connected by means of a conducting body (or even if the stump was completely removed and substituted by a small flesh piece or other conductive material). Along a similar line, Galvani accounted for the contractions of the other leg by assuming that the electric circuit occurred in an opposite direction.

For Galvani, similar explanations could account for the results of the other experiments visualized in the two plates illustrating the *Memorie* (Figs. 7.4 and 7.5). In one of

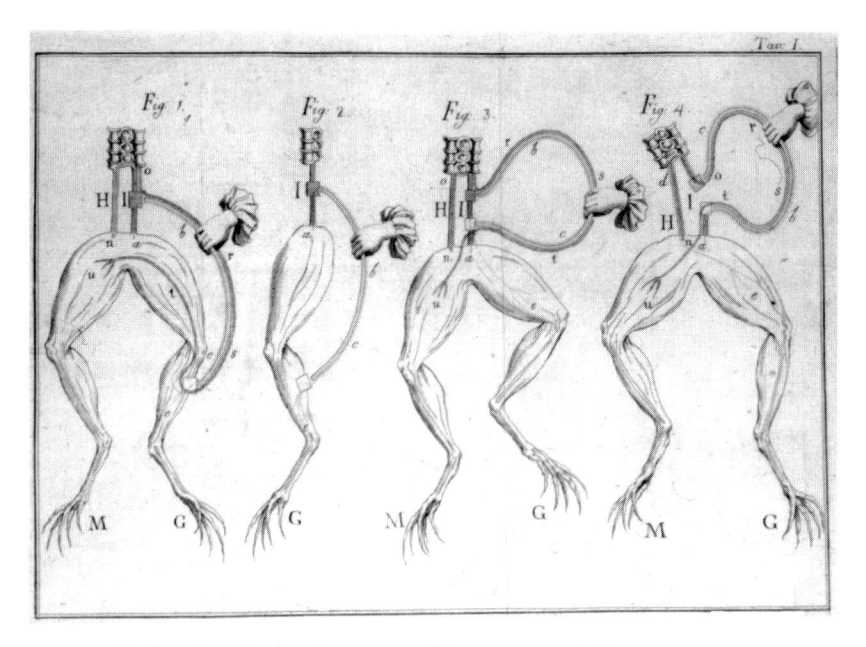

FIGURE 7.4. The first plate of Galvani's *Memorie sull'elettricità animale* (from Galvani, 1797).

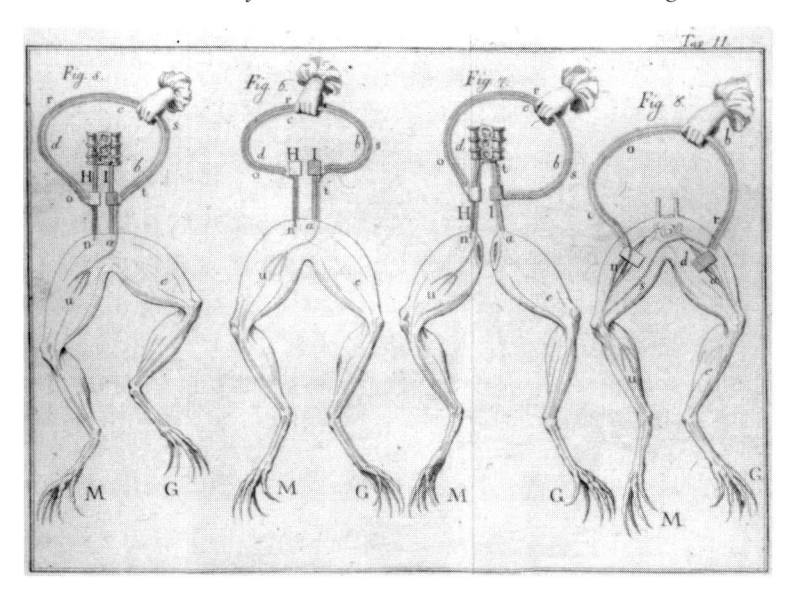

FIGURE 7.5. The second plate of Galvani's *Memorie sull'elettricità animale* (from Galvani, 1797).

these experiments Galvani illustrated the case of the contractions obtained by applying the armature exclusively to the muscle. This was the case that, as he overtly recognized, might have represented a strong objection to the idea of an obligatory circulation between nerves and muscles. He accounted for these contractions in this case by assuming that the electric circuit involved the thin nerve branches ramifying inside the muscle body. As he remarked, the idea that the electric fluid could circulate along the tiny nervous ramifications had been admitted by Volta himself.

According to Galvani, moreover, the idea that nerve fluid could circulate along the nerve branches embedded inside the muscle substance could account for a series of observations that Volta had interpreted against the animal electricity theory. These included the contractions obtained by Volta in living animals, and, moreover, the sensations brought about by the application of metal on the eye or tongue.

The doctor of Bologna was probably aware that the way he was interpreting this type of result could be considered an ad hoc explanation. This is why he made a series of experiments in which he tried to provide evidence to his contention. To this purpose he prepared a frog and ensheathed both sciatic nerves inside the thigh muscles of a second frog so that they were completely covered by the muscle substance. The application of the armatures and the closure of the circuit with a metallic arc to the surface of the muscle ensheathing the nerve brought about the contractions as would occur with naked nerves. With this experiment, which shows his extraordinary planning creativity, Galvani was providing evidence in support of an old idea that he had expounded only as a useful conjecture in the *Saggio sulla forza nervea* of 1782, but that had remained then only at the stage of a conjecture.

With the eight figures included in the *Memorie sulla elettricità animale* Galvani tried to account for all the crucial experimental situations that had marked his controversy with Volta: the arc between nerve and muscle (*Fig. 1–2* of Fig. 7.4), the arc on the nerve alone (*Fig. 3–4* of Fig. 7.4), the arc between the two nerves (*Fig. 5–7* of Fig. 7.5), and the arc between the two muscles (*Fig. 8* of Fig. 7.5). With their internal variations, for Galvani these four categories represented all possible combinations resulting from the arrangement of the three experimental parameters, namely the prepared animal, the armatures, and the arc. The persuasive force of these illustrations was also given by the fact that they provided simple and exhaustive graphical explanations, and, at the same time, they represented an ideal model of real experimental situations. Even though they were the results of a multitude of experiments, these figures put before the reader's eye the fundamental circumstances needed for the production of a particular phenomenon and this accounted for their importance. Galvani was aware of this when he wrote:

> For an easy and clear intelligence of the phenomena of animal electricity, it did not suffice, in my opinion, to have proved in a general way that all muscle movements excited until now depend on a particular circulation of that same electricity brought about by the muscle through the nerve and the arc. It seemed to me necessary to put under the eyes this circle, and show it in every particular case. (GEA, p. 31)

Even though drawn with formal elegance, the figures of the *Memorie* appear indeed to be graphical schemes suitable for clearly illustrating the electric flow that Galvani believed he had demonstrated in his experiments. In that regard, they are different from the plates of the *De viribus*, which, as we have remarked, were intended to convey to the reader an impression of realism and truthfulness with relation to the experiments illustrated. From a certain point of view, the figures of the *Memorie* represent a subsequent step in a pathway evolving toward the didactic attraction of the images illustrating the scientific work of the next century.

On the basis of the general analysis of the circuit of animal electricity demonstrated by the experiment, Galvani summarized his theory in a series of points:

> I. That the electricity which induces muscle contractions is already singularly gathered and cumulated in muscle [...] in a condition of imbalance [...].
> II. That this imbalance is the basis for the circulation of the electricity in the muscle itself.
> III. That this circulation consists of electricity leaving the muscle and returning to the same muscle by an inviolable law [...]
> IV. That this circulation occurs in the following way: The electricity leaves the muscle by the way of the nerve; it rushes to the place of that nerve to which it is attracted by the force of the armature, and arc; its comes out from the nerve in this same place, being drawn by the same forces; then it enters into

the arc and by means of it returns eventually with all the power to the muscle from which it left.
- V. That because of this tendency, and of the effort that electricity uses in order to come back to the muscle, it chooses always the shortest route to arrive there [...]
- VI. That nerves [...] are the natural and specific conductors of that electricity [...]
- VII. Finally, that the nerves exert this duty with their intimate and medullar substance. (GEA, pp. 30–31)

In its general principles this theory was analogous to that proposed 6 years before in *De viribus* (see Chapter 5.4). Contractions of the frog muscle were due to an electric current flowing from muscle to nerve and from this again to the muscle. For Galvani this depended on "a particular organization of these animals" and namely on a "particular machine." This had the specific function of maintaining electricity in a condition of imbalance, which was made possible by a "structure and organization of parts." In this conception, already asserted in the *Trattato*, Galvani distanced himself from the purely mechanical conception that had dominated medical and biological thinking throughout the eighteenth century. In the more particular context of the discussion on the role of electricity in physiological phenomena, Galvani distanced himself from Volta, who did not recognize any specific machine or any particular organization in the animal as the framework on which electricity could play its role in excitable fibers.

Undoubtedly Galvani's fundamental model of the muscle as an animal Leyden jar expounded in *De viribus* remained substantially unaltered in *Memorie sulla elettricità animale*. Moreover, some of the arguments elaborated in the new text, such as those concerning the nature and mechanisms of the arcs involved in the electric circuit of animal electricity, were connected to the discussion of the matter developed in the *Trattato*. Galvani reasserted with particular force the importance of the surface of contact between the different elements of the circulation pathway involved in animal electricity. In his view this was at the same time a surface of connection and also a surface of separation, that is, a zone capable of modifying the electric flow in a significant way. He remarked, for instance, that the contractions were more easily elicited by an arc made by two pieces of different metals put in contact than by the same metals "soldered—as it is said—and united together by means of the fusion of the metals." In his view, the relative inefficacy of this last type of arc was due "to a passage of the electric torrent from one metal to the other made exceedingly easy and almost insensible, due to the lack of the interruption and division of one metal from the other which is present when the two divided metals are put together in contact on purpose." For Galvani, despite the pressure to put tight contact between the two metals, there was, nonetheless, something like an "interruption [...] due to some tiny layer of air" in the zone of contact. This interruption would necessarily bring about a small resistance to the electric flow, which resulted in an increase of "the impetus and strain" of this flow. In this way electricity succeeded eventually in forcing the "vincible" resistance and thus produces the contractions with its increased energy (GEA, p. 52).

This way of reasoning based on the idea of an obstacle "minimum and vincible" drew on the discussion largely developed in the *Trattato* to account for both the particular force of the heterogeneous contacts and the efficacy of the "mutations of contacts" in the artificial and natural arcs. We remark here en passant that the idea that electricity could jump across a "tiny layer of air" was congenial to a scientist of the Enlightenment, accustomed as he was to phenomena resulting from very high electric potentials, as, for instance, the sparks (Chapter 3.2).

On this occasion, however, Galvani did not limit himself to old arguments or results, but he presented instead new experiments to support his interpretation. In one of these he prepared the arc with "two small heterogeneous strips," by cutting them at the tips to be put in contact "longitudinally so that they thus come to be toothed" with "not a few teeth, and well separated, and divided one from another." By putting together these two metallic strips the contraction appeared not only "to their first fitting, but they were renewed at any minimal pressure made over any one of these teeth." This was evidently due to the multiplicity of contacts that could be established with this arrangement. If instead some trick was used to tightly fix together the teeth of the two strips, there were no more contractions, whatever pressure could be exerted on them. This was for Galvani the sign that, in the absence of the resistance produced by the minimal air interval, "the electric torrent [...] flows placidly, and as silently, without exciting the contractions and give the minimal indication of itself" (GEA, pp. 53–54).

Another experiment that, according to Galvani, revealed the particular force of the contact involved the use of two armatures either heterogeneous or homogeneous when they were connected by an arc. There was, however, a particular proviso: One of the two armatures should be connected to the arc through "a drop of water which should be rather prominent." In this way the contractions were obtained "at the moment of the contact with the given drop, that is, at the moment of making of the circuit." When, afterward, the tip of the arc was immersed deeply into the drop, "no motion is excited as long as the contact with the water is maintained, as before." When however, on making the immersion deeper in the liquid of the drop, the tip of the arc eventually touches the armature below, "at once with the touching here again arise the previous movements, and usually even more vigorously." Movements are renewed by lifting the tip of the arc in such a way that "a new very thin water layer would intervene between the same surfaces," and again afterward bringing the tip of the arc again into direct contact with the metal of the armature (GEA, pp. 56–57).

In conclusion, even if in the *Memorie* Galvani recognized the particular power of the contact between metals, and notably the efficacy of the heterogeneous arcs, he continued referring to interpretative principles distinct from Volta's theory of the electromotive power of metals. We remark here that in many of the experiments by which he attempted to counteract Volta's theory, Galvani remarked on the particular efficacy of the temporal variation of the stimulus in the excitation of nerve and muscle as evidence. This is a fundamental aspect of neuromuscular electricity, which would need to await the new developments of modern electrophysiology in order to be fully explained (see Chapter 10).

As we have seen, in the fourth memoir Galvani dealt with the problem of the heterogeneous arcs by substantially reasserting the thesis expounded in the *Trattato*. He concluded with considerations of a physiopathological character that were also related to the discussions in the *Trattato*, even though the reference framework was less dominated by humoral theory, and the accent was more on the alteration of the solid components of the organism.

7.3 GALVANI'S RESEARCH ON ELECTRIC FISH AND THE VARIOUS FORMS OF ELECTRICITY

The fifth memoir stands out from the other *Memorie sulla elettricità animale* for the specificity of the theme, which concerned the study of the properties and mechanisms involved in the shock of the torpedo. As already mentioned, Galvani had long been interested in the study of electric fishes, and in 1795 he carried out a series of experiments on the torpedo that he decided to summarize and discuss in this memoir. His main intent was to compare the properties of animal electricity discovered by him with that, generally recognized, of the torpedo, and to find new evidence for his theory. Spallanzani was a pioneer of this research, because of his studies carried out about 15 years before "and new and interesting experiments" he had published.

Galvani's investigations on the torpedo published in the fifth memoir are of great interest not only for their importance in the comparison between the "stormy and lightning" electricity of the fish with the calm electricity of common animals but also for the specific results Galvani had been obtaining. Of particular interest was that concerning the integrity of the innervation of the electric organ for the production of the shock. He could also demonstrate the existence in the torpedo of "minimal degrees of electricity...suitable to stimulate the nerves of frogs and to excite in them muscle motion," even when not producing shocks sensible to the observer.

Given the very short time spent on the Adriatic coast, it is really amazing that Galvani was able to obtain many important results in his study of the fish. This shows clearly that, in his previous research, he had been acquiring an extraordinary skill as experimenter. This applied particularly to the ability in designing and realizing experiments capable of giving sound responses to his queries with great rapidity. Among the experimental setups he used in his torpedo experiments there is one that is especially remarkable—the one in which Galvani used his prepared frog as "an animal electrometer, and as one of the most exquisite" in order to reveal the electricity of the fish. Some decades later, this arrangement would direct Carlo Matteucci in his important experiment on the "induced twitch." This is the phenomenon by which the electricity associated to the contraction of a frog's leg might stimulate the nerve of another preparation (and thus produce contractions in this other frog: see Chapter 9.1). Matteucci's experiment differs from that of Galvani only because the galvanoscopic frog served to detect the electricity associated with the excitability of the muscle tissues of another frog rather then that of the torpedo. In this respect it is somewhat paradoxical that in his experiments Galvani tried to prove that, in contrast to the torpedo, the electricity associated with the neuromuscular excitation of a frog is not capable of exciting the nerve of another frog. It is also to be noticed that similarly to Galvani, in his experiments Matteucci also used prepared frogs to reveal the electricity of a torpedo. In contrast to the Bologna doctor, however, Matteucci had available physical devices capable of revealing the torpedo shock even without any recourse to the prepared frogs. As illustrated in Figure 7.6, in Matteucci's experiments the discharge associated to the fish shock, besides exciting frog contractions, also moved the galvanometer needle and produced electrochemical effects.

In the fifth memoir, after rapidly alluding to the practical aspects of his journey to the Adriatic coast, Galvani reminds the reader that in the torpedo had been recognized "two bodies said to be electric bodies or organs," very much similar to a series of "small magic

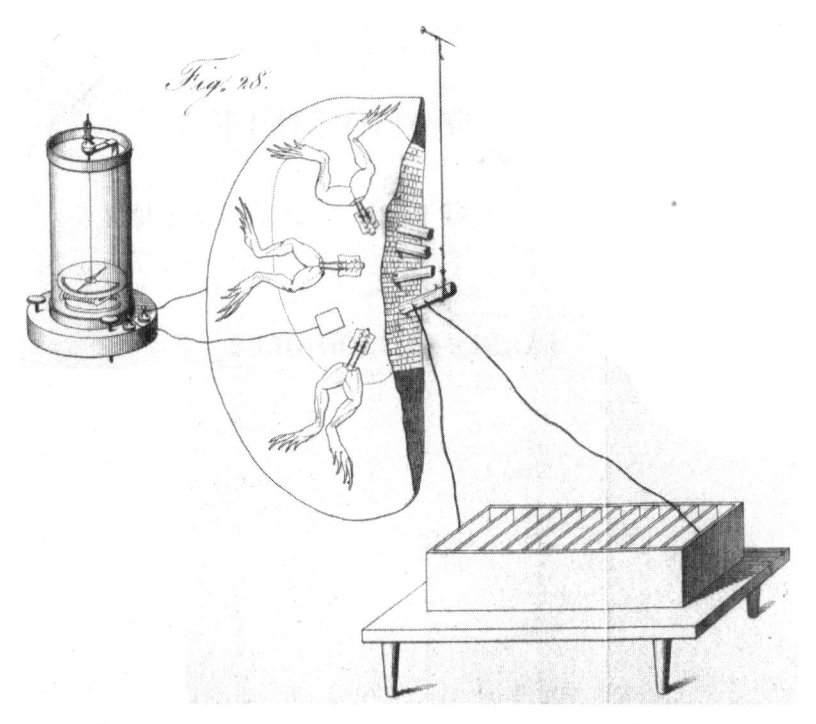

FIGURE 7.6. The experiment in which Carlo Matteucci used prepared frogs to detect the shock of the torpedo and, at the same time, measured the current with a galvanometer and obtained electrochemical effects from the fish discharge (from Matteucci, 1844).

squares," stacked one above the other, and that into these bodies there were the terminal arborizations of "an endless abundance of nerves." He considered next that brain could be the origin of these nerves and therefore that the brain "is the laboratory and reservoir of this electricity and nerves the conductors" (GEA, p. 65).

As mentioned in Chapter 3, the existence of a possible relation between the electricity of the torpedo shock and the nerve fluid emerged first in the studies of John Walsh and John Hunter. In particular, in 1773 Hunter was the first to point to the excess of innervation of the electric organs, compared to the other tissues of the organism. After considering the two possible functions of nerves, that is "sensation and motion," Hunter wrote:

> If it be then probable, that those nerves are not necessary for the purpose of sensation, or action, may we not conclude that they are subservient to the formation, collection, and management of the electric fluid; especially as it appears evident, from Mr. Walsh's experiments, that the will of the animal does absolutely control the electric powers of it's [sic] body; which must depend on the energy of the nerves. (Hunter, 1773, p. 487)

He then concluded with a statement that seemed to foretell the evolution of subsequent research on nerve physiology, which would indeed begin with Galvani's studies initiated a few years later: "How far this may be connected with the power of nerves in general, or how far it may lead to an explanation of their operations, time and future discoveries alone can fully determine."

To verify the possible existence of a relation between nerve fluid and the electricity of the torpedo (electric) organs, Galvani isolated and removed one of the two organs from the animal. He could detect no sign of electricity in this organ, whereas the other organ, which remained in situ and with intact innervation, continued to produce shocks. In another animal he sectioned the head, thus separating the brain from the rest of the body. In this case neither of the two organs was capable of producing "even the least sign of electricity, neither the biting nor the shocking one." Similar results were obtained if the brain was extracted rapidly from the skull so that the animal could maintain its viability and "blood circulation would suffer the least possible alteration." Despite these precautions, it could be that the result depended on the general attenuation of "life forces" caused by the experimental lesions rather than from the interruption of the flow of the electric fluid from the brain to the electric organs. To make the point more clearly, Galvani extracted the heart from the animal, and he could still see evident signs of electricity after this maneuver. If he cut the brain, "the said signs of electricity ceased, and it was no longer possible to excite any" (GEA, pp. 65–66).

The results of these experiments appeared so clear that Galvani remarked in his notebook, on May 17, 1795, the following corollary: "If therefore any electric force perishes after the separation and excision of nerves from the brain, it seems safe to conclude that electricity is administered from the brain to nerves."[4] With the ensemble of his experiments on the torpedo, Galvani believed that he had established an unequivocal relation between the nerve fluid and the fluid responsible for the torpedo shock, which was then unanimously considered electrical. Having thus proved that the brain was the source of the electric fluid responsible for the fish shock and the nerves were only conductors, he passed on "with the supply of analogy and of new observations to demonstrate that the same happens with common animal electricity." He applied metallic armatures and arc to the torpedo's muscle and found out that they contracted in a similar way to that of frogs or other animals. In that regard, he described a phenomenon that had "surprised him not a little," and which could be used to argue against the electromotive power of metals. He found that heterogeneous armatures applied to the nerves going to the electric organs of the torpedo, or directly placed on the organs themselves, did not give "any minimal increase of force and energy in the electricity of the said organs, and of giving shocks or in producing the biting sensation." For Galvani this meant:

> Metallic armatures are not responsible for the imbalance of electricity, nor do they provide some electricity of their own, but are only apt to increase a little bit the electric force when this is weak. This increase does not easily manifest itself when electricity is by itself strong, and vigorous, as is that proper of torpedo, but manifests itself in a plain way in the common electricity of animals, because this is much weaker and is perhaps modified in a different way. (GEA, pp. 68–69)

Galvani's surprise on that point depended only partially on the unexpected difference between torpedo and common animal electricity. It was also related to the information he had derived from some unidentified source (possibly Spallanzani?) that "the difference of metal played a rather important role in favouring or preventing the electric explosion in the torpedo," as he wrote in the unpublished memoir of 1786 (see Chapter 5.3).

The demonstration that in torpedo there were two forms of electricity, "one proper and particular" and the other "common to all animals," and a further observation that the nerves of both the electric organs and muscle originated in the brain and had the same "form and substance" suggested an important analogy to Galvani:

> It is thus one and the same the origin of the nerves going to the mentioned organs compared to those going to other parts and muscles; it is also the same structure and substance of the ones and of the others: it is thus appropriate to say that it is also the same use and office, and by consequence even the same should be the fluid they receive from the brain, and bring to the parts where they terminate. Concerning

---

[4] The experiments involving the ablation of the brain or the heart made by Galvani in order to ascertain the role of the nervous system in the control of the torpedo shock present various points of analogy with those made by Galen in the third century A.D. With his experiments Galen endeavored to prove, against the cardio-centric theories of Aristotle, the role of the brain in the control of motor functions (see Manzoni, 2001).

those which go to electric organs, it is demonstrated that these are conductors of an electric fluid. It should therefore be the same thing concerning those which bring to other parts, and those which branch in muscles. (GEA, p. 70)

The study of electric fish allowed Galvani to bring a new argument in support of his theory of animal electricity. The similarity between nerves of the organs and nerves of the muscles was reinforced by the observation already made by Michele Girardi, an anatomist who had been a student of Giovanni Battista Morgagni in Padua (and confirmed by Galvani), that some of the nerves going to organs had side branches which terminated in muscles. In a rhetorical way Galvani asked: "Should we then believe that the same nerve contains two very subtle fluids of a different nature, and that these two fluids flow simultaneously through the same nerve?" (GEA, p. 70). "No" was the obvious response.

In *De viribus* Galvani had tended to remark on the correspondence between the animal electricity of common animals with that of electric fishes rather than with that of the ordinary electricity of inanimate bodies. In the *Memorie*, on the other hand, he believed that he had definitely proved that fish and common animal electricity were carried by the same fluid, and he emphasized the similarity of fish electricity with the ordinary physical electricity. The apparent difference between the two forms of animal electricity was simply due to "a bigger accumulation of this fluid made in the mentioned organs, due to the extraordinary abundance of nerves which go to them, and due to the surprising multiplicity of surfaces therein, depending on the numberless plates which, as we have said, are made up by the prisms composing those organs." Galvani added other differences, pointing to the importance of "the different structures and organization of the parts" involved in the reception and elaboration of electric fluid between torpedo and frog. This is another instance of his special attention to the aspects of structure and organization that were, in his view, a characteristic feature of the animal machine (GEA, pp. 72–73).

Various experiments now pointed to the similarity between the animal electricity of torpedo and common physical electricity, particularly for those aspects concerning the strong efficacy of torpedo shock in stimulating contraction in prepared frogs. In one of these experiments Galvani placed a prepared frog "above the back of a Torpedo, extracted out of the water, and situated on a table" and noticed the strong convulsion of the frog placed above different parts of the torpedo's body. Afterward he put several prepared frogs on the fish body and they soon started contracting. He described it in these words: "It was for me a joyful show to see them all move at once, and, I would say to jump [*saltellare*]." Interestingly it was possible to obtain contractions even if frogs were not in direct contact with the torpedo, but communicated indirectly "through the water layer by which the table had been moistened at the moment that the torpedo was put above it" (GEA, pp. 74–75). By giving an unequivocal demonstration of the transmission of torpedo's shock through water in this way, Galvani solved a point that had been debated in the studies of this fish. It also had great relevance to ethological aspects (what would be the use of the shock for a fish if it did not spread in the liquid habitat of the animal?).

In his experiments with torpedo and prepared frogs, Galvani noticed that the contraction could occur even when he could not perceive any sensation corresponding to the torpedo shock and remarked that, in this way, it was possible to detect "those minimal degrees of electricity, which remained occult until now." In them he identified an expression of an electric fluid circulation that he supposed to be present in the fish even in resting conditions. In his view, this represented the counterpart in the fish of the continuous course of electricity, "calm and uniform" to which he had referred in *Trattato dell'arco conduttore* as the fundament of neuromuscular physiology in all animals. In the next chapter we shall see that Volta himself attributed great significance to these "beautiful experiments" in which Galvani demonstrated the existence in the fish of a continuous flow of electricity. In the views of the physicist of Pavia these results had a special importance in supporting his own hypothesis of the production of electricity in the fish.

By summarizing the experiments in which he compared the electricity of torpedo to common physical electricity, Galvani wrote by addressing himself to Spallanzani:

> From all these facts you see well, if I am not wrong, put in full light and in safety, the stimulating faculty of the torpedo electric fluid to be totally similar to that of common electricity; and also, moreover, confirmed the analogy of the electricity of the torpedo with that of a magic square. (GEA, pp. 74–76)

The analogy was supported by subsequent experiments in which Galvani put on the back of the torpedo some isolated frog muscles soon after their separation from the animal, as well as an isolated frog heart that had just stopped beating. In contrast to what had happened when the contact was made through the nerves of the prepared frogs, in this case there were contractions only when the torpedo gave a full shock, sensed by the experimenter, after being stimulated. In other words, there were no contractions present in the prepared frogs in the absence of clear shocks. Moreover, when the fish was relatively weak, at the production of the shock there were contractions in the skeletal muscle, but not in the isolated heart. Galvani discussed this relatively poor excitability of the heart (compared to common muscle) in light of the debate some years before in the scientific community with relation to the difference between the electric excitability of various organs and tissues of the body (see Chapter 3.3 and Chapter 6.2). To support his contention about the similarity between the common physical electricity and the electricity of the torpedo, he set up a new experiment. He arranged an animal chain made up of various pieces of excised skeletal muscle connected in series, and inserted in the chain an isolated heart. By stimulating the chain with a weak electric stimulus (either directly from the electric machine or through a Leyden jar), he noticed that the contraction of the heart lagged behind that of muscles. By decreasing further the stimulus intensity, only muscles appeared to contract, while the heart was seen "to remain totally still."

Let us remark here en passant how in these experiments with "chains" made up of different elements and excited with stimuli of different efficacy, Galvani used arrangements often adopted by Volta (see Chapter 8.3).

On the basis of his experiments, Galvani eventually asked himself if the animal electricity of the frog had a stimulating action similar to that of both the torpedo and artificial electricity, this being a matter of great importance in the controversy with Volta. The physicist of Pavia, as we have already discussed, explained the experiments with the metallic arc applied to two points of a nerve as evidence of an action of the metallic electricity on the "nervous force" in the section of nerve between the tips of the arc. Among Galvani's experiments aimed at ascertaining this point one is particularly significant. He applied two heterogeneous armatures (one of zinc and the other of silver) to the nerve and muscle, respectively. Afterward, he tied a moist hemp thread to the zinc armature terminating in another piece of zinc and connected it to the other armature so that the arc was completed. Lively contractions were soon produced. If, on the other side, the nerve of another prepared frog was connected to the thread of hemp (or metal), there was no contraction in this last frog, wherever the nerve was collocated.

For Galvani, on the one hand, these experiments made clear that animal electricity could not act as an external stimulus in exciting muscle contractions. On the other hand, they showed that heterogeneous metals could not increase the force of animal electricity to such a point that it became capable of exciting the nerve. In his view, the arc produced contractions only because it made possible the flow of electricity naturally unbalanced by physiological mechanisms. They did not act, as Volta pretended, by producing electricity capable of stimulating the nervous force in the nerve segment to which they had been applied.

An important point that needs to be considered here is why Galvani insisted on the fact that the animal electricity of the frog could not act as a stimulus for the excitation of nerves. As we have mentioned (and will discuss further in Chapter 9), about 40 years after Galvani's *Memorie*, Carlo Matteucci would provide evidence that the opposite is true, with his experiment on the "induced twitch." In this experiment the nerve of a prepared frog is laid over the thigh on another frog. When this last one contracts, also the first preparation moves. It is difficult to imagine that in the numberless variations of his experimental arrangements, Galvani never put the nerve of a frog on the muscle of another and observed the effects of electricity on them. We can hardly help from supposing that he was strongly conditioned by the needs of his controversy with Volta and he thus overlooked results which could show that even the "nonexplosive" animal electricity of a frog can be an effective stimulus for the nerve.

In the fifth memoir Galvani develops the theme of animal electricity in torpedo with relation to that of common animals and to ordinary artificial electricity; it is a crucial and recurrent theme all through his electrophysiological research. Many of those who have written about Galvani have often remarked that his conception of muscular motion is based on an essential distinction between animal electricity and the other forms of electricity known at the time. They have also remarked that this was the main point of distinction with Volta who attributed galvanic phenomena to common physical electricity. A certain historiographic tradition has even gone much farther by saying that Galvani's conceptions belonged to a vitalistic tradition which precluded any assimilation under the same laws of the forces acting on inanimate matter with those specific to the living world (see Chapter 1 and Chapter 5.3).

Volta himself contributed at least partially to this tradition. In his memoir *L'identità del fluido elettrico col così detto fluido galvanico*, published by his pupil Pietro Configliachi in 1814 (but written by Volta several years before), the physicist of Pavia repeatedly insists on the idea that for Galvani the electric fluid in animal bodies is modified and "animalized" in some way such as to become something different from ordinary electricity. Volta had expressed similar views in some unpublished notes added to his *Memoria terza sull'elettricità animale* of 1795, where he says that Galvani's animal electricity is an electric fluid "more or less deprived of its properties and characteristics, and dressed with others, *animalized* in a certain way, to which [Galvani and his supporters] gave the name of *electric-animal* fluid" (VEN, I, p. 303). Indeed, Volta varies somewhat in this critical attitude against his antagonist because in some of his writings he accused Galvani's collaborators (and particularly his nephew Aldini) of having favored the interpretation of such "animalized" fluid, which was not advanced by Galvani himself.

In concluding the analysis of Galvani's *Memorie sulla elettricità animale*, his last published work, we think it is now appropriate to linger for a while on the theme of the relation between animal electricity and common physical electricity that is evident all through the research trajectory of the Bologna doctor.

Galvani had been comparing the properties of the "nervous fluid" with those of the electric fluid since the first phase of his electrophysiological research in the 1780s. In a series of unpublished manuscripts of 1782, he remarked on the correspondence between the two types of fluid by pointing to their similar phenomenological properties. He was adhering to his conception of animal electricity in a subsequent memoir of 1787 (also unpublished). Galvani insisted on the similar behavior of animal to that of artificial electricity, both with relation to conductive bodies "like metals, humidity and water," and insulating substances like "vitreous, resinous or oily bodies." The phenomenological similarities between the two types of electricity were reasserted and extended in the third and fourth part of *De viribus*. He noted in particular that the force of both animal and ordinary artificial electricity was increased "with the help of the so-called armature, made of the same metals which physicists usually employ to coat vitreous and resinous bodies." Galvani's conclusion was then unequivocal in sanctioning the full electric nature of the "animal spirits": "Thus being these things, after our experiments, certainly nobody would, in my opinion, cast doubt on the electric nature of such spirits" (GDV, p. 402).

At the moment he was writing *De viribus*, Galvani was clear that the agent responsible for muscle contractions was of an electrical nature, even though it did not manifest some of the typical signs of "common" artificial electricity. These included the "sensation almost of a slight aura, the attraction and repulsion of light bodies" and the capacity for producing "in the electrometers invented until now, the minimal signs of movements." On the other hand, these signs were lacking in the case of the torpedo and other electric fishes, but nobody doubted that true electricity was implied in their shock. It was specifically the similarity with the animal electricity of torpedo and analogous fishes which, for Galvani, gave strong support for the genuine electrical nature of the electricity involved in the neuromuscular phenomena of frogs and common animals. Among the similar characteristics of frogs and electric fish electricity in *De viribus*, Galvani listed the fact that it did not need to be "provoked by previous artifices" and, moreover, that it needed to flow along a circuit to be effective.

In the *Trattato dell'arco conduttore*, Galvani seemed to oscillate between two attitudes. The first was an extreme idea that animal electricity is "an electricity totally distinct from common electricity"; the second, more moderate idea, was more in line with that expressed in *De viribus* and based on only some dissimilarity with common electricity. Such oscillation clearly reflected the difficulty he had in viewing the evident phenomenological differences that existed between the two and which were difficult to account for in a simple way. In that regard, it must be noticed that the *Trattato* was written in order to counteract Volta's contention that animal electricity did not exist and frog contractions were due to a common form of electricity. Galvani's' decision to distinguish clearly the two forms of electricity is thus comprehensible in view of the controversy with Volta, and corresponds in part to the comparable sharp distinction he made in *Memorie sulla elettricità animale* between the electricity of torpedo and that of common animals. Accordingly, it is interesting to note that in his *Taccuino*, where he recorded the protocols of the torpedo experiments, he did not insist on the difference between torpedo and frog electricity. On the contrary he underlined their similar nature, as for instance when (on May 14, 1795) he noticed that the torpedo shock seemed capable of recharging the animal Leyden jars present in the frog muscle with its electrical surplus.

On the one hand, Galvani aimed at maintaining the electrical nature of the fluid involved in muscle contraction, despite the obvious difference between animal and common artificial electricity. On the other hand, against Volta, he needed to note the specificity of the physiological mechanism involved in the electrical phenomena of animals, the salient aspect of his discovery.

Under the mask of the anonymity in the *Trattato* Galvani made it explicit that Volta's conception implied the following conclusion:

> [...] that all the action of animal electricity would be reduced to a pure mechanical stimulus, and by consequence would be removed any law proposed by Galvani of a circulation of the same electricity depending on the animal machine and structure. This electricity would thus lose all the new dignity acquired by the labour of the same author; and, quite the opposite, it would be reduced to having its office common with the simple and ordinary bodies, like wood, iron, and salt. (GAC, pp. 107–108)

Galvani never cast doubt (in the *Trattato* or in his other writings) on the genuinely electrical nature of the agent responsible for the contractions. Indeed, he preferred to call it "animal electric fluid," rather than "fluid analogous to the electric," because "animal and common electricity are in their characters [...] one and the same thing" (GAC, pp. 61–62).

Besides noting the similarity between the animal electricity of frogs and electric fishes, in the *Trattato* Galvani made a sharp distinction between his and the presumed animal electricity invoked by authors like Bertholon and Gardini (see Chapter 5). In his view, as well as in those of Volta, this was simply common electricity manifesting itself in living organisms rather than in inanimate bodies.

As we have often remarked, the fundamental aspect of Galvani's conception of animal electricity and of his distinction between this form of electricity and common electricity is the importance he gave to the operations of the particular machine he supposed to be present in living organisms. This aspect has often been overlooked by historians, which has led to confusion about the importance attributed by the doctor of Bologna to the agent involved in neuromuscular electricity. Together with the theory of the electrical nature of nerve conduction and muscle excitation, the notion of a specific machine responsible for the specific properties of the electricity involved in bioelectric phenomena is to be considered one of the fundamental achievements that physiology owes to Luigi Galvani.

As we have remarked, this conception only makes partial reference to the mechanistic views of the seventeenth century. These were based on purely mechanical interactions and on the importance of a deep knowledge of the anatomical organization of animal structures—particularly of their microscopic organization, as was the case for Malpighi (and for his teacher Giovanni Alfonso Borelli: see Borelli, 1680–1681). For Galvani the reference to structure and organization is more functional and dynamic than anatomic. Even his model of "the small animal Leyden jar" (which has a clearly defined structural dimension) is not developed primarily on the basis of an investigation of the minute anatomy of the neuromuscular complex. It derives mainly from physiological investigations and physico-chemical reflections, even though considerations about the tiny structure of the neuromuscular complex were afterward decisive for the elaboration of the final model. We can thus say that with Galvani, physiological investigations went beyond the seventeenth-century mechanistic theories. This occurred particularly because he was able to propose an interaction between classical mechanistic views and the new Hallerian approach to living phenomena together with the instruments and devices of the electrical research of the Enlightenment.

In the next century Claude Bernard made frequent recourse to the notion of the animal machine; he was a scientist who asserted the applicability of physical laws to the study of biological phenomena, but he tended to underline the organizational complexity typical of living machines and their specific operative functionality. Georges Canguilhem (1970) remarked that Bernard discovered the role of liver in glucose metabolism by following the concentration of this sugar on the various circulatory routes and not on purely anatomical considerations (see also Grmek, 1968). For both Galvani and Bernard the reference was to machines that do not reveal the principles of their functioning from a simple knowledge of their structure and arrangement of elements alone.

Like other members of the Bologna Academy of Sciences, Galvani was scarcely inclined toward explicit reflections on his scientific methodology. Nevertheless, with his research he offers an instance of the scientific approach which was at the forefront of his time. To him we could perhaps extend Mirko Grmek's assessment of Spallanzani. For Grmek the latter was a scientist who, because of his rare theoretical discussions, has long been ignored in the historical analysis of the evolution of the scientific method, despite the extraordinary novelty of his methodological approach to investigation in various fields of the sciences of his times (Grmek, 1982).

With regard to lessons on the scientific method given by Galvani *malgré lui*, we can provide an example from the fifth and last of his *Memorie sulla elettricità animale*, which was his last scientific publication. There is no explicit concern with general principles and theoretical discussions, but instead with practical recommendations on the appropriate way to carry out experiments. The proviso—he says—is intended not only for the experiment he has just described (that of the nerve placed in contact with the hemp thread) but also for "all the experiments reported in these few Memoirs." After alluding to the insulating flat surface on which the animal is placed, Galvani warns that:

> For the happy outcome of this experiment [...] it is on the other hand of extreme importance that the portions of the said surface, which remain under it [i.e. the nerve] is thoroughly dry, and free from any damp; otherwise either the phenomenon would not happen, or—even though it would happen—it would be doubtful and uncertain what was the cause from which it originates. (GEA, p. 85)

This exit in a minor tone is a clear indication of Galvani's stature as an experimenter of rank and, with reference to the panorama of life sciences at that time, it contains an important lesson concerning method. Nonetheless, this probably appears more so for scientists than for philosophers and theorists of the scientific method.

7.4 THE CONCLUSION OF THE GALVANI-VOLTA CONTROVERSY

Volta reacted badly to reading the *Memorie sulla elettricità animale*. In all probability he felt himself acutely stung by the passages accusing him of being inconstant and having often changed his opinions. In April 1798 he wrote two letters to Giovanni Aldini with an anonymous (but recognizable) signature (*cittadino N.N di Como*: meaning *unnamed citizen of Como*) which were published in the *Annali di chimica e storia naturale* of that year. He accused Galvani of not taking into account, or even of not knowing, his writings of the last 2 years. He expressed his resentment as follows:

> I confess to you that I have been strongly surprised and even scandalized by the assertiveness with which Galvani, in face of all that, still declares his opinion quite irreconcilable, and opposes, right at the beginning of his recent work (p. 3 of the Memoir) some of his propositions and many of Volta's, in the following terms— "He wants electricity to be the same and common to all bodies: myself one particular and proper to animals; he places the causes of the imbalance in the used artifices, and namely in the difference between metals; myself in the animal machine; he establishes that cause as accidental and extrinsic, myself as natural and internal; in short he attributes everything to metals, nothing to the animal; myself all to this, nothing to those, as afar only the imbalance is concerned." Now this *in short*, this rejecting any virtue to metal whatever or power of unbalancing the electric fluid; all that has hurt me, by contradicting not only so many arguments and proofs multiplied by Volta on the subject of Galvanism, but also to facts, and direct experiments, independent on Galvanism. (VEN, I, p. 525)

The last sentence of this passage alluded to the research carried out by Volta in the most recent period, research that—in Volta's opinion—Galvani did not know about. This was likely because the results of this research had been reported and discussed by Volta in letters addressed to Carl Gren and published in both in *Gren's Neues Journal der Physik* and in the *Annali di chimica e storia naturale* of 1797, the same year of the publication of Galvani's five *Memorie*. Apart from their importance in the controversy on animal electricity, Volta's results represented a new important contribution to the history of physics. The use that he gives to the word *Galvanism* (which starts appearing just in this period) is testimony to the turning point which is occurring in his research. The title of the letter to Gren *Sul Galvanismo, ossia sull'elettricità eccitata dal contatto de' conduttori dissimili* (*On Galvanism, i.e., on the electricty excited by the contact of dissimilar conductors*) indicates Volta's need to distinguish the new phase of his investigations from his previous experiments based on the use of frogs' legs to reveal the electricity of metals. He did not need the biological preparation any longer to detect the new electricity he had discovered. Although he still praised Galvani's "admirable discovery" of the contraction that could be excited by "suitable conductors of electricity," he reproached him for the inability to go further, as he himself had been able to do, particularly with "the experiments outside Galvanism, namely in those which do not involve any animal body or organ" (VEN, I, p. 525).

As to Galvani's arguments expounded in the *Memorie sulla elettricità animale* supporting the possibility that contraction could be obtained in the absence of any heterogeneity of the conductors, Volta reiterated his argument about the impossibility that one could be completely sure that two different conductors were completely homogeneous. But what was more important were the new experiments "independent of Galvanism," which had convinced most of the "community of physicists and notably the Transalpines (*oltramontani*)" to adopt the general theory of contact electricity (VEN, I, p. 539).

In a somewhat mirror-like way to that in which Galvani had excluded metals from his working bench, Volta had succeeded in demonstrating his metallic electricity without recourse to frogs or other animal preparations. He had succeeded in that by profiting from various devices capable of increasing the electrical potential of the metallic electricity by a change in electrical capacity. One of the tools capable of doing was based on his newly designed "condensing electroscope," a compact device based on a sensitive electroscope associated with the *condensatore*, the instrument that Volta himself had invented about 15 years earlier. The operations were as follows (see Fig. 7.7): The weak electricity generated by the contact of different metals is transferred to the lower, fixed disk of the *condensatore* while the capacity of the device is enormously increased by connecting the upper disk to earth; afterward the metals are removed and the earth connection broken so that the electric capacity decreases. The use of the condensing electroscope (or of other devices based on analogously functioning devices and particularly the "multiplier" of Nicholson which made the operation automatic) was especially necessary for the combination of different conductors that had a weak electromotive force. This happened with the combination of a metal and a humid body, and even more so with two different humid bodies. In these cases, it was necessary to repeat many times the manipulations aimed at increasing the detectability of the tiny charges separated at their contact.

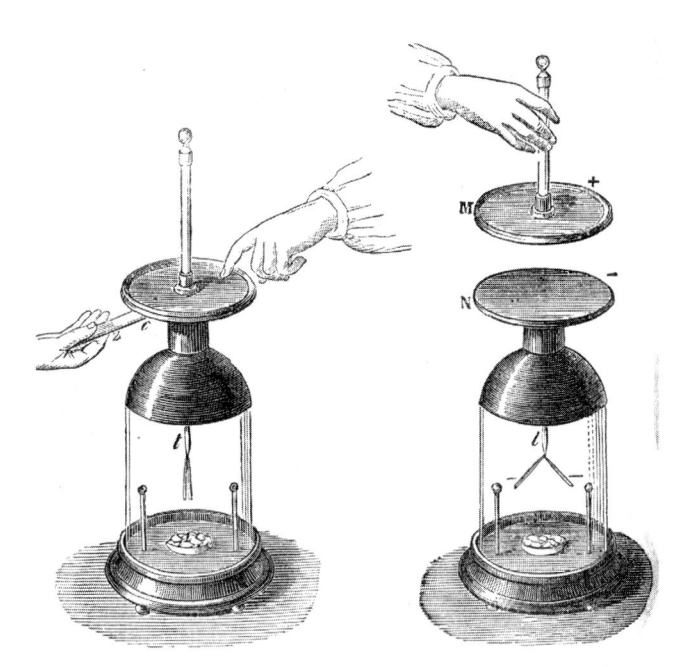

FIGURE 7.7. One of the tools used by Volta to detect metallic electricity, based on the association of his *condensatore* with a sensitive electroscope (from Ganot, 1859).

These experiments in which, as remarked, no animal preparations were used thus confirmed Volta's contentions that dissimilar metals (i.e., "conductors of the first class") were much stronger motors of electricity than humid bodies ("conductors of the first class"). Volta could thus conclude by saying:

> Having now not only proved the thing, but made it visible by the above described experiments with the metallic discs, and even with the non-metallic ones, which becomes sensibly electrified, and give signs to the electrometer, by only fitting them together, what are you going to say, my dear Aldini, and what Galvani himself? (VEN, I, p. 555).

Unfortunately, Galvani could not respond because in April 1798 he had problems of another kind. He had refused to take the oath for the new *Repubblica Cisalpina*, established in the north of Italy after Napoleon's armies had taken over the various Italian states, including that of the Pope, who ruled over Bologna. Galvani's refusal had resulted in the loss of all his public functions and their relative salaries. Moreover, his health started deteriorating and he died within a few months at his brother's home in Bologna, on December 4, 1798 (see Bresadola, 2011a).

Galvani's death was the immediate cause of the end of the controversy on animal electricity, at least of the controversy as it had developed over the previous 10 years. A second, important factor was Volta's invention of the battery, which will be announced by the Pavia physicist with a letter of March 20, 1800. It was addressed to Sir Joseph Banks, president of the Royal Society of London, the scientific institution of which he was fellow and from which he had obtained the prestigious Copley metal for his research on galvanic phenomena in 1794.

In the next chapter we will consider some of Volta's final research, and we will analyze his many contributions to life sciences, which are an almost forgotten aspect of his scientific work.

[this] immortal work
which inaugurated a new epoch in the whole field of science,
is a most brilliant illustration of the extreme fruitfulness
of an intimate combination of the exploration
of the laws of inanimate nature
with the study of the properties of living organisms.
—Bohr (1937)

8
The Electrophysiological Work of Alessandro Volta

IN VARIOUS PARTS of this book, and particularly in the first chapter, we have discussed the extensive way in which collective imagery, at both popular and erudite levels, has been pervaded by the idea that Galvani and Volta were in many respects two antithetical scientists. On one side was Galvani, who was depicted as a representative of old medical thinking, strongly anchored to vitalistic conceptions, and who fortuitously became (somewhat *malgré lui*) a pioneer in an unexplored field. He would open new paths of discovery, notwithstanding his incapacity to progress with the assurance of the rigorous and systematic investigator, thus losing the chance to pick the mature fruits of the research he had started. On the other side Volta was seen as the expression of modern science, with its rejection of immaterial entities and the taste for accurate experimentation and precise measurement. Volta was thus represented as the physicist capable of bringing to completion, through laboratory investigations, the research program that the Bologna doctor had only been able to begin.

As already mentioned, this cultural tradition started during the period of the Galvani-Volta controversy, and it has been perpetuated through a conjunction of unfortunate historiographic circumstances. Among them was an insufficient consideration of the historical context in which the research of the two scholars flourished, as well as the sparse attention paid by historians to the scientific problem at the heart of the controversy.

In recent times, the opposition between Galvani and Volta has been seen as an expression of two antithetical viewpoints: the medico-physiological of Galvani and the physical of Volta. These were considered as irreducibly opposed one to the other, and thus incapable of interaction without mutually excluding and nullifying one another, like the two views of an ambiguous visual image (Pera, 1986/1992).

In parallel with this, and implicit to the aforementioned opposition, another view has taken root in the collective imagery. By inventing the battery, Volta granted to physics an ultimate, crowning victory over physiology so that animal electricity was relegated to the periphery of scientific thought.[1] We have already discussed the difficulty of using labels such as "physiologist" and "physicist" in relation to the scientific endeavors of "natural philosophers" of the eighteenth century, and we will return to this theme. In the previous chapters we have provided what we hope are useful arguments for reconsidering Galvani's scientific work, outside the simple categories cherished by certain traditions in the history and philosophy of science. We want here to focus our attention on Volta by noting from the outset that the superficial opposition between physics and physiology has contributed not only to an incorrect evaluation of the work of the Bologna doctor but also to a disregard of many aspects of the scientific endeavors of the scientist of Como. In particular, there has been little attention for Volta's interest in the physiological and medical implications of electricity in animal organisms. This interest played a fundamental role in Volta's invention of the battery. As we shall see, moreover, in some important phases Volta's research was stimulated and guided by Galvani's studies.

Without doubt Volta made substantial contributions to the development of physics between the eighteenth and nineteenth centuries with the invention of his battery (and many other instruments) and also with his theories. On the other hand, the scholar of Pavia was a man of the Enlightenment and, together with the literati of his era, shared interests in the fundamental themes of "natural philosophy" and "natural history" outside the field of physics (in modern acceptance). In addition to electricity, his studies delved into the fields of "airs" (he discovered the "flammable air of the marsh," i.e., methane), of meteorology, and of geology. He was also deeply interested in chemistry. One of his great regrets was that, because of the lack of an adequate quantity of quicksilver, he could not perform before Lavoisier the experiment on the formation of water from an explosion of a mixture of "inflammable metallic air" (hydrogen) and "dephlogisticated air" (oxygen). Volta's interest in animal electricity was not limited exclusively to the possible role of electricity in animal economy but concerned other fundamental themes of the period. We will quote here a letter that he addressed to Lazzaro Spallanzani in 1773, in which he said that he had carried out an experiment of "leg reproduction" on a terrestrial salamander, similar to those made by the famous naturalist. Volta wrote that "the reproduction did occur, although rather slowly," and concluded with the following words:

> Forgive me my unskillfullness in these matters, and provide me, through research and your writings, with better lights on this, and on the other parts of Physiology

---

[1] In Chapter 1 we listed some of the obsolete concepts to which animal electricity is assimilitaed in the view of some scholars: the "heavenly spheres" of the Aristotelian-Ptolemaic cosmology; the "humours", "virtues", and "sympathies" of classical medicine; Descartes's vortices; and Stahl's phlogiston.

and Natural History. A singular taste for these is stimulated in me by reading your works, and the works of those, who, like you, endeavour to put them within the reach of great numbers of amateurs. (VEN, VI, p. 4)

Volta was 26 years old when he wrote this letter and his "singular taste" for physiology and natural history accompanied him throughout his scientific life. As to his experiment on the terrestrial salamander, it is perhaps worth noting that it did not have the impact like the experiments he carried out about 20 years later on Galvani's frogs. This is yet another instance of the unpredictability of the paths of scientific discovery.

8.1 VOLTA AND LIFE SCIENCES

One of the myths rooted in the collective imagination is that Galvani acted only as an initial catalyst for Volta's interest in the experimental study of the effects of electricity on animal organisms. The writings of scholars of different epochs and nationalities, like Emil du Bois-Reymond and Giovanni Polvani, give the impression that, once Volta's interest in animal electricity had been sparked off by reading *De viribus*, the scientist of Como progressed alone through the unexplored paths of the new electric science. It is as though he was supported exclusively by the cleverness of his own thought and by the rigor of his experimental methodology, leaving behind the efforts of the "pioneer" Galvani (or the other researchers who got involved in the matter).

By reference to the most fruitful years of Volta's investigations (those spanning from 1792 up to 1800, the year in which he announced the invention of the battery), Polvani expresses himself with these words, certainly full of emphasis and rhetoric:

> Of this wonderful world Volta is the true dominator. This is not only because the other scientists, both Italian and foreigners, ask him for explanation, listen to him, admire him, and—if his antagonists—are afraid of him, spy on him, escape from him. This happens also because it is he, and only he, who creates the circumstances for the new events. It is he who directs, and pushes and brings where they should lead, as if he knew already the goal and the value of his work. His research is not therefore that sequence of fortuitous and fortunate events which revealed to Galvani, attentive and perspicacious as he was, one of the most elusive and hidden secrets of life. It is instead a continuous passage from one position to another, with planned and definite moves, as with the rock climber who foresees, perceives and by him alone creates the path whereby, sloping more and more upwards he gets to the top: alone, there yonder, eventually he rests, more than surprised at the peak he has scaled and of the immensity he can dominate. (Polvani, 1942, p. 265)

If, on one hand, Volta did not limit his interest exclusively to the physical aspects of the electrical research on animal organisms, on the other, in his path toward the battery he was certainly not alone and far from any other, as immersed in the solipsism of his "wonderful world," as Polvani is pleased to say. On the contrary, he was particularly attentive to the investigations of other scholars, and especially to those of Galvani. In his path of discovery Volta was indeed confronted by new scientific questions, which induced him to broaden his experimental and conceptual horizons, also toward others' research, thus continuously reconsidering the working hypothesis he was elaborating, and—at the same time—providing him important, and sometimes decisive, cues.

In particular, Galvani's studies bring back anew, in a continuous way, at the heart of Volta's attention, animal preparation, that wonderful "animal machine" which, since the beginning, had surprised him for its extraordinary capacity of reacting to the weakest electric influences, an apparently winning challenge of animate nature against the most sensitive humanmade artificial electrometers (which were then the thin-straw device of Volta and the electrometer with gold leaves of Bennet). Because of his background as a natural philosopher, and also for the stimulation of other scholars' researches (above all Galvani), the animal preparation was to be a constant reference point in Volta's research, up to and beyond the battery.

It is perhaps worth remembering here what Volta wrote in a memoir published in 1802, 2 years after the invention of the battery. In trying to understand the principle underlying the functioning of his new device, he asked himself if an assembly of three conductors of the same class could be effective. After acknowledging that until then he had been unsuccessful in his attempts, he was inclined to be hopeful for the possibility of such an achievement. He wrote: "Even if not art, nature has found the way of achieving the goal, in the electric organ of Torpedo, Trembling eel (Gymnotus electricus), etc., made

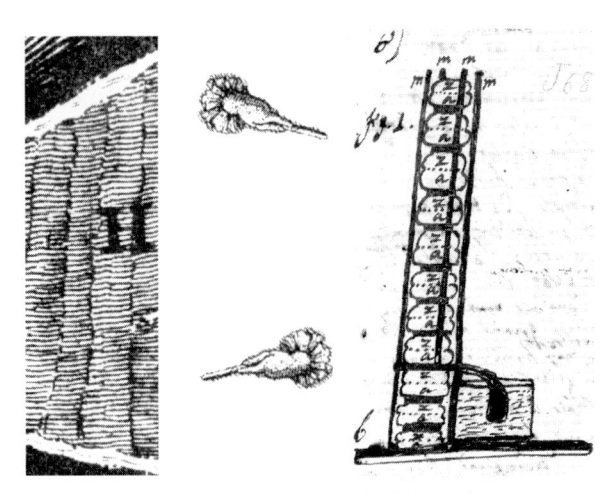

Natural electric organ   Artificial electric organ

FIGURE 8.1. The electric organ of the torpedo and the battery of Volta (image from Piccolino, 2000; original material from Hunter, 1773, and from the autograph of Volta's rough copy of the communication to Banks on the invention of the battery existing in the archives of the *Istituto Lombardo, Accademia di Scienze e Lettere* of Milan).

up exclusively of conductors of this second classes, i.e. humid conductors, without any of the first class, i.e. without metal; and perhaps we are not far from achieving it also by art" (VEN, II, p. 62). For Volta nature (specifically the animal kingdom) could open unsuspected possibilities to physics that he should try to reach with his "art" (see Fig. 8.1).

Volta's attention to the animal world was not limited to the observation and reflection on the particular contrivances that nature had developed in living beings, but it also assumed a variety of aspects, sometimes with the characteristics of a genuine interest in the specific field of electrophysiology and medicine. From the beginning of his research on animal electricity Volta was aware of the possible medical applications of what he was observing. In the *Memoria seconda sull'elettricità animale*, published on May 1792, after only 2 months of research in this field, he reached conclusions of great relevance, both theoretical and therapeutic, concerning the effects of electricity on organisms. Starting from the need for circulation of electricity through animal tissues in order to obtain contractions, he wrote:

> If on the other hand, the electric fluid is not placed in such course that it could pass across the mentioned nerves and muscles, but only will accumulate or thin out, it will not produce such effects. Let us put, for instance, the frog, either prepared or intact, on the Conductor of the Electric machine, so that it would acquire the strongest electricity. The frog will stay motionless until one would not draw from it a spark, or else, increasing and pushing excessively, the electricity would not spring up from some part of the frog's body, e.g. from the tip of a hanging leg, as a strong electric jet.
>
> Therefore the electric status *plus* or *minus*, that is the excess or scarcity of electric fluid in all body frame, an electricity whatsoever of the stagnant type, or (as it is called) a simple *electric bath*, does not affect our extremely sensitive small animal, even when prepared. This type of electricity does not affect it such as to excite nerve sensibility and muscle irritability, such as to produce those contractions, which occur just when the electric fluid just flows and passes across them, even in a small dose.
>
> If one considers the matter, he will see well that, at least as far muscle movements are concerned, the electric status of the atmosphere can have a small or no influence on animal economy. And thus, in this respect, we can expect little or nothing from the method of electric application by simple bath to the treatment of diseases. In order to obtain some sensible effect we must push the electric fluid instantaneously, or in a shocking way, from one side to the other of the animal, direct its flow, in a way that it stirs nerves and fibres etc, that is by operating with the other known methods of *Medical Electricity*. We must on the hand confess that this practice has not made the progress that, amid so many supporters, it seemed to promise. (VEN, I, pp. 48–49)

By reasoning about the physiological implications of his first studies of electric stimulation of muscle motion, Volta undermines the attempts to use static electricity for treatment of diseases as they were widely diffused among the *electrifying* physicians of the eighteenth century. These attempts were generally based on the immersion of the patient in a bathtub full of water charged by contact with an electrical machine. In *De viribus* Galvani himself had considered the possible effects of the *bath application* of electricity and thought it was useful in treating diseases like rheumatisms, in which it was necessary "to accumulate in muscles a larger amount of electricity" in order to attenuate "the stagnant humours" (GDV, p. 413).

On the basis of his considerations Volta downplayed the possible impact of atmospheric electricity on human health. These influences were considered of great significance in the medical systems elaborated by Pierre Bertholon and Francesco Giuseppe Gardini. In some way they represented a return, in an electrical form, of the idea of atmospheric influences on animal organism; it had been one of the characterizing aspects of classical medicine (see Chapter 5.2). Volta's conclusions were therefore fully framed within the ongoing medical debate of the period and went in the direction that medical progress will take in the following century.

By comparing Volta's first medical deductions with the "conjectures and corollaries" expounded by Galvani in the fourth part of *De viribus*, we are far from correctly concluding that Volta adopted an exclusively electrophysical perspective. On the contrary, we could say that paradoxically medicine profited more from 2 months of research carried out by the "physicist" of Pavia, than by 10 years of research by the "doctor" of Bologna. This conclusion can, however, considered valid only with respect to the short period mentioned. Despite the importance of Volta's medical and physiological results, it is with Galvani's work and his notion of animal electricity that the course leading eventually to modern electrophysiology starts. As we now know, this proved to be a science with abundant consequences for medicine.

In his writings Volta explicitly acknowledged his interest in physiology and medicine, particularly after the invention of the battery. In a letter addressed to Marsilio Landriani in September 1800, he wrote that he was then working for "9 months" on the new apparatus:

> both varying its construction, and trying to improve it, both extending the experiments, aimed not only at perfecting electric theories, that is to better and more deeply understanding the newly discovered laws; and also, to a great extent, to make out of it useful applications to Physiology and Medicine. In that respect my research has offered me many results curious and interesting at the same time, that I can qualify as true electrico-physiological discoveries. These I will communicate to you on another occasion. (VEN, II, pp. 10–11)

In the draft of a subsequent letter to the same Landriani, he described the "trough" version of his new apparatus (the version *a corona di tazze*), which in addition to producing movements "causes a more or less intense burning on the skin of sensible body parts, if the contact is continued; and eventually excites in the organs of taste, sight and hearing, the sensations proper to each one of these senses, of flavour, light, sound." By addressing himself to his correspondent, he then added: "On these effects I have discovered some very interesting and curious phenomena. I mean both on the entire living bodies and their organs, and on the excised members or parts, in which there remain somewhat of a vitality, sometimes even when one would be led to assume that any vitality is completely extinct" (VEN, II, pp. 148–149).

In the springtime of 1801, in a letter addressed to an unknown correspondent (probably the Turin abbot Anton Maria Vassalli), Volta complained about the fact that the research stimulated by the invention of the battery concerned almost exclusively the chemical effects of the new instrument. Scholars—he wrote "seem to take into little account the other effect, to be called electro-physiological, which, on the other hand, are in no way less singular and surprising," whereas "on the contrary, I did occupy myself to a large extent with them from the beginning" (VEN, II, p. 16).

Volta's attitude was not exclusively a personal view but, at a general level, did reflect a general vision of an era in which it was difficult to define rigid barriers between different research fields. This comes out also on the occasion of the most prestigious award he received, the *Prix Bonaparte* from the Institut de France (an award that was considered by many as the final sanction of his triumph over Galvani and, in some way, to reflect the victory of physics over physiology).

In the report accompanying the award (founded by Napoleon "for the discoveries concerning electricity and galvanism"), in the séance of the Institut de France of "17th Messidore of the year 10" (June 29, 1802), the history of electricity is distinguished in two phases: the first one characterized by friction electricity, and the other by metallic electricity (i.e., galvanism), championed, respectively, by Franklin and Volta. In the following the report alludes to "the chemical effects offered by the new apparatuses" and continues by saying:

> But it is mainly in the application to animal economy that it is important to consider the galvanic apparatuses. We know that metals are not the only matters which put electricity in motion with their contact. Metals share this property with some liquids, and it is probable that it extends, with diverse modifications, to all natural bodies. Is it not true that the phenomena exhibited by the torpedo depend on a similar actions likely exerted between the different parts of their organization? And does not this action exist with a less sensible intensity degree—and yet not less real—in a much more considerable number of animals than has been until now believed? The analysis of these effects, the complete explanation of the mechanisms which determine them and their similarity to those presented by Volta's battery, would probably provide the key to the most important secrets of animal physics. (VEN, II, p. 123)

We will consider soon the main discoveries and the main "electrico-physiological" effects to which Volta was directing his interest, and we will allude, moreover, to the possible medical applications he envisioned and was planning on the basis of his electrical researches. On the other hand, the feature of Volta's research that is particularly difficult and at the same time attractive is the distinction between those aspects that are more strictly physical from the physiological (and medical); far from adopting exclusively the "electrophysical" point of view, in his studies the scientist of Como operated a continuous and fruitful exchange between the electrophysical and electrophysiological perspectives. In that respect he adopted an attitude similar to that of Galvani and of the other great "electricians" of the eighteenth century (like Henry Cavendish, for example).

Let us consider the experiments on the tongue, which played a crucial role in distancing Volta from Galvani's conception of animal electricity and in directing Volta's subsequent research (see Chapter 6.2). With this experiment carried out in 1792 Volta succeeded in determining the polarity of the currents produced by the metallic arcs, 4 years before being able to measure this current with a physical instrument. He noticed that the acid sensation produced by applying the tin of a bimetallic arc made by tin and silver to the tip of the tongue was similar to that elicited by the positive polarity of the electric machine. If, on the other hand, the tongue tip was touched with the silver side of the bimetallic arc the sensation was weaker and of an alkaline character. This type of sensation could be reproduced by applying to the tongue the negative pole of a charged electric machine.

Considered from the point of view of physics (in the modern sense), this experiment appears as an instance in which a sensory experience is effectively exploited in order to ascertain an electric parameter (the polarity of a current), in the absence of any adequate physical system of measurement. However, the experiment can also be considered from the point of view of sensory physiology in which case it is one of the first demonstrations of the polar nature of the nervous excitation produced by an electrical stimulus on a given sensory apparatus, that is, of the dependence of the physiological response on the current polarity. Volta's observation was to be confirmed as leading to "the law of polar excitation" formulated about half a century later by German physiologist Eduard Pflüger.

In his tongue experiments Volta was particularly cautious in trying to exclude possible causes of errors. In the case of bimetallic arcs he needed to show that the sensation was due to current circulation and not simply to the application of metals per se. To this end, he first showed that the taste sensation is produced only by closing the circle between the two metals used for the stimulation, while no comparable sensation is produced by simply putting the metals in contact with the mucous surface of tongue and mouth. He went on, moreover, to set up various arrangements designed to confirm this conclusion. He showed, for instance, that the same acid taste is elicited even without a direct contact between tongue and metal: It is enough to dip the tongue in a glass full of water, which is, in turn, in contact with the positive electrode of the bimetallic couple (usually tin or zinc). It was possible, moreover, to produce the same acid sensation by putting "two metallic spatulas or spoons, one on the flat part and the other on the tip" both of silver on the tongue; the sensation is produced if each of the two spatulas is connected to, respectively, a silver and a tin lamina "separated from one another" and immersed "in a jar full of water," so that their extremities protruded from the liquid surface (VEN, I, p. 140).

In its experimental "virtuosity," Volta succeeded in performing a remarkable, and somewhat paradoxical, experiment in which a weakly alkaline solution was used to establish the contact on the positive side of the bimetallic arc. At the very moment that the tongue tip was immersed in this solution, he perceived an acid sensation that he attributed to the passage of a current. This sensation transmuted afterward into an alkaline taste, as the solution diffused and reached the tongue surface.

The polar nature of the effects induced by electric stimulation emerges in Volta's other electrophysiological experiments. In the case of the pain sensation felt with particular intensity when metallic electricity was applied to wounded surfaces or sores, it was more easily excited by applying the negative pole than the positive one. This difference was more marked when the powerful electricity of the battery could be applied after 1800. In a memoir published in 1802 Volta expressed his surprise at this strong asymmetry of the pain sensation induced by stimuli of opposed polarity with these words:

> It is difficult to understand the reason why the burning sensation is more intense and lively when excited by the negative head or pole of the battery than by the positive one, despite the same degree of electric *tension*. The difference is very large, as is the burning sensation excited by a battery of 20 couples applied in the first manner in comparison to that by one of 60 couples applied in the second manner. (VEN, II, p. 77n)

Volta showed that even muscle contractions in frogs were excited in a different way by stimuli of different polarity. As also Galvani before him, Volta observed that contractions were produced both at the onset of the current application (i.e., at the closure of the circuit) and at its offset (i.e., at the circuit opening or breaking). In a letter addressed to Tiberio Cavallo in 1793, he realized that these last contractions appearing "at the break of the circuit" were produced less easily and were normally weaker compared to those produced at the current onset. In another letter addressed to Francesco Mocchetti 2 years later, Volta returned to the subject with reference to the physiological effects induced by the electricity of a Leyden jar in chains formed by both humans and frogs, these ones having been prepared "at a variable degree." He drew attention to the efficacies of the two different current polarities in the "completely prepared" frog:

> It is relevant here to observe that the direction of electric current is not indifferent. I have indeed already found (from the springtime of 1792 the year in which I published my first Memoirs on Animal Electricity) that the current is much more effective if the direction is from torso or spinal cord toward the lower limbs, i.e. when it passes from the trunk of nerves to their ramification than vice versa. Let's first find

the minimal electric charge which can induce full contractions in the prepared frog when the discharge is directed from sciatic nerves towards thighs and legs they innervate. If one applies afterwards the same charge in the opposite sense it will not get the least motion; nor will it be achieved with a charge two or three times stronger; it needs to be usually 5, and sometimes 6 or 8 times stronger; and almost of that strength to shake a frog that is not completely prepared. (VEN, I, p. 364)

Similarly to Galvani, Volta had remarked that, by using metallic arcs, contractions were obtained transiently at the instant in which the contact was applied to the animal (and sometimes, as noted above, even when it was suddenly removed): Normally, after the first shaking, the preparation remained totally still despite maintaining the contact, and consequently the circuit. To obtain repeated contractions, and thus induce a muscle spasm in the legs (similar to that associated to tetanus), it was necessary to establish and break the circuit in rapid alternation.

In principle there were two possible explanations for the intermittence of the contractions induced by the maintained application of the arcs (or, according to modern physiological terminology, for their "transient" or "phasic" character). On the one side, it could be a characteristic related to the physiological mechanisms of nerve and muscle excitation induced by electric stimuli. On the other side, it could depend on a totally different cause, of physical type, namely from the possibility that the currents produced by metals were of inherently short duration, and produced uniquely at the onset and offset of the application of the arc, with no maintained phase.

The problem was of great importance for both physiology and physics. Volta was able to provide an important and exhaustive answer for both scientific domains. He showed that the current produced by a maintained metallic contact was physically continuous ("perennial" or "indefectible" according to his own terminology), and the transient character of muscle contractions was due to the specific physiological mechanisms of the neuromuscular system. As a matter of fact, the tongue experiment convinced Volta that electric fluid "when pushed continuously in a given direction according to the prevailing force should circulate without pause." This is because the application of a bimetallic arc, tin-silver or tin-gold (with the tin extremity applied to the tongue tip) produces "a sensation of a well-defined sour taste which continues and goes on in an increasing way as long as the communication between the two metals is maintained" (VEN, I, pp. 138–139).

On the basis of an electrophysiological experiment Volta concluded that "the action of metals is not a momentary electric discharge followed by an equilibrium phase [*repos d'équilibre*]. But a maintained action, the expression of some type of force expressed by metals on the electric fluid of the bodies contacted, without pause, with more or less energy, everyone according its own nature" (VEN, I, p. 219). This was a fundamental conclusion for physics because an entirely new branch of electrical science started to emerge: electrodynamics.

In most of his "electrophysiological" studies Volta seems to move with extraordinary perspicacity and skill to select a series of impressive successes in a decidedly difficult field. Nonetheless, his course was not without thorns and in some circumstances his results were contradicted by the research of other scholars. One example is the presumed ineffectiveness of electrical stimuli in producing contractions in those muscles "the movements of which do not depend on the action of will," and particularly the heart and bowels. Volta dealt with this subject in the *Memoria seconda sull'elettricità animale* and in a letter addressed to Martinus van Marum in August 1792, where he said he was unable to induce muscle contractions in many invertebrates. Among the animals with muscles supposed to "rebel" against electrical stimulation, the scientist of Como listed "leeches, slugs, snails, earthworms, and others," on which he claimed "to have made trials in every possible way," by both using metallic arcs and artificial (i.e., friction) electricity "strong enough to produce sparks and give a sensible commotion to the arm." Not all invertebrates appeared, however, refractory, and contractions were easily excited in "shrimps, beetles, grasshopper, butterflies and even in flies," for example (VEN, I, pp. 124–126).

The presumed electrical insensitivity of the heart asserted by Volta could not but arouse interest in the scientific milieu of the time, mainly because the heart was one of the intensively studied organs in research on irritability. It was not by chance that Felice Fontana, one of the strongest advocates of Hallerian irritability, was among the first to enter the arena by showing that metallic arcs could provoke contraction in cardiac muscles similar to voluntary muscles (Fontana, 1792).

Informed of Fontana's studies, Volta hurried to repeat the experiments and he confirmed the results obtained by the "physicist" of the Grand Duke of Tuscany. Volta noticed that it was indeed possible "to accelerate heart movements, and to awake them once they had ceased" by using bimetallic arcs. He added, however, that it is necessary "in order to succeed with heart, particularly in warm-blooded animals, to take more care and attention regarding the conditions than is required for the other muscles endowed with voluntary contractions," and, particularly, it is necessary "to use the bimetallic couples more efficacious in their electromotive action." To support "the difficult excitability of the heart by electric stimulation in comparison to voluntary muscles," Volta also added the fact that "these experiences on the heart" succeeded only for a short time after the preparation of the animal "whereas those on flexor and extensor muscles, all in short the muscles subservient to voluntary motion, succeed for an incomparably longer time" (VEN, I, pp. 160–161).

Volta returned to the same subject on various occasions. An instance is evident in the second letter to Cavallo written in 1792, where he remarked on the sparse or absent excitability to "a weak current of electric fluid" in the case of those "muscles on which the action of will has no effect, such as the stomach, the bowels etc, and not even the heart, despite its strong irritability" (VEN, I, p. 191).

Volta's attitude in this circumstance seemed to be analogous to that he held in the controversy with Galvani concerning the efficacy the monometallic arcs and the contractions without metal: tactical argumentations, ad hoc moves, largely inspired by his desire to not fully acknowledge his own errors of both experimental and theoretical character. However, even in that which, at first view, can appear an error or an oversight, comprehensible perhaps for the feverish rhythm of his research in the years of his most productive research, Volta exhibited great powers of observation and critical acumen. In a letter addressed to Joseph Banks in 1795, he pointed out a very important difference between the electrical excitability of the heart compared to skeletal muscles. The application of a weak electricity localized to the nerves innervating these muscles—he wrote—elicited powerful muscle contractions. In the case of the heart the situation was different. First it was necessary to use a stronger electricity, like that produced by the couples made of the most active metals. Moreover, and importantly, it was necessary that "one of two metals of the couple is applied to the proper substance of heart, or, at least, that the heart is situated in such a way that a considerable part of its body would stay within the circuit, and consequently would be permeated by the electric current excited by the two metals." "In brief"—Volta concluded—"it is needed that the electric *stimulus* influences the muscle *in a direct way*" (VEN, I, p. 252).

With these words Volta was perhaps the first to point out that the stimulation of the heart nerves is not per se sufficient to elicit contractions in the cardiac muscle, and that electricity needs to act directly on the muscle fibers of the heart. As we now know, Volta's observation corresponds to the physiological reality: The contractions of the heart are normally generated in specific, auto-excitable muscle fibers situated in a particular zone of the atrium ("sinoatrial node"). Heart nerves exert only a modulator action on these intrinsically rhythmic fibers. Specifically, the excitation of the nerve fibers belonging to the vagus nerve leads to a reduction of the pulse frequency and to a weakening of the contractile force. On the contrary, the excitation of the sympathetic nerves increases both pulse frequency and contraction forces. Neither of the two heart nerve systems can, however, bring about the contraction of the heart muscle directly. An analogous situation also holds for the muscles of the stomach and bowels, where the contractions arise as a consequence of autoexcitability phenomena and also in response to mechanical stimuli. Also in these muscles the nerves exert exclusively a regulatory action, which can be either of an inhibitory or excitatory type; the main difference with the heart is that the vagus nerve has a stimulatory action on the stomach and bowels, rather than an inhibitory one.

The explanation of the mechanism of voluntary movements, and the difference between these and involuntary movements, represent a constant theme in Volta's "electrico-physiological" research. In a letter dated October 19, 1798 and addressed to Luigi Valentino Brugnatelli, the editor of the "Annali di chimica," Volta wrote:

> For the next volume of the Annali di chimica I am going to have a small dissertation concerning some conjectures of mine on the power of will in animals to move the electric fluid in the brain, that is at the origin of nerves that go to voluntary muscles; and, with such a stimulus, to excite the *nervous energy*, meaning that unknown

virtue through which the contraction in those voluntary muscles is eventually effected. I have communicated by letter such opinions [...] now for several years to more than one of my correspondents. (VEN, I, p. 559)

Among the correspondents alluded to by Volta were Cavallo in 1792 and Van Marum, Mocchetti, and Orazio Delfico in 1795. This was evidence of Volta's recurrent interest in the origin and mechanism of voluntary movements (see Chapter 6.2). In that respect it is worth noting the relationship between these dates and those of the publication of Galvani's main works: *De viribus* in 1792, *Trattato dell'arco conduttore* in 1794, and *Memorie sulla elettricità animale* in 1797. As we have already said, and will have the occasion to discuss more in detail later, this is an indication that Galvani's works provided Volta with important themes of reflection, contributing in a significant way to define the direction of his research.

In the letter to Brugnatelli, Volta explicitly alluded to the "Galvanians" and declared to his correspondent that he had in his mind a hypothesis of "conciliation" between his own ideas and those of the Bologna colleague. This is witness of a new attempt by one of the two main protagonists of the controversy on animal electricity to come to an accommodation with his antagonist, by admitting, at least in part, the validity of the opponent's interpretation of the phenomena (see Chapter 7.1). In contrast to the proposal of a few years before, Volta did not criticize the way the experiments were performed. As already mentioned, Volta had then assumed that the electricity was of the "galvanian" type (i.e., of animal origin) when contractions were brought about by using homogeneous metals, or when they were obtained by a direct contact between the animal tissues, whereas, in the other circumstances, the electricity involved was of his own type, deriving from the electromotive action of the contact between different conductors. He argued instead on the nature of the phenomena themselves involved in the experimental manipulations. He wrote to Brugnatelli:

> This is how even I admit a true and proper Animal Electricity, after having done so much to demonstrate the non-existence of a pretended Animal Electricity, i.e., that which is excited in the cut limbs with the artifice of metals. The true electricity that I admit is that of the Torpedo and Trembling Eel and in the other fishes which give the shock; I am now inclined to attribute a similar electricity, in the above mentioned way, to all animals; and so much animal as it depends on the *anima* (i.e. soul), i.e. it obeys to the will, and does not properly extend itself—or only poorly—outside its place. (VEN, I, pp. 560–561; italics is ours)

We will return to Volta's important allusion to electric fishes, which was a fundamental element not only of his hypothesis on animal electricity but also of the research endeavor leading him to the invention of the battery (see Section 8.4). Let us concentrate now on the meaning of what Volta proposed to Brugnatelli in this letter, and that he developed further in another letter also written in the autumn of the same year and addressed to Johann Peter Frank, a celebrated physician and Volta's colleague at Pavia University. A first consideration concerns the distinction put forward by Volta between the voluntary contractions in living animals and those obtained in the dead frog, prepared "in Galvani's manner," or in its parts separated from the animal body. In the letter to Frank this distinction is made clear and explicit:

> I say that in the experiments of Galvanism, i.e. in those made with cut and prepared limbs, or even with entire—but dead—animals undergoing the external application of conductors, that is to say, of motors of electricity; on the other hand, in the live animal, endowed with brain and will, I am myself inclined to another explanation for the motions of voluntary muscles, for which the electric stimulus applied to the respective nerves is the most effective (in contrast to what occurs in the heart and other involuntary muscles, more excitable by mechanical and chemical stimuli applied directly on the muscles themselves). For the voluntary muscles, as I say, I believe that the motions are determined and produced at their origin by a movement of electric fluid in the nerves going to the said muscles; and I think that this movement is made at the command (*ad nutum*) of the will where the will has its seat at the beginning of the specific nerves. (VEP, III, p. 414)

According to Volta, there are therefore two essentially different categories of phenomena. On the one hand, there are those involved in the experiments of galvanism, which are produced by the contact between different metals on the prepared animal or in its dismembered parts. In these cases, and in Volta's stimulation of the tongue and on other sensory experiments, the mechanism is purely physical and corresponds to the general theory of contact electricity. On the other hand, there are the experiments on live animals. According to Volta, in these experiments the contractions are brought about by the electricity contained in nerves and moved by the action of the will. Volta's distinction is different (as we know) from the previous one concerning the difference between experiments with metals and without metals; it emerged in 1796 after his success with the measurement of metallic electricity obtained without any recourse to the frog preparation or to other biological detectors.

The disappearance of animals from his experimental bench marked an important turning point in Volta's research, allowing him to separate his own experiments from those of Galvani and of galvanism more sharply. This did not mean, however, that Volta stopped being interested in the physiological processes underlying organic life, as we shall see in the following sections. In particular, he elaborated his idea of the role of electricity in voluntary movements on the basis of various factors. First of all, there was the experimental observation that very weak electricity applied exclusively to the nerve suffices to produce contraction in the corresponding muscle. This was a form of electricity that was—in Volta's words—"the minimal physical, or material, action that can be attributed to the immaterial will, because finally we need to attribute it one such action." Moreover, the electric fluid can flow through the conductive matter of nerves without "the need of any peculiar instrument of a secretory or other nature, by which it is prepared or extracted, because it is nice and ready, as occurs in all other conductors, waiting only for an action capable of putting it in motion" (VEP, III, p. 415).

In Volta's view, voluntary movements were thus produced by the electric fluid present in nerves (as in any other conductors) and put in motion by an "impulse" of the will. This was a type of electricity very different from that of Galvani, inasmuch as it was not linked to any particular "machine" of the body and did not involve any peripheral mechanism for its functioning, as was the case for Galvani's "animal Leyden jar." Volta was not inclined to go further in his proposal of reconciliation vis-à-vis Galvani. This appeared clearly in his letter to Brugnatelli:

> If the Galvanians are content to reduce animal electricity within these terms, I will be very happy to agree with them. It is possible, on the other hand, that they refute this way of reconciliation, that I am pleased to offer them; that they still pretend that Electricity is excited by a purely organic force, meaning that the electric fluid which is prepared and works in the brain, is accumulated within these last ones, or inside the interior faces of muscles, and is put in whatsoever condition of unbalance, and because of this unbalance, once discharged, stimulates the muscles themselves. It can be, I say, that they continue on by asserting such kind of electricity produced, as they pretend, by a purely organic mechanism, even in the cut limbs or muscles [...]; If, in sum, they do not surrender to such a proposal of reconciliation, I might even withdraw it, that is not even grant that other Animal Electricity, depending on and moved by the will in the entire and living animal. (VEN, I, p. 561)

When Volta was writing these words, Galvani (if not the Galvanians) was already surrendering. This was not, however, due to the reconciliation proposed by his adversary that he had already refused, although in other terms, 1 year earlier. It was due to illness, which would, a little more than a month after the date of the letter, reduce Galvani to death. One should also say that it would have been difficult for Galvani to be fully acquainted with Volta's proposal because the physicist of Como never published his "*dissertanzioncella*" promised to the editor of the *Annali di Chimica*. As a matter of fact, in the autumn of the year 1798, Volta was already fully immersed in the line of investigation that would bring him, within the space of 1 year, to the invention of the battery.

8.2 VOLTA'S RESEARCH ON SENSATIONS

It has emerged from the previous chapters how, in his experimental research on animal electricity, Galvani did not show any real interest in sensations and in the physiological mechanisms underlying them in his research on animal electricity. From the beginning of his electrophysiological investigation, he refused to study these arguments, explicitly declaring that sensations were beyond analysis with the available scientific tools. Galvani did not doubt the importance of sensations and sensory function for human life, which was comparable to that of motor function. It was rather the difficulty of investigating sensations properly, according to the scientific standards of the experimental science of the epoch. While muscular motion "makes itself sensible to the observer," sensations not only "exceeded the boundaries of human sight and human understanding" but were in addition "untrustworthy" and "inconstant," as he remarked in *Trattato dell'arco conduttore* when criticizing Volta's experiments on taste sensations (see Chapter 6.2).

Even though he declared that he would come back to sensory mechanisms after clarifying the problem of muscular motion, neither in *De viribus* nor in the following works (nor in the unpublished manuscripts) did he deal in a systematic way with this theme, which was a crucial aspect of the natural philosophy of his century. In addition, it is possible that this was also due to the type of methodological reasoning which restrained a "Galilean" scientist from penetrating the "obscure labyrinths" of the senses. It is noteworthy from that viewpoint that the subsequent experimental study of sensation has been developed mainly by physicists. Consider, for instance, the fundamental studies brought about in the nineteenth century by Ernst Mach on the psychophysics of perception. In the field of physiology the scientific study of sensation has followed that of motion with a delay of about a century, emerging in its full form only in the middle of the twentieth century.

Compared to Galvani, Volta displayed a different attitude. As mentioned in the previous section, in the study of the time characteristics of the electric action generated by bimetallic contact, he attributed more value to the continuous or "tonic" properties of taste sensation brought about by the metallic arc applied to the tongue than to the transient character of the contractions excited in the frog preparations. On Volta's side logical intuition undoubtedly played a part in this choice because "in the matters of nature" one is led to accept more easily that a continuous agent would produce a transient effect rather than the opposite (that a momentary cause would produce a continuous effect). Moreover, there was the acumen of the shrewd observer since it is likely that Volta happened to observe that by applying a continuous mechanical stimulus he would obtain only a transitory contraction in the prepared animal. Volta's conclusion also benefitted from the common physiological culture of the natural philosophers of the Enlightenment, and particularly the conceptions connected to the notion of Hallerian irritability. This pointed to the transient and phasic character of muscular responses elicited by continuous stimuli (and notably by chemical irritations).

Intuition, observational acumen, and scientific culture do not, however, provide a secure shelter for the natural philosopher against possible experimental errors, particularly when engaged in the study of complex phenomena like those in the physiology of living organisms. This is why Volta soon thereafter carried out a control experiment in order to confirm that a continuous electric flux is necessary in order to bring about a taste sensation analogous to that produced by a prolonged application of the metallic arc to the tongue. To produce taste sensations in this control experiment, he used the artificial electricity of the electric machine. In this way he could show that, in order to obtain sensations comparable to those generated by the maintained application of a metallic arc, the machine must not be discharged on the tongue in an instantaneous manner. It is necessary instead to rotate the disk of the machine in a continuous way with the precaution of avoiding any excessive accumulation of electricity on the conductor of the machine (which might result in a sparking discharge of the machine) (VEN, I, p. 127).

With relation to the taste sensations, Volta remarked: "This continuous current [...] without accumulation, or charge, without tension appearing at the electrometer, is much more similar to the one which is induced by the simple armatures or metallic contacts in the experiments we are dealing with." This observation is important because it also pointed to the typical character of the electricity excited by metallic contacts, that is, a flow of a large amount of charge but at a low tension. This characteristic contrasted with that of the electricity proper to the electrostatic machines, which were usually charged at very high tension, although with a small quantity of electric charge, as Volta afterward noticed:

> It can be supposed that the action of such metals in the site where they fit together is that of moving and putting in circulation not a small but a great amount of electric fluid, and yet with a mild course; so great that the ordinary action of the electric machine does not move a bigger one, although it does that with a more vigorous impetus and tension. (VEN, I, p. 210)

Volta had already noticed these characteristics of the current excited by metals ("a small, but mild and placid, current" that displays "such a small force or tension") in a letter addressed to van Marum in 1792. He will come back to them in many subsequent works, among which was another letter, also addressed to van Marum and written between 1796 and 1797. In this letter he wrote that the current excited by dissimilar metals can only partially be assimilated with that produced by a large Leyden jar charged "to a very small tension"; this was because the discharge of the jar is temporary, whereas metallic electricity is "a current which continues without pause as long as the circle is not interrupted etc." According to Volta, metallic electricity "must be compared not to any electric discharge whatsoever, but instead to this electricity which, more or less abundantly, flows placidly, and without interruption, from the prime conductor of the machine, to the Conductor of the Cushion with which it communicates, comes back to the prime Conductor etc, continuing its circulation as long as the action of the machine is maintained" (VEN, I, pp. 488–489).

It is, however, after the discovery of the battery that Volta would insist on the singular properties of metallic electricity. These properties led several scholars to assume that the agent involved in the new apparatus was not the common electricity of the electric machines, but a new type of electric fluid, more similar to the electric fluid assumed by Galvani in his experiments on frogs (this accounting for the phrase "galvanic fluid," somewhat ironically used for a long while to designate the new electricity discovered by Volta).

Before returning to the electrophysiological aspects of Volta's works, we notice here en passant, that besides a fruitful exchange between physics and physiology there was, in his experiments that have just been mentioned, an effective interaction between the study of the electricity of metals and that of the "artificial electricity" (the expression used for the electricity of the friction machines).

On the basis of his experiments, Volta thus arrived at establishing the continuous character of the action of dissimilar metals in moving the electric fluid, and attributed to specific physiological mechanisms of the neuromuscular system the different time characteristics (transient or phasic) of the motor responses elicited in frogs by the prolonged application of bimetallic arcs.

On the basis of the initial experiments on taste sensations he could then approach the analysis of a second series of sensory stimulation studies. After succeeding in inducing taste in the tongue experiments he applied bimetallic couples near the eyeball. He described these experiments in some letters addressed to Tiberio Cavallo in 1793 (parts of which were published in the *Philosophical Transactions of the Royal Society*). In one of these letters Volta said that he applied near his own eye "a small sheet of tinfoil" and partially introduced in his mouth "a small silver spoon": "every time that these armatures will be brought to a reciprocal contact, or will be put in communication by means of another metal, the eye will be struck by a more or less lively light sensation, which disappears at once like a flash, [even] if the communication continues" (VEN, I, p. 218).

This light sensation (a "transitory flash" in the English translation of Volta letter's made by Cavallo; see, for instance, in VEN, I, p. 207) produced by the prolonged application of the metallic arc was therefore quite different in its time course to the taste sensation brought about by the same stimulus applied to the tongue. If, following the application of the metallic arc, one wished to obtain a new luminous sensation in the stimulated eye, it was necessary to separate and reunite the metal again in a repeated way, so that the circuit was closed and opened again and again:

> If, instead of leaving the things so at rest, one amuses himself by breaking and re-establishing the communication between the two metals alternatively with more or less rapidity, the flashes come again in shocks; one has the sensation of an undulating or blazing light, and eventually of an almost continuous light, if things are made in a way that separations and reunions are made with the greatest swiftness, through the succession of small strokes, or by rubbing one lamina against the other in such a way to make it trembling or hopping. (VEN, I, p. 218)

In this experiment Volta achieves the temporal fusion of a luminous sensation due to an alternating stimulus, according to a mechanism that is referred to in modern sensory physiology as "flicker-fusion." If we consider that he arrived at this effect simply by using a piece of tinfoil and a silver spoon, we cannot but remain astonished by his extraordinary experimental ability in studying physiological processes.

There is more. Despite the discontinuous character of the visual sensation brought about by the maintained application of a metallic arc, Volta knew, on the basis of the tongue experiments, that the electric flow excited by metal is of a continuous nature. He attempted therefore to see, by keeping the metallic arc in contact with the ocular tissues, if some residual form of visual sensation might remain after "the initial live flash." Indeed, by putting himself in a completely dark room he succeeded in perceiving a very weak luminosity during the prolonged application of the arc to the eye. The sensation was, however, so weak—he said—that "one was aware of it because of the following more complete obscurity which dropped on the eyes as a dark veil at the moment the metals are separated." Moreover, "it seemed, in addition to that, that at the instant itself of the metal separation another light flash would occur, but very weak, followed by the just mentioned, more complete obscurity" (VEN, I, p. 219).

With these experiments Volta succeeded in providing evidence, more than two centuries ago, of the complex time characteristics of the visual responses excited by the a prolonged electric stimulus to the nervous elements of the eye; this consists of a prevalent transient sensation at the onset and offset of the stimulus and a very dim light sensation during the maintained phase. This is indeed an achievement of extraordinary physiological significance. Although in the circumstance of his experiments this behavior emerged with an artificial stimulus (electricity), it is indeed the expression of an important property of the physiological response of the visual nerve cells in their natural condition (and also of other sensory systems). Indeed, the visual system has evolved to signal with a greater efficacy the environmental stimuli that are informative, among which are the rapid changes of ambient light. In most conditions, at least under natural circumstances, these changes are brought about by objects moving in the visual field or the eye itself moving. For an animal (and, of course, for humans) a moving object is more biologically relevant than a still object or a stationary background. As a matter of fact, the moving object could be a potential predator or prey, or a con-specific animal that should be indentified and possibly approached for social or adaptive purposes.

All animals are aware of the particular characteristics of visual objects conferred by their motion. This is, for example, the reason why we move our limbs or hands when we seek somebody's attention or remain still (or move very slowly) when we do not wish to be seen. This does not happen as a consequence of purely psychological characteristics of the visual system, devoid of any biological substratum. It is because the temporal variations of retinal illumination (like those produced by moving objects or by an intermittent light) represent a more effective stimulus for the nerve cells and circuits of the retina (and of the more central stations of the visual system) compared to a steady level of light (like those produced in the presence of stationary objects or backgrounds).

The first indication of this eminently transient character of nerve cell responses to electrical stimulation in the initial stages of the visual system came indeed from experiments carried out more than two centuries ago by Volta. This happened with the use of "a small sheet of tinfoil" and "a small silver spoon." This was indeed a great achievement, especially considering the simplicity of the means used to achieve it.

Even in his experiments on "visual electrophysiology" Volta followed his procedure of varying his experimental arrangements in thousands of ways. He thus applied metals not only in close proximity to the eyeball but also to the eyelids, inside the mouth, or into the nostrils, succeeding in producing more or less intense visual sensations in many instances. He assumed that in some of these circumstances, and particularly when the two tips of the arc were applied into the nostrils and mouth near the root of the tongue, the electric fluid might stimulate the optical nerve directly (an indication that he was not ignorant of the anatomical implications, which is comprehensible when we consider that Antonio Scarpa, one of the greatest anatomists of the period, was among his colleagues at Pavia). During his trials, Volta became aware that various circumstances "can have an influence on the greater or lesser intensity of these sensations, as the greater or lesser application of the metallic lamina to the eye bulb, the smoothness of the applied surface, the place and extension of this application, the more or less abundant fluid bathing the eye." Although he noticed that "there should be, moreover, some differences among the eyes of different individuals which would make them more or less sensitive to this type of stimulation," he noticed:

> On the other hand, among the persons enjoying good sight, I have not found any who could not perceive a flash, when the experiment is properly conducted. That is, when the metals are well chosen, well applied, and brought all together at once to reciprocal contact.

And added:

> I am by the way convinced that the experiment would succeed also on persons blind as consequence of cataract, or for any other fault whatsoever, other than the insensibility or paralysis of optical nerves. Here is then the possibility that these trials could be of some utility, allowing us to establish if such fault would exist. Who knows then, if some relief of the paralysis might derive from a well administered electrical stimulation, both at the beginning of the process or in more or less advanced phases. But let the matter rest with the physicians. (VEN, I, p. 222)

Despite the closing sentence, this passage shows, in addition to Volta's cultural and physiological intuition, his great attention to medicine, and—in particular—an interest directed to the possible diagnostic and therapeutic applications of electricity. It seems likely that he never used electricity for such purposes in the ophthalmologic field, but he did make recourse to it in an attempt to treat deafness. This happened in 1802 when he applied electricity "to a girl of about 15 years, deaf since her birth." The patient underwent electric treatment in both ears with shocks repeated every second, for a duration of 10 minutes. Even though he recognized not having great success in his therapeutic attempt "on the other hand, it cannot be denied that the patient has acquired the hearing sense to the point to notice some sounds, sometimes not very intense, at distance of some feet." He made the curious observation that "she could perceive better, and also in more initial phases, the obtuse and hollow sounds" (VEN, II, p. 181).

Volta made this attempt at the therapeutic application of electricity by using his battery, even though from his allusion to the administered "shocks," it seems that the new device was used in the same way as the traditional electrical instruments (electric machines and Leyden jars), that is, by producing sudden and short-duration discharges. On another occasion he made an explicit distinction between the electricity "mild, but active, and continuous" of the battery and that powerful and instantaneous of the "wearing ordinary machines," by suggesting that "better successes seem to be expected from the first" compared to the second. Moreover, by speaking in the third person, he wrote that "some of these experiments have been already undertaken, and some admirable successes obtained with the electro-motor apparatus of Volta [i.e., the battery] are already claimed in various parts of Europe, and most of all in Germany, particularly in the treatment of deaf persons" (VEN, II, p. 313). As a matter of fact, from the beginning of the nineteenth century the Voltaic battery progressively substituted the other devices that had been used until then for the therapeutic application of electricity; the battery also gave a new impetus to the investigation and practice of medical electricity (see Fig. 8.2).

In the letters to Cavallo of 1793, Volta alluded to some of his attempts to stimulate the touch sense with electricity, although he described them as having "only limited success." In these experiments he tried to produce what we would better define as pain sensations. This is done by applying the bimetallic arc to "the eyeball, internal rim of the eyelids, and most of all to the lachrymal glands." He notices that in order to excite these sensations it is necessary "to use stronger means," that is, "the metals which have more activity when they are put in opposition, silver or gold on one side, tinfoil or better zinc on the other" (VEN, I, p. 245).

By applying the stimulus near the eye he could elicit both pain and light sensations, and he could thus make an important distinction between the excitability characteristics of the two sensory modalities, which revealed some important differences. As to the visual sensation, it appeared immediately "at the instant of the contact" of the metals, whereas for the other sensations he noticed that:

> Assuming of course that everything remains in its place and the communication is not interrupted, a sensation of pain and burning of the touched eye parts comes

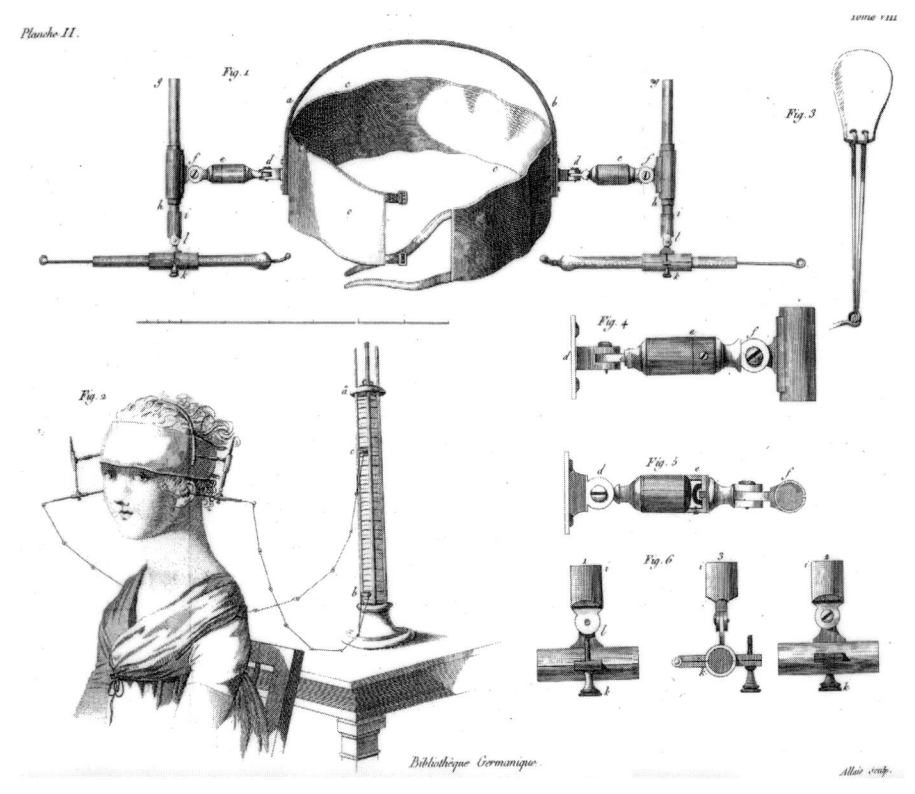

FIGURE 8.2. The medical application of Volta's battery to the therapy of some diseases of the ear (from Grapengiesser, 1802).

out little by little; this sensation increases not only for a few seconds, as happens for the taste sensation in similar experiments on the tongue, but instead for one or two minutes, up to the point of becoming insupportable, and even produce a little local inflammation. (VEN, I, pp. 245–246)

Volta was quick to exclude the possibility that these phenomena were due to some form of mechanical irritation by showing that "this sensation of burning pain" disappears when, with the separation of the metals, the passage of current is interrupted, even though the contact with the surrounding ocular tissues is maintained. The other method he used to demonstrate the necessity of a current flux in order to achieve the pain sensation made apparent another significant difference between this sensory modality and the visual (or taste) sense. First of all, he showed how no sensation is produced by using an arc made by two laminas of the same metal (for instance, "silver and silver, zinc and zinc, tin and tin") and "that very little is produced if, instead of using gold and silver on one side, and zinc and tinfoil on the other (which act more efficaciously when put in opposition), one uses silver and gold on one side, and copper or iron on the other, or even iron and zinc, copper or tinfoil."

Soon afterward he remarked:

> These less efficacious combinations of metals do not fail in evoking the light sensation in the back of the eye, of taste on the tip of the tongue, although these are much weaker than the better matched combinations, that is those of zinc and silver and gold; if therefore they have almost no effect on the external parts of the eye, we should say that the touch sense is not that excitable, at least for the action of electric fluid, compared to the sight and taste sense. It is, moreover, not so readily so, even compared to the taste sense, which, on the other hand, is itself much slower than the sense of sight. (VEN, I, p. 246)

With these experiments, still based on the use of dissimilar metals, Volta therefore established the different electrical excitability (the different "threshold" in the words of modern sensory physiology) of the sensory mechanism underlying pain sensations (indicated by him as the "touch" sense), compared to sight. Moreover, he made clear the different latencies and time courses of the sensory response (short latency and transient character in the case of sight, long latency and maintained response in the case of pain sensation).

As to the characteristics of electrical excitability, and for the time properties of the response, the pain sense seemed to correspond more to the taste than to the sight. Volta had the opportunity to verify this correspondence in another type of experiment, in which he made use of the electricity produced by an electrical machine. In previous experiments he had noticed how easy it was to stimulate visual sensations with instantaneous electrical discharges (normally associated with spark production) of an electrical machine or of a Leyden jar. With these kinds of stimuli, Volta was not able to produce pain sensations analogous to those induced by a bimetallic arc. As in the case of taste sensation, he was obliged to use the electricity of a machine that is rotated continuously with an application capable of insuring a moderate and sustained electric flow, without production of sparks.

With this method it was possible to reproduce both the burning sensation caused by the maintained application of metallic arcs and the mild inflammation that arises when the arcs are applied for a particularly prolonged time. Thus, it appeared clearly to Volta that, in order to produce a taste or a burning pain sensation, it was necessary to have "a continuous torrent of electric fluid" like that produced by heterogeneous metals or by an electric machine, rotated in a continuous way. On the other hand, in order to produce a visual sensations (and also to produce muscular movements), the instantaneous discharges of friction electricity are more effective, while the metallic arcs are efficacious only at the onset and offset of the contact (VEN, I, pp. 246–247).

With the invention of the battery Volta introduced a new and more powerful device for stimulating sensory systems. In the letter in which he communicates his invention to Banks, in addition to the sensation of continuous pain, of slow onset and ingravescent character, and of burning (*cuisson* in the French original), there was also a different pain sensation. This was of rapid onset and had a pricking character (a *piqûre*). These two types of pain sensation identified by Volta in his experiments correspond to two different forms of pain sensation (or "nociception") in modern sensory physiology: They correspond respectively to the delayed or "slow" pain, and to the initial or "fast" pain. Besides their different perceptual characteristics, the two modalities can be distinguished in terms of the different cutaneous receptors that are activated by the natural stimulus, and, moreover, also for the different diameters of the nerve fibers transmitting the sensory signal to the nervous centres. In the case of slow pain, these are small-diameter fibers, lacking the insulating cover of myelin (C-type), with a very slow conduction speed (less than 0.5–2 meters per second). In the case of fast pain, on the other hand, the fibers are of larger size and endowed with a myelin sheath. They conduct at a speed ranging between 4 and 9 meters per second. The different conduction type accounts for the slow or fast type of the resulting pain sensation, respectively, which were described by Volta in his experiments of sensory electrophysiology.

In 1793, by alluding to his research on sight and touch-pain sensations, Volta claimed he carried out experiments aimed at producing hearing and olfactory sensations, too. He never succeeded in exciting the olfactory system, either by using the electricity of bimetallic arcs or that of the battery. With the new apparatus he was able to produce sensations of a type that he interpreted as sounds. It is on the basis of this result that in 1802 he made his attempt at the medical application of electricity for the treatment of deafness (see earlier). The experiments on excitation of auditory sensation were first described in the 1800 letter to Banks announcing the invention of the new instrument. After speaking of the other type of sensations (and of muscular movements) that he obtained with his device, he wrote:

> I have now only to say a few words on hearing. This sense, which I had in vain tried to excite with only two metallic plates, though the most active of all the *motors*[2] of electricity, *viz.* one of silver or gold, and the other of zinc, I was at length able to affect it with my new apparatus, composed of 30 or 40 pairs of these metals. I introduced, a considerable way into both ears, two probes or metallic rods with their ends rounded, and I made them communicate immediately with both extremities of the apparatus. At the moment when the circle was thus completed I received a shock in the head, and some moments after (the communication continuing without any interruption) I began to hear a sound, or rather noise, in the ears, which I cannot well define: it was a kind of crackling with shocks or effervescence, as if some paste or tenacious matter had been bubbling. This noise continued incessantly, and without increasing, all the time that the circle was complete, &c. The disagreeable sensation,

---

[2] Interestingly the 1800 English translation of Volta's original communication on the invention of the battery, published in *The Philosophical Magazine*, the French term *moteurs* was rendered with the word *exciters*.

and which I was afraid might he dangerous, of the shock in the brain, prevented me from repeating this experiment many times. (VEN, I, p. 580) (English translation from Volta, *Phiłolosophical. Magazine*, September 1800, p. 308, revised)

Besides the results concerning some peculiar physiological effects, there is a general principle that emerges from Volta's experiments on electric stimulation of senses and muscular contractions. This represents one of his greatest contributions to neurophysiology. It is the principle whereby the specificity of the physiological effect depends not only on the type of agent used for the stimulation but also on the type of nerve stimulated. In physiology this principle is now known as the doctrine of the "specific nerve energies" and is ascribed to Johannes Müller, the great German physiologist who formulated it in a rigorous and general way in the first half of the nineteenth century. According to this doctrine, if, for instance, the eye or the optical nerve is stimulated in a variety of ways—like mechanical or chemical irritations, light or electricity—the outcome will always be a sensation of light. The same holds for the other senses—taste, smell, hearing, and somatic sensations.

The conception that underlies this doctrine had been already foreseen and formulated by Volta in the first phase of his electrophysiological investigations, soon after his experiment on stimulation of the tongue. In *Memoria seconda sull'elettricità animale* (1792), Volta wrote:

> It, therefore, becomes manifest that according to which nerve is stimulated and to what is its natural function, such is the effect that ensues correspondingly, that is to say as regards sensation and motion, when that nervous virtue is activated on subjecting it to the influx of electrical fluid. (VEN, I, p. 63)

In Volta's formulation, this law of the constancy of the effects brought about by the stimulation of a given nerve is considered only in the context of the action of electricity. It can easily be extrapolated to other kinds of stimulating agents, in that a key role in the production of the physiological effect is ascribed to the specificity of the involved "nervous virtue" (*virtù nervea*), rather than to the stimulatory action itself. It is to be noted that, in announcing his principle, Volta also took into account motor nerves. As a matter of fact, at the moment he was writing the words just quoted, he was relying uniquely on the comparison between the physiological response brought about by stimulation of motor nerves with the sensory effects induced by stimuli applied to the tongue. As he advanced further in his investigations by producing other kinds of sensations, he would reassert his principle and extend its reach and validity.

One of the main results achieved by Volta in his research on sensations was the specification of the two fundamental classes of physiological responses: on the one hand, those of transient, or "phasic" type (muscular movements, visual sensations), which are better stimulated by rapid electrical stimuli of impulsive character; on the other hand, those of sustained or "tonic" character (the sensations of taste or the burning type pain), which usually require prolonged stimulation. This last type of sensation was easily produced with bimetallic arcs. This is because, as Volta showed, the bimetallic arcs generate a continuous flow of electricity, whereas it was possible to produce a maintained electric flux by using traditional electric machines only when particular precautions were taken. To this purpose, it was necessary to rotate the disk of the electric machine in a uniform and uninterrupted way, while avoiding the accidental discharge of sparks from the prime conductor of the machine.

On the other hand, to excite a sustained physiological response of a phasic or discontinuous type (vision, muscle motion), it was necessary to break and close the circuit in rapid alternation when using metallic arcs. This procedure, in which Volta had become particularly skillful, was referred to for a while as the method of "Voltian alternatives," from the phrase *alternatives de rupture* used by him in his 1793 letter to Cavallo (VEN, I, p. 244; see Chapter 6.3).

Similarly, when using the electrical machine with the disk in continuous movement, it was possible to produce sustained excitation of vision and muscle contraction by making recourse to tricks unlike those necessary for producing continuous taste or pain sensations. It was necessary to manipulate the apparatus to produce small, repeated discharges. This could, for example, be brought about rather easily by creating a small gap in the discharge circuit. If, on the other hand, electricity flowed uniformly, and without discharges (i.e., according the situation more effective for exciting taste and burning pain sensations), then there were normally no contractions or visual sensations (except at the onset and at the offset of the circuit as occurred with the bimetallic arc).

These observations of Volta were similar to those previously made by Galvani, when he remarked that frog contractions were produced more easily with a sparking type of electricity. This happened even if the preparation was located at some distance from the electrical machine; this condition could be more effective than that corresponding to a direct connection between a continuously operated machine and the frog. Galvani interpreted these results as evidence for the need of an impulsive action of electricity, in the form of a sudden impact or a rapid vibration (see Chapter 4.2). He assumed that muscles became insensitive to prolonged, non-sparking electricity, because the continuous flow of electric fluid modified the excitability of animal tissues.

On his side, Volta ascribed the phasic character of muscle contractions or visual sensations to the particular physiological properties of the mechanisms involved in these phenomena. He expressed his views on this theme in a particularly expressive way in a letter addressed to van Marum in the period 1796–1797:

> All this comes to say that, in order to shake or stimulate substantially the muscle or the excitable nerves, it is necessary that a considerable portion of electricity passes across them in a very short time, in an almost indivisible instant, in such a way that, in rushing there, it would produce a small push. […] In order to achieve this, a

larger flood [of electricity] is required. That is that the Machine should deliver at every moment, and put in motion, a larger amount of electric fluid; or that this one would go on being accumulated on the prime conductor, and subsequently leap onto another conductor separated from it. It would thus cross, due to the accumulated tension, the dividing gap, and proceeding with such an impetus, it would go further and invade the said animal organs etc. (VEN, I, p. 481)

If we integrate this reasoning with what Volta had written in 1793 to Cavallo concerning the way "nerves accommodate in a certain manner and become unexcitable in the presence of a continuous electric stimulus" (VEN, I, p. 234), we see how he was coming near to grasping, at least at a phenomenological level, the fundamental properties of nerve excitability, as clarified by modern neurophysiology (see Chapter 10).

In the continuation of the letter to van Marum of 1796–1797, Volta tried to provide a quantitative estimation of the "very short time" needed by the electric fluid to pass across the nerves in order to stimulate the contractions effectively. He would return to this problem in a memoir written in 1802, after the invention of the battery that we will analyze in detail later because it represents one of the most profitable moments of interplay between the electrophysical and electrophysiological perspectives in his scientific endeavor.

8.3 SENSATION AND MUSCULAR MOTION IN VOLTA'S "CHAIN" EXPERIMENTS

All through his research Volta showed a particular dexterity in setting up experiments characterized by a strong demonstrativeness and ready reproducibility. These assisted in illustrating his conclusions and often integrating a variety of results within a single experiment. We have already mentioned the experiment he carried out in 1793 in which he was able to produce simultaneously both visual and taste sensations by applying a metallic arc between the surface of the eyeball and the tip of the tongue. A variation of this experiment was based on the use of "chains" of people forming a circuit according to a procedure that had been previously used widely by Walsh and Galvani (and that was common to other electricians of the age). In comparison with the experiments made on a single person (generally Volta himself), he remarked that the sensations produced in members of the chain became attenuated as the circuit of the current became longer and more difficult (with the multiplication of the members of the chain, or when the contact was made in a less tight way). If the experiment was carried out correctly, the sensations could still be perceived when the chain was made up of 8–10 people. After the invention of the battery, Volta was able to produce physiological effects in chains of 20 people (VEN, I, p. 399).

The chain experiment can be varied in "thousands of ways," and also one or more prepared frogs can be within the electrical circuit. When the circuit is completed, there will appear "simultaneously the convulsions of frog legs, the flash in the eye, the taste on the tongue tip; i.e. the movements and sensations which correspond to the proper office and functions of the nerves that the electric fluid, put in motion by the action of metals, encounters on its way" (VEN, I, p. 225; see Fig. 8.3). It is to be remarked that, although Volta uses the word "simultaneously," the order in which he listed the effects corresponded precisely to their increasing latency (as he later declared explicitly in the description of comparable experiments).

With the invention of the battery Volta succeeded in combining the diverse physiological effects on himself, benefitting from the great power of the new apparatus. In the famous letter to Banks of March 20, 1800, after describing the way the various sensations or contractions can be produced, he wrote:

> But the most curious of all these experiments is, to hold the metallic plate between the lips and in contact with the tip of tongue; since, when you afterwards complete the circle in the proper manner, you excite at once, if the apparatus is sufficiently large and in good order, and the electric current sufficiently strong and in good course, a sensation of light in the eyes, a convulsion in the lips, and even in the tongue, and a painful prick at the tip of it, followed by a sensation of taste. (VEN, I, p. 580)

Returning to the arrangement with chains in which—according to Volta— "experiments succeed in an exceedingly nice and pleasant way," these could be used not

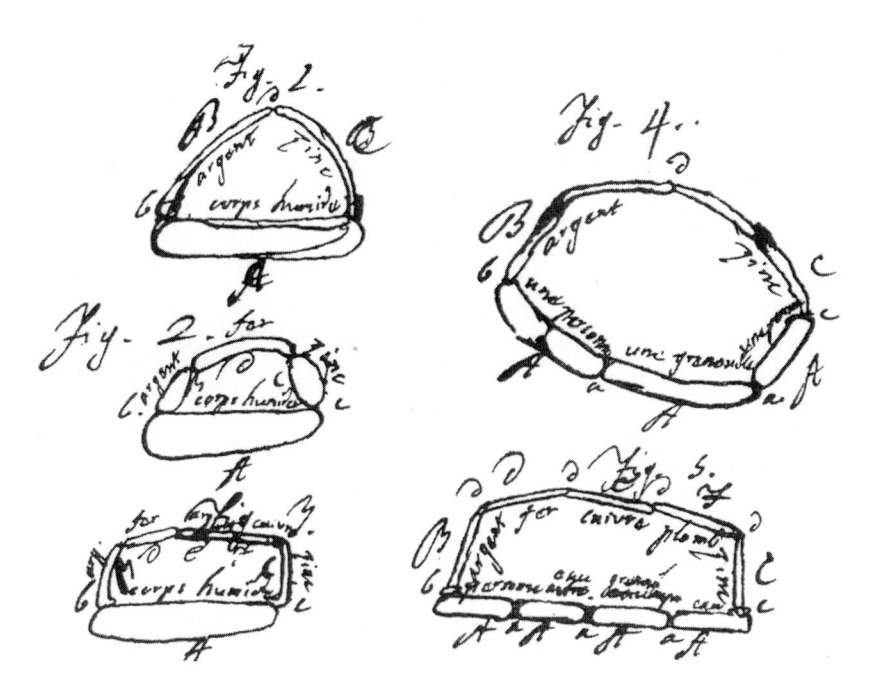

FIGURE 8.3. Various schemes of the complex chains contrived by Volta for his experiments on metallic electricity (from Volta 1918–1929, vol. 1, p. 228; and from Volta's manuscripts at the *Istituto Lombardo, Accademia di Scienze e Lettere* of Milan, J 14).

only to combine various physiological effects but also to perform other types of analysis, both "physical" and "physiological." We have already mentioned that, in the experiments of the joint stimulation of the visual, taste, and pain sensations, Volta noticed that the less powerful metal combinations succeeded in stimulating vision and taste but did not produce pain. In this experiment he used the chain in order to make evident the different excitability of the various sensory modalities, and he thus obtained a result of definite physiological interest (VEN, I, p. 256). It is worth noting here that the difference of the various metallic couples with relation to their electromotive power had been established by Volta mainly on the basis of the different efficacy of the couples in exciting the taste sensation or in producing the contractions in frogs. Experiments of this type, based on the use of a physiological response in order to evaluate quantitatively a physical action, were afterward verified by Volta by comparing the physiological effects brought about by the various metallic couples with those induced by the various degrees of the artificial electricity produced by electrical machines and stored in Leyden jars.

In another experiment involving chains, Volta introduced various frogs in the electric circuit prepared in Galvani's manner, noticing the greater ease of contractions compared to the stimulation of the various sensations. However, with great physiological acumen he remarked that the result could not be interpreted simply as evidence of a greater intrinsic excitability of the motor nerves compared to sensory nerves or sensory receptors. He supposed that the difference could be due to the conditions of the experiment: In the prepared frogs the current could be applied directly to the motor nerve, whereas in the experiments of sensory stimulation (made on humans and thus involving anatomically intact subjects) only a part of the current passed across the sensory structures to be stimulated, a large amount being shunted by the surrounded tissues (VEN, I, p. 240).

In an experiment that we could define as a control, Volta showed that, in the completely prepared frogs, the contractions are easily excited even by the less powerful metal combinations, whereas "in the frogs in which only the viscera are removed, and the loins are left in situ, the contractions are excited with much more difficulty, with more difficulty than the taste on the tongue and the flash in the eye" (VEN, I, p. 241). On the other hand, in the case of a frog in which only the viscera had been removed, the contractions could not be produced if the frog was put within a chain of two people, even when using the most effective metallic couples.

Volta confirmed these (and other) conclusions in experiments based on chains including frogs (in addition to persons) by using as a stimulus the electricity accumulated in a Leyden jar charged at progressively lower tensions (measured with a Bennet type electrometer). He noticed that "the charge of a jar not arriving to 1/8, 1/10 degree of the said electrometer" is effective in producing contractions only if the frog is prepared in the most satisfactory way, with the nerves denuded and placed within the chain in such a way that the "only the crural nerves are left as exclusive communication between the legs and the trunk" (VEN, I, p. 243). In his great inventiveness (and in his love for chains), Volta could not miss the opportunity of making all these experiments involving the stimulation of frogs prepared at various degrees (and also human subjects) at once. This is what Volta wrote in a letter addressed to Francesco Mocchetti in 1795:

> This is a beautiful way to do the experiment. Let us make a chain of persons, some of whom establishing a direct contact through their hands, others communicating through the interposition of frogs; one here totally intact, another there skinned and disembowelled, and, between two other persons, a third frog completely prepared; for the good success of the experiment the persons should have moist hands, so that the weak electric current does not find an impediment to pass across this way: moreover they should not be on a moist floor, to avoid the current deviating through this other path. Having thus arranged the things, the first person of the chain holding with his hand a small Leyden jar should touch with the hook [i.e., the conductor of the jar] the last person of the chain: if the charge arrives to one degree of Cavallo's pendulum electrometer; or, else, [to one degree] of that with thin straw, which is even more sensitive, then the person would not perceive any shock; this is because the path offered to the passage of the electric current is too long; but the frog would perceive it, even the half-prepared and the unprepared one. Let us now repeat the trial with the small jar only to ¼ of a degree; there would not be contractions in the intact frog; but only in the two others. Finally let us decrease even more the intensity of the discharge (for instance by dividing that of one degree over another jar of the same capacity, and the residual electricity over another, and subsequently this electric quarter over another one etc) to such a point that only 1/20 or 1/30 of a degree would remain. By making the trial, only the completely prepared frog would be strongly shaken, in no way the others. (VEN, I, pp. 363–364)

In this experiment the scholar of Como demonstrated the different excitability of the various frog preparations in a single trial. The less excitable frogs (i.e., those prepared "to a less perfect degree") served as a control (of a negative type) with reference to the more excitable ones (which show contractions with that particular stimulus). Similar experiments—Volta says—can be carried out, also using metallic electricity by exploiting the different electromotive power of the various metal combinations.

Coming back to the experiments described in the 1793 letters to Cavallo, Volta noticed that with the less charged Leyden jar (i.e., "not so charged as to arrive to 1/8 or 1/10 of the gold-leaved Bennet electrometer"), which was capable of producing contractions in the totally prepared frogs, it was no longer possible to excite the contractions "if many of these frogs were put in the circuit, either one after the other, or intertwined to persons." This was—for Volta—an evident consequence of the obstacles encountered by the electric fluid in its course, because of the greater length of the paths, and, moreover, because "it must pass across a greater numbers of such extremely narrow trails [the nerves]." In experiments using a single frog the same effect could be obtained, if the current flow was made more difficult in various ways, like "the interposition of a bad-conductive body, no matter how thin it could be, of a thin cloth, of a membrane, of a thin paper, which are not well moistened"; and also "if the hands that are connected together, or those which hold the frogs, are not wet; or if the contacts are made only in a few points etc." Volta comments on these results by saying:

> Here is how the experiments with ordinary artificial electricity serve to clarify the others, where the play comes from a kind of electricity produced by the simple contacts of two metals of different species, just like we had observed in the other experiments, the contact between them, or with the conductors of a different class, which are the humid conductors. (VEN, I, p. 243)

We could note how, in interpreting the range of these experiments, Volta had a clear—although intuitive—notion of the concept of electrical resistance (a resistance to the current flow) that he assumed increased by increasing the length of the chain, by interposing bad conductors, and by making narrow the contact surface, particularly in the case of humid bodies (and also in the presence of other types of obstacles). He had also the notion of the partitioning of the current flow, which played an especially important role when the current had to pass across a spatially extended conductor as was the case for animal bodies. These concepts would be clearly expounded by Volta in the last scientific memoir he wrote, which was published under the name of his old student Pietro Configliachi, in 1814 (see later).

Concerning the obstacle opposed to the passage of electric fluids, Volta was well aware that this was much greater with humid bodies than with metals. In one of the 1793 letters to Cavallo he noticed that the physiological effects decreased drastically by augmenting the number of people in the chain, or by reducing the contact surface between moist bodies. On the other hand, "concerning the metallic arc, it can, on the contrary, be made up by any extension whatsoever of a very thin wire, and the pieces, multiplied at will, could touch each other even in small points (at the condition that these contacts are made with a certain pressure, and the metals are clean)," without any substantial diminution of the effects. He continued by saying: "The reason for this great difference is easily grasped if one considers how metals succeed as conductors over humid bodies and water itself" (VEN, I, p. 227).

The weak conductive power of solutions and moist bodies and the need to make recourse to wide contact surfaces were wisely exploited by Volta in one of the experiments described in the 1793 letter to Cavallo. In this circumstance he succeeded in varying the physiological effect of the metallic electricity, simply by varying the extension of the contact surface. To achieve this goal, Volta made a very simple experiment that was not based on the use of chains of conductors and was easily reproducible (both by the natural philosopher and amateurs of his time and by modern readers). This is the way he described the experiments:

> I take a cup of zinc mounted on a silver pedestal; having filled it with water, I dip my tongue tip inside the water; in this position I touch the silver pedestal with one or

two fingers of the previously moistened hands; a weak sensation of undefined taste: I hold this same pedestal, and I clench it strongly with the hands; the taste becomes vivid, I perceive a well defined, and much more lively, acid taste. (VEN, I, p. 237)

The modern physiologist can see here an experiment in which, by changing the electrical resistance of the stimulating circuit, one changes the voltage of the stimulus and consequently the biological effect; one difference is that the variable resistance is not that of a specially designed physical device (the potentiometer) but simply that of the contact between the experimenter's hand and a silver pedestal.

This is an example of how Volta made an intuitive use of the concept of electrical resistance, a concept that he evidently seemed to master in the execution of his experiments and in their interpretation. Concerning the concept of diffusion and spatial partitioning of the current in physiological experiments, it is already present in a clear way in the letter to Baronio of April 3, 1792, and it would be returned to in *Memoria seconda sull'elettricità animale* (VEN, I, p. 48) as well as in many other subsequent writings, until it would be expressed with particular evidence in the already quoted memoir of 1814 (see Section 8.4).

Volta's experiments with chains were also fundamental within the framework of his analysis of the polar nature of the physiological responses to electrical stimuli (see Section 8.1 and Fig. 8.4). In the second letter to Mocchetti of 1795, after describing the experiment on the tongue showing the difference of the effects produced by stimuli of opposed electric polarity, Volta went on to detail the results of other trials in which he showed that frog contractions at the closure of the circuit were induced more easily with a given polarity of the stimulus (i.e., with the trunk of the preparation—which is the part richly endowed with nervous tissue—facing the zinc of a zinc-silver arc), while those that appeared at the cessation of the stimulus were better produced by the current of opposite polarity (the muscles of the leg facing the zinc). Eventually, in his experimental fantasy aimed at creating experiments where various effects can be seen at once, he described one in this way:

> The show is more beautiful and gracious by experimenting with two frogs at once, dipping them in the glasses but in opposite ways: this is because at the making of the circuit with the above mentioned arc of two metals, one contracts—that is

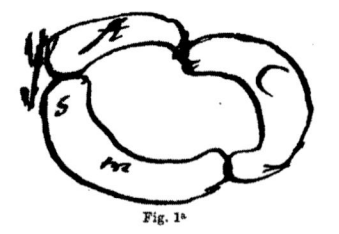

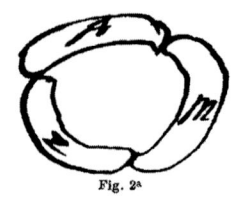

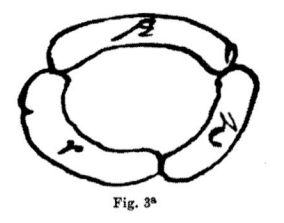

FIGURE 8.4. Three different circuits with the frog and metal used by Volta in the experiments described in the 1795 letter to Francesco Mocchetti (from Volta 1918–1929, vol. 1, pp. 383–384; and from Volta's manuscripts at the *Istituto Lombardo, Accademia di Scienze e Lettere* of Milan, E 33 g).

that with the trunk facing the zinc—while the other remains quiet; this last one leaps afterwards at the break of the circuit—on the one or the other side—while the first frog remains quiet; and things continue in this way in alternation. (VEN, I, p. 385)

For the modern physiologist it is impressive how Volta, working in a period in which it was very difficult to quantify the electrical stimulus, could articulate a fundamental law of nervous excitability, by using simply devices (Leyden jars, metallic arcs, metallic cups or glasses filled with water): the responses induced at the onset of the electrical stimulus (ON-response of the modern terminology) are better elicited by stimuli of a given polarity (negative polarity on the nerve), while the responses at the stimulus offset (OFF-responses) are better elicited by opposite-polarity stimuli. This law is well known to modern electrophysiologists who, however, generally ignore the fact that it was first formulated by Volta.[3]

We have repeatedly remarked how, even before the invention of the battery, Volta wondered why the electricity excited by the contact of different metals (according to the "decidedly singular law" he had discovered) was so powerful in producing physiological effects, whereas "it did not produce either sparks, or any of the other electric signs" observable with the electricity of the ordinary electric machines (VEN, I, p. 303). He correctly accounted for these differences by assuming that metals put in motion a greater quantity of electric fluid, but at a very low "tension" compared to the electrostatic machine. A tension that he estimated to be "lower than 1/4 or 1/10 of the most sensitive jar electrometer" by comparing its effect with those induced by weakly charged Leyden jars. Eventually, starting from 1796, he could measure directly the tiny tension produced with metallic contacts by using the "duplicator" or "multiplier" device invented by William Nicholson or his own *elettroscopio condensatore* (Chapter 7.2).

---

[3] As we shall see in detail in Chapter 10, the flow of current in a given direction, such as to "depolarize" the membrane (i.e., reduce the potential difference between the interior and the exterior of the fiber) tends to excite the nerve fiber because it favors the opening of a class of membrane channels (sodium channels) responsible for the discharge of the nervous impulse. Rapidly, however these channels close because they enter a state called "inactivation." This induces in the fiber a condition similar to that described by Volta with the phrase "the nerves accommodate and dispose themselves to the rest." This inactivation (or "accommodation" in Volta's terminology) is the main reason why the fiber becomes unexcitable in the phase that follows the application of a "depolarizing" stimulus. The other reason is the opening of a different type of membrane channel (potassium channels), which eventually leads to an increased negativity of intracellular potential. When the current is of opposite polarity (such as to increase the intracellular negativity), it does not excite the membrane because it brings the potential farther away from the threshold for the generation of the electric impulse. If, however, this hyperpolarizing current persists over time, it brings about an increase of membrane excitability. This happens because a prolonged hyperpolarization removes a component of the inactivation of the sodium channel that exists even at the resting potential (and for other reasons, too). This increased excitability can be such as to reduce the threshold of excitability to a point that the sudden interruption of the hyperpolarizing current becomes a stimulus powerful enough to produce the discharge of the electric impulse. As Volta correctly noticed, when compared to the responses arising at the onset of the stimulus, these responses produced by the interruption of the stimulus (OFF-responses) appear less frequently and are usually less intense.

The problem of the difference between common electricity of electrical machines (friction electricity or triboelectricity) and that of metallic electricity came back again after the invention of the battery. This was because the new electricity exhibited a great power in stimulating physiological effects, and also in producing other actions (like those of chemical or thermal nature), but still it appeared weak (compared to friction electricity) in producing sparks, attracting light bodies, and in giving other electrical manifestations. As already mentioned, this brought many scholars of the age to suppose that the fluid put in motion by the battery was not common electricity, but a new fluid, endowed with different electricity. This "new" fluid was referred to as "galvanic fluid" or, more often, simply as "galvanism," an attribution that underlined its similarity to the fluid involved in Galvani's experiments. The problem of the "galvanic fluid" annoyed Volta for a long while and obliged him to intervene in the discussion by writing several memoirs. Their titles obsessively contained the phrase "on the identity of the electric fluid with the galvanic fluid."

We return to this problem in order to note how acutely Volta was able to face the difficulties arising from the apparent differences between ordinary electricity and the electricity of the battery. As already anticipated, this was one of the circumstances in which there was a fruitful interaction between the various perspectives potentially on the stage of his research in Volta's work, among which he was able to move with great freedom and fantasy. With the intention of solving an eminently physical problem, the scientist of Como addressed physiological experiments. While doing so he arrived, as we shall see, at evaluating an important parameter characterizing the temporal aspects of neuromuscular excitability with astonishing precision.

In an analogous way to what he had done in the first phase of his research (when he was dealing with the comparison between metallic and ordinary electricity), Volta also showed that the apparent differences between the electricity of the new device and that of a Leyden jar charged at the same electrometric degree (or "tension") decreased progressively as the capacity of the jar increased, and it became almost unnoticeable when using many big jars connected in parallel (according to an arrangement indicated as "battery" according the terminology of the age). With this aim, Volta compared first the shock produced by a single jar of "one square foot of armature" with that produced by big jars or by batteries of jars "with armature of 60 square foots" or more, and eventually with the shock produced by the his newly invented electromotor device.

The arguments he developed to interpret these observations in *Memoria sull'identità del fluido elettrico col fluido galvanico* (*Memoir on the Identity of Electric and Galvanic Fluids*), published in the *Annali di Chimica* of 1802, start from the consideration that the time of the discharge of Leyden jars of different capacity, charged at the same tension, increases with increasing the capacity; consequently, compared to a small jar, the discharge of a big jar can be considered as a temporal series of succeeding discharges similar in amplitude to that of a small jar. In his experiments of physiological stimulation Volta had become aware that there is a fundamental physiological duration, very short but finite, during which temporally subsequent stimulations can sum their exciting effects on the nerves. This meant that a big Leyden jar was able to force a larger quantity of fluid across the nerves, because of the longer duration of its discharge, as if this was the sum of a temporal series of short stimuli. Accordingly, the shock it could induce would be stronger.

This occurred, however, as far as the duration of the discharge was shorter than the physiologically useful time characterizing the excitability of the nerve; this is an important parameter of both motor and sensory physiology currently indicated as "integration time" or "summation time." Volta was able to evaluate this time with his knowledge of the discharge time of various jars of different capacity. He succeeded in this task by noting the value of the discharge time of a large Leyden jar (or a battery of jars) over which there is no further increase in the physiological response when increasing the size of the jar (or the number of jars in the parallel assembly). As we shall see, in the attempt to solve an evidently physical problem, Volta arrived at physiological conclusions of great interest. As a matter of fact, he estimated the physiological value for the summation of shocking sensations brought about by stimuli of different durations with an amazing precision: *un minuto terzo* ("one third of a minute," a term which, in the sexagesimal system used for indicating short times, meant 1/60 of a second, that is about 16.66 milliseconds, a value astonishingly similar to the ca. 17 milliseconds measured with modern techniques; see, for instance, Kimura, 1989). Volta achieved this when the best chronometers were not suited to resolve times shorter than about 1 second (although scientists, and especially astronomers, had developed psychophysical methods to divide the minimum measurable time of a chronometer in the order of tenths of a second; see Chapter 9.1). The details of how Volta arrived at his conclusion cannot be dealt with here; it was based on the combination of extremely skilful experimentation and acute guesswork. Let us just say that he started from the assumption that the "effective" discharge time of a Leyden jar charged at a given electric tension—that is, the time needed to make ineffective its physiological action—decreases by decreasing the capacity of the device, and, next evaluated this time in a very large assembly of Leyden jars (a relatively large, and thus measurable, time). It was by considering Leyden jars with smaller and smaller metallic coatings (which means progressively decreasing capacities, and thus shorter discharge times) that Volta eventually ascertained the useful summation time for the physiological responses with the astonishing precision noted earlier.

8.4 VOLTA'S RESEARCH ON ELECTRIC FISHES AND THE INVENTION OF THE ELECTRIC BATTERY

Even though we cannot go into all the other aspects of Volta's research on the fruitful interplay between physics and physiology (for instance, the experiments in which he made use, although in an intuitive way, of the concept of the "internal resistance" of an electrical device), we need to consider at least the main aspects of this interplay. These concerned the research leading to the invention of the electric battery, and particularly the importance in this context of Volta's reflections on electric fishes. The letter announcing the invention of Volta's electric battery, written in French and dated March 20, 1800, was sent to Joseph Banks, president of the Royal Society of London. It was read to the Society in the séance of 26 June of the same year and soon afterward published in the *Philosophical Transactions* (see Fig. 8.5). The publication was in the original French language, with only the title in English: *On the Electricity Excited by the Mere Contact of Conducting Substances of Different Species.*

Starting from the first lines of his letter, Volta made clear how the extraordinary novelty of his device was to produce an electrical disequilibrium by containing at its interior exclusively conductive materials (in contrast to the other devices of the time capable of providing a source of electricity). He noticed, for instance, that the Leyden jars necessarily had at their interior "one or more insulating layers." The special characteristic of producing an electric flow with no insulating body inside made—in Volta's view—the new device similar to the special organ by which the electric fishes produced their powerful discharges. For this and other reasons, he proposed to call his new instrument *organe électrique artificiel* (artificial electric organ). This is the way Volta expressed himself on this point in the English translation of his letter published in September 1800 in Nicholson's *Philosophical Magazine*:

> To this apparatus, much more similar at bottom, as I shall show, and even such as I have constructed it, in its form to the natural electric organ of the torpedo or

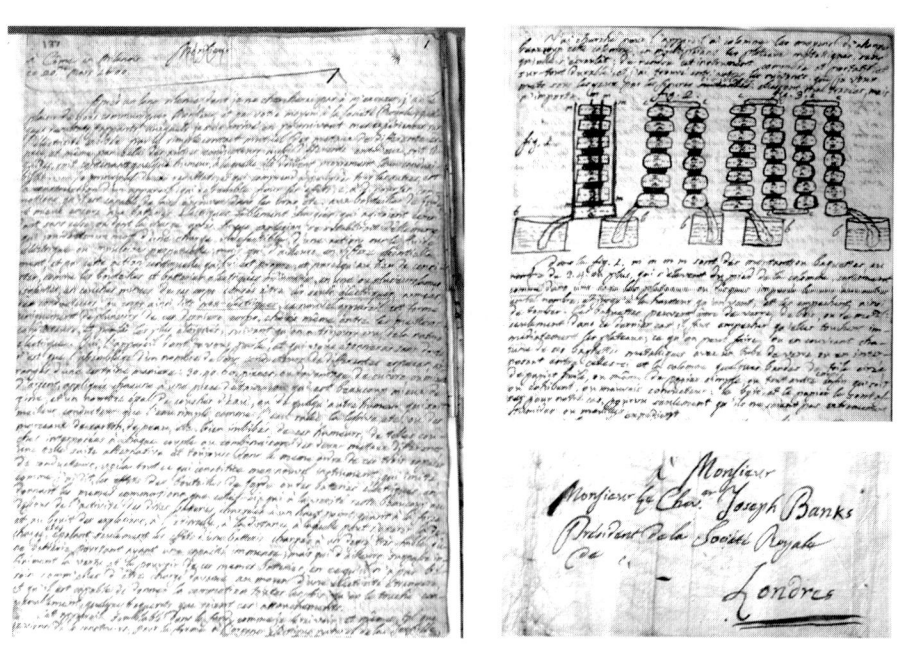

FIGURE 8.5. Pages from the original letter addressed by Volta to the president of the Royal Society on March 20, 1800, announcing the invention of the battery (from the Archives of the Royal Society of London).

electric eel, &c. than to the Leyden flask and electric batteries, I would with to give the name of *the artificial electric organ*: and, indeed, is it not, like it, composed entirely of conducting bodies? Is it not also active of itself without any previous charge, without the aid of any electricity excited by any of the means hitherto known? Does it not act incessantly, and without intermission? And, in the last place, is it not capable of giving every moment shocks of greater or less strength, according to circumstances shocks which are renewed by each new touch, and which, when thus repeated or continued for a certain time, produce the same torpor in the limbs as is occasioned by the torpedo, &c.? (Volta, 1800, pp. 290–291)

As a matter of fact, being made of an assembly of disks of two different metals (generally copper and tin, or silver and zinc), stacked one above the other, with interposed moist disks, the new apparatus bore a morphological resemblance with the electric organ of the fishes, also made of disks staked one above the other (Fig. 8.1). To make the resemblance of the artificial device with the electric eel (*anguille tremblante* in the French original, i.e., "trembling eel"), Volta even went on to suggest that the columns of the artificial organ "might be joined together by pliable metallic wires or screw-springs, and then covered with a skin terminated by a head and tail properly formed, &c" (Volta, 1800, p. 302; see Fig. 8.6).

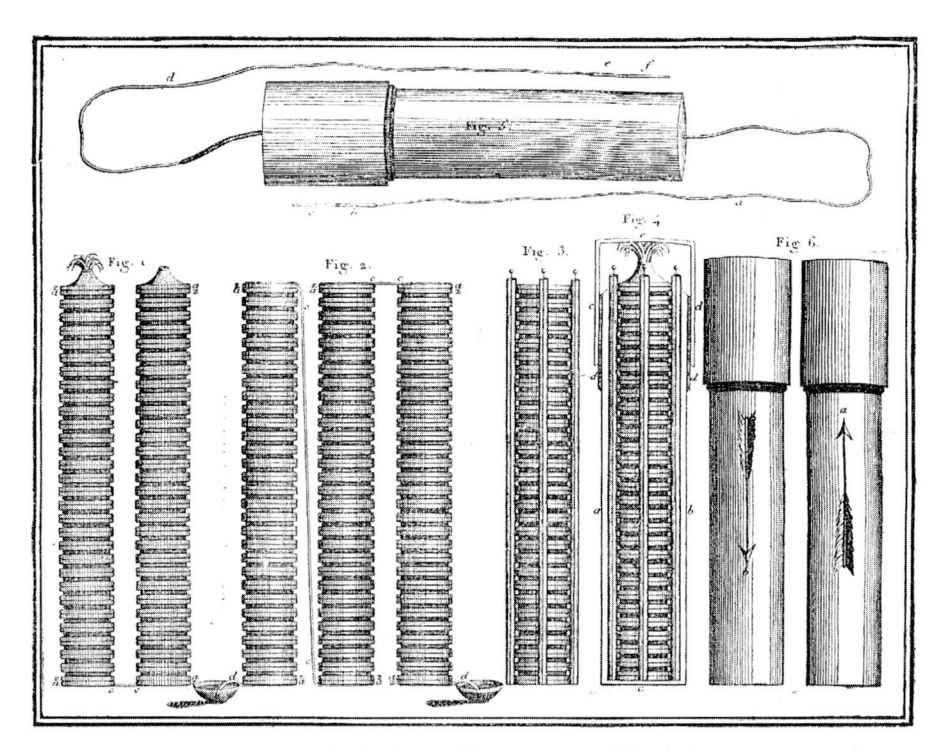

FIGURE 8.6. A "mobile" version of Volta's battery (from *Description de la pile électrique*, 1801 in Volta, 1918–1929, vol. 2).

Although the morphological similarity was evident and important, Volta did not consider that it was the essential element of the similarity between the natural and artificial "organ." In his writings he repeats the idea that the two organs are similar "at bottom," "in their essence," in the way they are built. The expression *organe électrique artificiel* is similar to that of the "artificial torpedo" used in 1776 by Henry Cavendish; it designated an artificial device capable of imitating the effect of the torpedo fish by discharging the electricity of an assembly of Leyden jars. Volta showed that with his new apparatus it was possible to reproduce the effects of the fish, similar to (and even better than) what occurred with Cavendish's artificial torpedo, and among these effects there was also the diffusion of the shock at distance in the water.

In Volta's view, it is indeed the comparison between the two devices, his own and that of the "celebrated English physicist" ("Lord Kavendish," according to a spelling that recurs in his writings) that showed the differences and underlined the importance and the novelty of the instrument he invented. In the 1814 memoir, with reference to the apparatus contrived by Cavendish, he wrote that "this small machine [*macchinetta*] which he [Cavendish] was pleased to call *Artificial torpedo*," even if it imitates the effects produced by the natural torpedo "is not similar, however, to the natural one in an intrinsic way, nor for any virtue, or action that this [i.e., the fish] possesses" (VEN, II, p. 267n). Moreover, on speaking of "the experiments and observations of Kavendish, [...] very beautiful and instructive, which make a great honour to the sagacity of the celebrated English physicist," Volta claimed (under the veil of anonymity) that they, on the other hand "still lacked of that much, that their completion was added by our Italian [i.e., Volta himself] with the discovery of the Electromotor, and with the application of this wonderful instrument to the true explanation, or at least very likely, of the phenomena of the Torpedo, and moreover, to the most perfect and definite imitation of these phenomena" (VEN, II, p. 268n).

Such "imitation" was indeed not limited to the capacity of producing the shock nor to the characteristic proper to Volta's apparatus (but absent in Cavendish's artificial torpedo) of doing that by an intrinsic force, being "always charged and always ready to give shocks" (in contrast to the Leyden jars which were the source of power in the device of the English physicist and needed to be charged again after every shock).

With reference to the natural organ of the fish, Volta added:

We are induced to believe that it is at bottom, i.e. as to the essential construction, the same [as the Electromotor]; namely that its virtue and activity comes from the general principle established by Volta. This is to say that different conductors, placed at reciprocal contact are also *motors* of electricity; keeping in mind that, how it is easy to suppose, the small sheets, or pellicles, stacked one above the other in great number in their small columns, or tubes, composing these electric organs, differ in such a way, that two or three dissimilar matters comes one after the other in alternation, with also some liquid humour between them; to sum up they are likely in the appropriate order, as indeed are arranged the double small metallic sheets, and separated by a third conductor, in the [electric] piles; we understand thus well why these would deserve well the name of artificial electric organs proposed by Volta. (VEN, II, p. 268n)

Afterward he continued by saying:

KAVENDISH did teach us nothing concerning the origin, that is the cause of this prodigious electricity of the Torpedo, and nothing about the role and function exerted therewith by its organs called electrical; nothing the other Physicists with their so many, and so various hypotheses, which serve more to confuse than to make the matter clear. (VEN, II, p. 269n)

In speaking of the other "Physicists" Volta had particularly in mind another Englishman, William Nicholson, who, following Cavendish's conceptions, in 1797 had proposed that the electric organ of the torpedo was composed of a multitude of minute capacitors or "electrophores." According to Nicholson, the natural electric organ could be modelled by an apparatus made of columns composed of disks of resin or mica stacked one above the other.

Volta had already vigorously opposed this idea in the last part of his 1800 letter to Banks. The reason is that with this letter, rather than simply announcing the invention of a new apparatus, Volta intended to announce the discovery of a new principle capable of transforming both physics and biology. He wished to propose the possibility of producing and maintaining an electric flow "with the mere contact of conducting substances of different species," without making recourse to any insulating material (as was the case with the electrical machines and capacitors of the age and was implicit in artificial organs models proposed by Cavendish and Nicholson).

Volta was aware, moreover, that his new device would not only revolutionize the physics of electricity but also conclude a chapter of the eighteenth-century science, by showing how an animal organism could produce and store an electrical disequilibrium inside the tissues of its body, made as it was of electrically conductive materials, and eventually how it could use this disequilibrium for the needs of animal economy.

Volta's interest in the torpedo and other electric fishes, and in particular in the possibility they offered of a genuine animal electricity (an electricity largely negated by him for the ordinary animals of Galvani's experiments), represents a fundamental aspects of the path leading the scientist of Como to the discovery of the electric battery. And also in this case Volta's antagonist, Galvani, plays a significant role.

We have already mentioned how Volta had been interested in electric fishes even before reading Galvani's *De viribus* and how he signified that the electricity of these fishes was decidedly different from that produced by rubbing the hair or the skin of an animal (a purely physical electricity—in his opinion—similarly excited in biological as in any physical body). In *Memoria prima sull'elettricità animale* (*First Memoir on Animal Electricity*) written in May 1792 he came back to these arguments by praising Galvani's discovery, which, in his opinion, illustrated the possibility that an electricity similar to that of the torpedo and other electric fishes could exist in common animals. Again he distinguished this genuine animal electricity from the electricity "excited by the rubbing of hairs in animals, of hairs and cloths in men," which was nothing other than the "the usual artificial electricity produced by rubbing" (VEN, I, pp. 18–19; see Chapter 6.1).

In the subsequent writings, the reference to electric fishes became rare until it exploded again in 1798, that is, in the year which immediately followed the publication of Galvani's *Memorie sulla elettricità animale* (see Chapter 7). This is a crucial period in the research of Volta, who, after measuring the electricity of his bimetallic couples with purely physical instruments (and without any recourse to Galvani's frogs), focused his attention on various combinations of conductive materials, capable of producing electromotive actions. The main goal was to make these actions more powerful and thus potentially useful also outside the limits of a *cabinet de physique* (see Chapter 7.4).

A significant reference to electric fishes appears in a letter written in 1798 and addressed to the famous physician Johann Peter Frank (Volta's former colleague at Pavia University). This letter has been mentioned earlier in this chapter with relation to the role that electricity could have on the contractions of voluntary muscles. Volta wrote:

> This empire of will in moving the electric fluid which exists naturally in nerves is manifest, and appears with effects both vigorous and wonderful in the *Torpedo* and *Gymnotus electricus* endowed as they are with the extraordinary power of giving the true electric shock, and even the spark. These animals are provided with an extraordinary and abundant apparatus of nerves, precisely called *electric organs* (described by HUNTER and others) because, not serving any other function of the common animal economy, they serve only the function of vibrating and put in motion a great quantity of electric fluid, capable of producing on those who touch them the known commotion: so true this is that, these organs being excised, the animal lives well, and for a long time, and it has only lost the faculty of giving an electric shock. These things have long been known. But Dr. GALVANI has discovered more, and published last year some letters to Prof. SPALLANZANI containing various and beautiful experiments made by him on torpedoes. Deprived of the brain, i.e. of the organ where the will resides, the torpedoes at once lose the virtue of giving the shock, although they continue to live apparently well, that is they remain lively for long time, swim as normal, and make all the other movements etc. On the contrary, when deprived of the heart, which makes them more languid and much beforehand abolishes the motion together with the other functions, they maintain nevertheless the virtue of giving more or less vigorous shocks until the complete cessation of motion with the extinction of life. Here is then how the impulse of the electric fluid,

we are dealing with, depends originally and exclusively on the will of the animal which has its seat in the brain. (VEP, III, pp. 416–417).

In 1798 Volta reformulated a possible involvement of electricity in animal physiology and made reference to electric fishes and particularly to the "beautiful experiments" on the torpedo of the Bologna doctor which pointed to the fundamental role of the central nervous system in the control of the fish shock. Already in 1792, with reference to Galvani's hypothesis of the "animal Leyden jar," Volta had noted the relationship between "the muscle movements caused by such a play of electric fluid, naturally imbalanced between the interior and the exterior of the same muscles" and the "the faculty of giving a true electric shock in the manner of the Leyden jar," a property of the "Torpedo" and of the "Trembling eel" (VEN, I, p. 17). In the letter to Frank he was extending to the common animals a small portion of that faculty specific to the fishes of moving, under the control of the will, an extraordinary quantity of electric fluid.

Galvani's experiments continued to be a stimulus for Volta, capable of redirecting his attention toward the animal world, in periods in which, with the physical demonstration of the electromotive power of metals, he seemed to have cut his dependency on Galvani's frogs.

It is difficult to retrace with precision, and without ambiguity, the path leading a scientist to a discovery, and we are certainly not attempting to do that in the case of Volta and his battery (see Heilbron, 1978; Pancaldi, 1990, 2003). Perhaps even Volta himself could not tell us precisely how "toward the end of 1799" he arrived "to the great step" which, as he wrote in a memoir published in 1802, "would bring him to build up the new shaking apparatus [i.e., the battery] etc; which has thenceforth caused so much astonishment to all Physicists" (VEN, II, p. 59). It is certain, on the other hand, that in the experiments he carried out since 1795, by varying the combinations of different conductors in order to excite the flow of electric current more powerfully, Volta quickly realized that he could not succeed in his aim by simply combining various metallic couples. Eventually he came to the conclusion that humid bodies must be interposed in the chain of metals. Volta's attention to humid bodies was to a large extent a response to Galvani's concomitant experiments. The scientist of Como, who after the initial observations on animal electricity, tended to attribute all the importance to metals, was obliged to concentrate on humid conductors (i.e., the conductors belonging to the "second class" of his classification). This was a consequence of the "contractions without metal" experiments carried out by Galvani (and also by Eusebio Valli and Alexander von Humboldt; see Valli, 1793 and Humboldt, 1797). In his fantasy and experimental compulsiveness, Volta used these humid conductors in various forms, that is:

> as bodies containing in a considerable abundance some humour, both in the liquid state, free and fluent, as soaked cloths, chords, leathers, earths, bricks etc., fresh and juicy animal and vegetable bodies; either in a solid state or in combination, as flesh,

tendons, nerves, membranes, and many other materials, and "glutinous humours," and also "saline liquors" of various types and composition. (VEN, I, pp. 369, 381)

The necessity of using humid bodies is also found in the letter to Friedrich Gren in which Volta communicates having been able to demonstrate the electricity of metals without making recourse to Galvani's frogs, or to other physiological methods. When he tried to combine (unsuccessfully at the beginning) numerous metallic couples in order to multiply the weak electricity of a metallic contact, the publication of Galvani's *Memorie sulla elettricità animale* directed his attention again toward animal physiology by advancing the torpedo as evidence that "electricity is not extraneous to animal economy." In particular, fish electricity has a relation with nervous phenomena as shown by "the great apparatus of nerves" of which the "particular organs" of the animal are provided, and the abolition of the contraction after "the brain has been extracted, or the big nerve trunks have been excised" (VEN, I, p. 560). These last quotations are from a letter written by Volta to Luigi Valentino Brugnatelli on October 19, 1798, where again he discussed the possible role of the will in moving the electric fluid along the nerves.

The strong electricity of the fish that comes out "of the border of the animal, and overflows, thus shocking and fulminating others" is for Volta (as he remarks in the same letter to Brugnatelli) an evident testimony that nature had already succeeded in the goal he personally was trying to achieve with his art: that is, to make powerful the weak electricity generated by the contact of two dissimilar conductors. The structure of the electric organs of the fish, made by humid disks stacked one above the other, thus becomes for Volta the mental image capable of guiding him (despite many unsuccessful attempts) to the decision of interposing, between the bimetallic disks stacked in column, the fateful disks (the *bullettini*) moistened with water or some other "saline liquor"; or "the circles of cardboard, of leathers, or of some other spongy matter, capable of holding a great quantity of water, or of the humour necessary for the good success of the experiments" (VEN, II, p. 18).

In some way, the organs of the torpedo (or of the "Trembling eel") became for Volta what the Leyden jar had been for Galvani. It is certainly a kind of historical paradox that a "physicist" like Volta was directed in his endeavor by a mental image derived from the animal world, whereas a "physiologist" like Galvani drew his inspiration from one of the most famous physical tools of eighteenth-century science.

Starting from the moment in which Volta communicated the invention of the electric battery to the Royal Society, the reference to electric fish became a recurrent motif in his writings, until it turned out to be one of the central themes of the last scientific memoir, that is, the text he published in 1814 under the name of his pupil Pietro Configliachi.

Before analyzing the way in which Volta's interest in electric fish developed after the invention of the battery, we think it appropriate to discuss here the rather assertive opinion on the relation between Volta and electric fish expressed by one of the most important experts on Volta, Giovanni Polvani, an opinion that has become a reference for many subsequent historians (see, for instance, Gigli Berzolari, 1993).

In his critical analysis of the letter to Banks, Polvani expressed his concern over the way Volta communicated his discovery, and particularly his omission of any information that could help illuminating the mental processes leading him to his invention. This is how Polvani expressed his astonishment:

> All this brings wonder, and this wonder is increased by the lack of Volta's writings preceding the letter to Banks capable of explaining this passage and of accounting for all these omissions. Moreover, the unexpected fact that Volta presents the artificial electric organ as the reconstruction of the electric organ of the torpedo and the gymnotus, adds wonder to wonder; this is because Volta had never dealt with the structure of the electric organs of those animals, as also he had never dealt with *ex professo* the electric phenomenology inherent to them. All that Volta had written was what he had expressed eighteen years before to Madame Le noir; after that, even during the debate with Galvani, Volta had spoken of torpedoes and gymnoti only rarely and incidentally, without entering into the details, and just such as to indicate the generic property possessed by some animals of creating the electric fluid in conspicuous doses, and moving it at their will. Why then invoke now the problem of the structure and functioning of the electric organ of the torpedo and present his invention as the solution of this problem? (Polvani, 1942, pp. 347–348)

Polvani's conclusion was that, with his communicative strategy, Volta aimed at making clear that the animal electricity supposed by Galvani (and clearly evident in the fish shock) was identical to the artificial electric fluid put in motion by the contact between metals. Thus, he considered as misleading the reference to Volta as "interpreter and emulator of nature," who had been meditating on the "power of the torpedo ray," put by Configliachi in the engraving portraying Volta as a senator of the Kingdom of Italy (see Fig. 8.7). It was so misleading for Polvani that he omitted this inscription from the same image used as facing-title page in his monograph on Volta (where it is substituted by Volta's signature). Polvani does not even consider the possibility that Configliachi's decision to write this inscription below the image of his teacher might be due to some confidence communicated to him by Volta, and thus might reflect the deep feeling of his teacher.

We have repeatedly said how electric fish attracted the attention of natural philosophers and of literati in the second half of the eighteenth century. The possibility that electricity could be subservient to some function of animal economy was one of great themes of the scientific culture of the age. To say—as Polvani does—that Volta had never personally dealt with electric fish in his research is misleading, also because in this way Polvani does not appreciate the importance of mental images in the path to scientific discovery. Despite not having studied the torpedo personally, Volta was surely very occupied

ALEXANDER . VOLTA .
IN . RE . ELECTRICA . PRINCEPS .
VIM . RAIAE . TORPEDINIS . MEDITATVS .
NATVRAE . INTERPRES . ET . AEMVLVS

FIGURE 8.7. Portrait of Alessandro Volta as a senator of the Kingdom of Italy with the inscription added by his pupil Pietro Configliachi to underline the importance of Volta's reflection on electric fish in his invention of the electric battery (from Volta 1918–1929, vol. 1; original engraving by Giovita Garavaglia).

with reflections on the structure and function of its organ at the moment of the "great step" that he made "toward the end of 1799."

In a similar way, we cannot pretend a study ex professo of astronomy by William Harvey, who wrote that he was guided in his discovery of blood circulation by reflections on the movements of planets. Nor we can pretend knowledge of herpetology or ancient cultures from August Kekulé who was guided in his discovery of the closed-ring structure of benzene guided by his daydream of the snake seizing its own tail (the mysterious Ouroboros of ancient symbology).[4] The examples could be multiplied and, even though one should be careful about the way scientists present the influences leading to their achievements, it is sure that mental images, metaphors, and analogies may play important roles in the discovery process. This is because they might act as catalysts for the mental processes involved in the crucial phases of the discovery. They might also allow scientists to figure out the processes, behind their discovery in a metaphorical and symbolical way; because

---

[4] Kekulé (1867); see Gillis (1973). On the value of the dream images in the path of scientific discovery, see Mazzarello (2000).

of their complexity and elusiveness, these processes might not be recordable in the form of an analytic account devoid of ambiguity.

Returning to Volta and electric fish, it is remarkable that Polvani does not mention in any way the influence played by Cavendish's famous paper (in 1776) on the artificial torpedo. This was an article that Volta knew well and which undoubtedly played a role in his formulation of the laws of capacitors.

Concerning the route leading to Volta's "great step" (resulting in the invention of the electric battery), the reference to electric fish is evident in his saying that the new apparatus is similar "at bottom, as I shall show, and even such as I have constructed it." Only a prejudiced mind like that of Polvani (and also of other modern historians), wishing to apply a strict distinction between physics and physiology could miss the significance of Volta's reflection on electric fish, and more generally on physiology. Such prejudice is poorly suited to eighteenth-century science, and surely inappropriate to the case of Volta.

We have already said that after the invention of the battery the reference to electric fish almost became a constant feature of Volta's writings. By following the way in which he developed the comparison between the artificial and the natural organ we can understand, among other things, how undefined he considered the boundary between the products of nature and those of "art." In addition, we can see how, being convinced of the principle underlying the production of electricity in electric organs, he used this principle in order to understand the operative mechanisms involved in the functioning of these organs.

In a letter addressed to "the citizen N.N." and dated May 18, 1801 (published in the *Journal de chimie et de physique*) after saying that the new apparatus was able to produce shocks which "resemble so much those produced by the Torpedo, and by the other fishes, which have assuredly been called electrical" and that "to the imitation of the Torpedo etc, and even more constantly and unfailingly are renewed at any contact," Volta wrote:

> In addition to the mentioned shocks, which—as I already showed you—announce themselves to be electrical, at least so much as those of the Torpedo and Trembling eel etc, that apparatus [i.e., the electric battery] offers to you other, non-equivocal electric signs, which have not been observable in the Torpedo; some very sensitive signs observable with the electrometer, and even the spark; and those signs can be seen not only by using a good Condensatore and using it in the most appropriate way (I am actually persuaded that, with the help of this instrument, also invented by me, and so useful and instructive in the subtle electric research, I would be able to obtain electroscopic signs also in the Torpedo); I say not only with the help of the Condensatore, but also without such help, with the only precaution that the Apparatus [i.e., the electric battery] is large enough, and well in order, and in experimenting one would observe the due cares. (VEN, II, pp. 26–27)

This passage is also interesting because it shows the attitude that Volta would retain in the comparison between his electric battery and the torpedo. It was a strategy, based, as repeatedly remarked, on a fruitful interchange between different viewpoints, as had occurred in his former electrophysiological research: (1) the artificial organ is similar to the natural one, and even (2), better than the fish organ, it exhibits some characteristics stemming from the common functioning principles, and, moreover, it manifests signs that also belong to the natural organ; therefore, (3) by refining the experimental approach, one should be able to obtain those signs also in the natural organs, despite the fact that until now this had not been possible.

In his subsequent research, Volta would attempt to point to the complete similarity between the organ of the torpedo and the artificial apparatus; he would call it electromotor or "pile," following the phrases *appareil à pile* or *pile galvanique* used by the French scholars to designate the most common variation of his battery, the *appareil à colonne* ("column apparatus"). Volta's comparison aimed not only at bringing forth evidence of the similarity between the two apparatuses (artificial and natural) but also at providing a functional explanation. Thus, the results obtained in the one of the apparatuses would serve to explain the functioning of the other. Moreover, it stimulated experiments aimed at showing properties still undetected of the one or the other of the two "organs."

In a memoir published in 1802 in the *Annali di chimica* of Brugnatelli, Volta discussed another similarity—that between the electromotor apparatus and Leyden jars charged at the same "tension" or "electrometric degree." He noticed that for low tensions, there was no need for a gap in the circuit, not even "the minimal interruption; otherwise neither the charge, nor the discharge could be made appropriately; this is because the charge that arrives only to 1 or 2 degrees of the thin-straws electrometer [i.e., about 40–80 volts in the nowadays electric units] is so weak that it cannot leap across the gap of 1/60 of a line [1 line corresponding approximately to 2.5 mm], and perhaps not even to 1/600 of a line" (VEN, II, pp. 71–72).[5] He wrote that "having built a small machine [a "spincterometer" as he would call it later] capable of measuring with precision the sparking distance during the discharge of an electric device," he found that this distance depended on the electrometric degree in an almost linear way, whereas it did not depend on the capacity of the Leyden jars.

On the basis of his measurements Volta explained why Cavendish, in his 1776 paper on artificial torpedo, had difficulty in making the shock of a weakly charged assembly

---

[5] In the Kingdom of Lombardy–Venetia, where Volta lived, a line (*linea*) corresponded to about 2.5 millimeters and thus the minimal estimated distance for a weakly charged Leyden jar was, in Volta's case, about 4 microns. Although it is difficult to compare Volta's values with modern measures (particularly because of the difficulty of knowing precisely the electric potentials involved in his experiments), Volta's conclusion on very short sparking distances at low tensions is fairly accurate, since—for example—the sparking distance of an electric battery charged at 40 volts is about 0.25 microns, so small that the spark would be undetectable without the help of a powerful microscope (on the theme of sparking distances with both electrostatic devices and electric batteries, see Shaw, 1904).

of Leyden jars pass across a metallic chain, if the chain was not stretched so as to make tighter contact between its links (Cavendish, 1776, p. 225; VEN, II, pp. 72–73).

As already mentioned, after the invention of the battery, Volta was obliged to concentrate much of his experimental and argumentative efforts on the attempt to demonstrate the perfectly electric nature of the fluid circulating in the new apparatus. He had to face numerous objections pointing to differences between the ordinary electric fluid and the fluid of the battery. Among the apparent anomalies of the behavior of the battery was the observation that its efficacy in producing electrometric signs was not much reduced by the presence of a humid external layer, as was the case for normal electrostatic devices and particularly with Leyden jars. Volta was able to account for this anomaly on the basis of the scarce conductive power of water and water solution by taking into account the research in which Cavendish had showed that "pure water is 400,000,000 times less permeable to electric fluid than metals." He wrote that in his own experiments he had found somewhat different values for the relation between the conductive power of water and metals.

He later made a series of experiments with electric machines, Leyden jars, and with his battery, in which he seemed to master the laws of the combination in series and in parallel of electric circuits, as well as the concept of spatial diffusion in an extended conductor. In the case of the battery, he wrote that if the apparatus is in order (and particularly if the humid disk interposed between the metallic couples are in tight contact with each other and well soaked in the conductive solution), the shocks are produced even in the presence of a humid layer surrounding the column of the battery (and potentially shunting the current). This happens—he said—because the current of the battery "being copious, even though pushed with weak tension [...], instead of following a unique path, as it is offered by a moist conductor [...] divides itself in various ways, although longer, in two distinct conductive arcs, and it flows through them, and arrives to its termination, encountering, along the ensemble of the ways, less resistance than along only the first one" (VEN, II, p. 101). On the other hand, in experiments of this type Volta became aware of the fact that, even though the battery was still effective when surrounded externally by a layer of water, its power nevertheless increased by drying the external surface and became particularly strong "by removing the external humidity as far as possible" (VEN, II, p. 104).

Volta was certainly mindful of these experiments when, some years later, in 1805, wrote these lines in a letter to Configliachi.

> The researches that are more at my heart are aimed at rendering, if possible, detectable by the electrometer this electricity of the Torpedo. It would be no value to try this by leaving the fish immersed in the water; it would be convenient to draw it out, and explore it with suitable means, exposed to the air and even a little dried so that it will not be dripping with water. (VEN, II, pp. 193–194)

Configliachi was about to leave for the Riviera, with the aim of carrying out experimental research on the torpedo in La Spezia, the place where these fish had been studied about 20 years before by Spallanzani (see Chapter 7; see also Piccolino, 2003a and Finger & Piccolino, 2011). This letter published in *Annali di chimica* with the title *Sopra esperienze ed osservazioni da intraprendersi sulle torpedini* (On Experiments and Observations To Be Carried Out on Torpedoes) revealed not only Volta's great interest in electric fish but also the extraordinary and free way in which he operated on two levels, involving both the natural and artificial organ, moving with efficacy and self-assurance from physics to physiology and vice versa.

In addition to drying the fish's body, Volta suggested to Configliachi (if it was difficult to obtain clear electrometric signs from the torpedo's discharge) to partially remove the skin, or even to separate the organs "completely from the other parts of the animal" so that "they will remain deprived of any adherence, bare and clean." If these maneuvers failed, he said that he would like to "unmake those organs, and recompose them on purpose, either as they were before, or varying, either the position of the small tubes, either the series of pellicles, or small disks they are made up with abundance, and by moistening them with various liquids etc., to see if, and by how much, they keep of their former action or force; this could perhaps even increase" (VEN, II, p. 200).

From these passages it is clear that Volta, besides applying to the study of torpedo the knowledge derived from the study of the electric battery, tended to consider the natural electric organ of the fish as a laboratory device, which can be manipulated in various ways. It can be disassembled and reassembled, and the arrangement of its components can be varied in order to understand better the mechanisms of its functioning. In an analogous way, dealing with the tricks to be adopted in order to excite with the torpedo electricity the taste sensation, he wrote:

> However, in order to separate the taste sensation from the sensation of the shock and the burning, it will be convenient to leave the action of those organs [of the torpedo] to become much weaker, as we do precisely for the same purpose with the batteries. (VEN, II, p. 196)

We cannot but be amazed at the way in which Volta assumed that the electric tension of the torpedo organs could be controlled in the same way it could be with an artificial instrument, his battery.

To establish the polarity of the shock produced by the organ of the torpedo, Volta suggested exploiting the polar nature of the pain sensation excited by the electric stimulation on a bare and delicate region of the skin, or even better on a sore surface. This is—as he wrote:

> since, with the artificial electromotors, the negative pole is the one which excites a much more biting and burning pain; therefore, were the same to happen with the

Torpedo, which we should consider as a natural electromotor, one might understand, from such an experiment, in which of the two parts is located the electricity *in plus* and that *in minus*, if the discharge happens from the back to the belly, or from the belly to the back; a thing that also would be important to ascertain. (VEN, II, p. 196)

We notice here en passant that the phrase "natural electromotor" corresponds symmetrically to *organe électrique artificiel* previously used to denote the electric battery.

It was also possible for Volta to use taste and visual sensation in a similar way to pain sensation in order to establish the polarity of fish shock.

Passing afterward to effects on inanimate substances, Volta believed that it would be possible to produce electro-chemical effects with the fish shock. These would consist of the "beautiful chemical phenomenon of the *oxidation* of a wire of silver, copper etc. and the development of many small bubbles of *hydrogen gas* from the other metallic wire," this being in his opinion "the best, desirable, completion in the essays of comparison between the Torpedo, which is a natural electromotor, and the battery, which is an artificial electromotor" (VEN, II, p. 197).

However, Volta declared that he did not expect that the natural electromotor would be able to produce a series of physical effects obtainable only with particularly powerful electric batteries; for instance, "the pleasant and amusing phenomenon of making red-hot and fusing the tip of thin wires and metallic leaves." In his opinion, this was because there were not enough "pellicles and thin layers" in the natural organ to produce these effects, which required a rather powerful electricity. A further reason was that humid conductors, like those present in the fish organs, are generally less powerful than metals in producing electric actions (VEN, II, p. 197).

Volta well understood that one possible reason for the difference between artificial and natural organs is that the fish discharges its electricity "in a very short instant," whereas the battery generates a continuous flux of electricity. This problem had a particular relevance in view of the possible measurement of electricity because the common electrometers and the condensatore could detect long duration currents much more effectively than instantaneous discharges.

It thus became crucial for Volta to try to make the torpedo produce electricity continuously so that a large enough charge was delivered to the measuring instrument for it to be detected. He assumed that the shock would be produced when, as a consequence of some mechanical effort of the fish, the contact between the disks composing its columns becomes tighter. In developing this line of argument, Volta wrote to Configliachi that a good way to make torpedo electricity continuous would be as follows:

> …by placing a weight, on the back of the Torpedo laid down on with its belly on the basin, or directly on the wet cloth, that might compress it enough, one might, I think, oblige those organs to act in a continuous way; then the Condensatore will

be able to draw the sufficient charge of electricity, as it draws it from an ordinary pile put in a good order. It would be a good thing, that the weight placed on the back in the site correspondent to the place of the said organs were of metal, and that the Condensatore would be connected to this weight instead of the naked back. (VEN II, 195)

Poor torpedoes, we are tempted to say, compressed under a metallic weight in order to make their electricity more evident to the instruments! Indeed, if such a reaction might be justified, especially given the increased sensibility toward animal pain and distress that characterises modern society, we can also see in Volta's attitude a further expression of his tendency to apply the same principles and laws to both the animate and inanimate world. For him the torpedo was rather like a physical instrument, a kind of electric machine, the functioning of which could be studied using a similar approach to that used to understand the functioning of "physical" devices.

Volta had another fundamental problem somewhat related to that of forcing the fish to produce a continuous discharge: understanding how the fish could give the shock intermittently, and control its production (unlike an ordinary electric device which, being charged, was always ready to produce the shock upon simple contact). In order to account for this, he posited particular importance to some small movements observable in the fish in the preparatory phase of the shock (which had been interpreted as evidence of the mechanical nature of the shock by previous scholars like Lorenzini and Réaumur: see Chapter 3.4). He assumed that these movements were expressions of the mechanical effort of the fish in order "to produce a wide and perfect fitting" contact between the disks of the organs or with purpose "of squeezing some humour, and dripping it in such a way that it would soak better the pellicles or small disks stacked one above the other [...] some humour—I say—either mucous or lymphatic, or some other which would happen to dunk more those pellicles, or small discs thus making them better conductors or motors, or to fill the interstices thus establishing the necessary communication, or make them more complete" (VEN, II, p. 195).

Volta's explanation, based on an imperfect communication between the "pieces" of the natural electromotor in resting conditions (that would become more tight and effective at the moment of the shock due to some mechanical effort of the fish), necessarily implies that there is always in the organ some small "passage [*trascorrimento*] of electric fluid from the back to the belly" due to some persistent, imperfect contact between the disks. Outside the delivery of the shock, this electric flow should be necessarily "very scarce passage though," as is the case for the artificial electromotors "purposely built" to weaken their electromotive power or with this power "weakened" spontaneously due to the loss of "moistened disks" or to other causes. In this regard Volta quoted explicitly Galvani's *Memorie sulla elettricità animale*, where the Bolognese scholar made use of his prepared frogs in order to reveal the electricity of the torpedo. "Having placed some of them—Volta wrote—on the back and on the sides of the fish laid down on a moistened cloth, he saw them shaking now and then, and sometimes even in a continuous way, even though the Torpedo was not irritated, or gave any sign of discharging the shock" (VEN, II, p. 202; see Chapter 7.3).

Galvani's observation on the production of a weak electricity by the torpedo outside the phases of the shocks fitted with Volta's mechanical model of the control of the shock. This was because—as Volta wrote—"it would bring to see the electric organs as always mounted, and in a state of action at a certain grade, since they would have between the different parts a certain communication." It could then be assumed that the supposed compression of the organs was due to the mechanical effort of the fish to tighten the contact and thus complete the communication in order to produce the strong electricity of the full shock.

Volta's approach to the torpedo and to the mechanisms of production of electricity and control of the shock in its electric organs was eminently physical and electro-mechanical. It could perhaps be invoked as evidence of the exclusively physical perspective underlying his attitude with animal electricity, thereby justifying somewhat the analysis by historians of science or philosophers like Giovanni Polvani and Marcello Pera. In principle, this could be considered a legitimate analysis of Volta's treatment of electric fish and could therefore serve to support the idea of an irreducible difference between "physicists" and "physiologists" in eighteenth-century science.

However, we find a very similar attitude in other scientists of the same period like, for instance, Lazzaro Spallanzani, Volta's colleague in Pavia and one of the most eminent biologists of his era. As noted in the previous chapter, Spallanzani was also interested in electric fish. In the period of 1783–1784 he published a series of experimental studies on the torpedo that contributed to bringing this fish and its electricity to Galvani's attention. Among the experiments he was planning during a scientific journey to La Spezia (the same place where Configliachi was later going to study the fish) there were some similarity to those suggested by Volta to his pupil, for instance, those intended to measure the electricity of the fish with "the condensatore of D[on] Alessandro" (meaning Volta himself: see Spallanzani, 1932–1936, IV, p. 122).

Spallanzani investigated the existence of a residual electric activity in the organs separated from the animal both anatomically and physiologically (for instance, by cutting the nerves) in a similar way to Volta's suggestions to Configliachi. Anticipating Galvani, he studied the electrical activity of a torpedo which appeared lifeless and he showed that the excision of the electric organs did not bring about the death of the fish (which kept swimming for a while after the maneuver).

For Spallanzani to show the existence of a residual electricity in apparently dead animals or in organs separated from its body meant the possibility of isolating, even in the case of fish shock, the component elements of a living machine. As a matter of fact, Spallanzani used the phrase *quelle singolarissime macchinette* ("those very singular small machines") to designate the electric organs of the torpedo in the letter sent to Charles Bonnet and published in 1784 in *Opuscoli scelti sulle scienze e sulle arti* (and also elsewhere). A similar attitude had guided Spallanzani in his other experimental researches, where he tried to show the persistence of functional capacities in organs or tissues separated from the animal or in amputated or desiccated bodies. This was the case for his studies on rotifers (or "wheel animals") and on other "revivescent" organisms. These studies showed that small animals, "dead" as a consequence of prolonged desiccation, could be returned to life by simple procedures such as rehydration. The same applied to Spallanzani's investigations of hibernation or in his attempts at physico-chemical animal fertility as well as in his experiments on animal regeneration. These highlighted the surprising properties of earthworms, salamanders, and frogs in regenerating parts of the body after amputation. Again a similar attitude had guided Spallanzani in his research showing that some "vital" processes could be studied outside the animal body (the digestion of small pieces of meat by gastric juice, tissue respiration in cut pieces of animal bodies) or for a while after an animal's death (for instance, with the luminescence of fireflies and jellyfish).

In his tendency to resolve and analyze separately the functional mechanisms of living organisms, Spallanzani was following the path traced by the anti-vitalist tradition of the seventeenth century. This considered the organism was composed of "machines," obedient to physical laws similar to those ruling over the inanimate world, and encouraged studying the functioning of these machines (without investigating the ultimate causes of the living phenomena or the principles of life). Spallanzani's general attitude in his biological researches revealed how poorly defined the boundaries between life and physical sciences were in the Enlightenment. In that respect, we should not forget that, in addition to his investigations of living organisms, he had been involved in research on volcanology, mineralogy, and the properties of gases. Like Spallanzani, another great scholar of the age, Alexander von Humboldt, shared similar broad scientific and cultural interests; he also studied animal electricity and particularly electric fishes (both electric eels in the Spanish Guyana and torpedo in Naples: see Finger & Piccolino, 2011 Finger, Piccolino, & Stahnisch, 2013a and 2013b).

The opposition between physics and physiology (and to some extent that between vitalists and mechanists) is extraneous to the experimental methodology of both Spallanzani and Volta. It is not, moreover, a useful approach to interpreting the research of Galvani himself, despite what has been suggested traditionally (see Chapter 3.1). In the case of physics and physiology, the distinction was also poorly defined from a linguistic point of view; in the language of the time there were expressions like "animal physics," and in Italian the term "fisico" (corresponding to the present "physicist" in English) was often used to indicate the doctor (a usage still present in the English "physician"), as well as the physiologist and sometimes the scientists tout court.

In the last part of his 1805 letter to Configliachi, Volta declared, undoubtedly with a certain pride:

> I have no doubt that as the more you will examine, and fathom the electric organs of the Torpedo, so you cannot but discover a greater resemblance with my batteries, or

even an essential conformity with those that I call of the third type [i.e., composed exclusively of humid conductors]. The singular construction of these organs was long a mystery for both Physicists and Physiologists, and it is still so for many, but it ended being so for me from the moment that I succeeded in building my motor apparatuses, namely the above mentioned batteries of the third type [i.e., those made all by humid conductors], that are, I endeavour to say, basically the same thing as those organs. The experiments and researches, that I proposed to you, are aimed at verifying and confirming that in all manners, such as to eventually convince those that might still have doubts, or move objections. If such experiments succeed well, as I hope they will, they will show how shocks could be obtained from Torpedoes outside water, or even from their electric organs alone, and besides the shocks all the other phenomena that my batteries present. On the other hand, I have already shown how reciprocally the batteries imitate perfectly the Torpedoes, also in water, shocking and benumbing the hand plunged in the water itself: shocking it even before it could touch the fish body; also at a considerable distance: a thing that was not understood, and that I explain, and confirm with other experiments, in a way that removes any difficulty. As imitation of the Torpedoes which throw shocks in their native element, I put in action these batteries of mine, and make them give similarly shocks even under water, and also to a plunged hand, which does not however come to touch them: in sum I reduce them to be true artificial Torpedoes. (VEN, II, pp. 202–203)

In this significant passage we can see the vastness of Volta's scientific project, which is also a reason for the success obtained in the extraordinary path going from his initial research on animal electricity to the invention of the electric battery. By reflecting on nature, on both the forms and the mechanisms it offers to the view of the acute observer, the "art" succeeded in creating new instruments and new powerful machines. Afterward, from the analysis of these machines, the researcher could arrive at understanding the mechanisms underlying animal physiology.

From an historical point of view, it is of secondary importance that the electric organ of the torpedo (and other electric fishes) works in a way substantially different from what Volta had supposed more than two centuries ago (see Keynes & His Martins-Ferreira, 1953; Bennett, 1961; Finger & Piccolino, 2011).

The investigations of both Volta and Galvani are linked to John Walsh's scientific work. It is perhaps appropriate to reflect on the fact that their scientific endeavors, at both conceptual and practical levels, have created the conditions for the birth of modern electrophysiology. Electrophysiology would eventually lead to our understanding of the mechanisms that underlie the extraordinary power of electric fish ("the untamed art of the extraordinary Torpedo," as Walsh wrote in his laboratory notebook on July 9, 1772 with an allusion to a poem on the fish by the Latin poet Claudianus: see Piccolino, 2003a; Finger & Piccolino, 2011). This development occurred in unpredictable ways, and often along routes very different from those envisaged by the scientists who started the process more than two centuries ago. It is true, however, as Walsh recognized in his article on the torpedoes of La Rochelle, that the subject was "not only curious in itself, but [was] opening a large field of interesting enquiry both to the electrician in his *walk of physics*, and to all who consider, particularly or generally, the animal œconomy" (Walsh, 1773, p. 461). Moreover, Walsh noted auspiciously that "as artificial Electricity has led to a discovery of some of the operations of the torpedo, the Animal if well considered would lead to a discovery of some truths in artificial Electricity which were at present unknown and perhaps unsuspected" (Walsh, 1772, p. 145).

From an historical perspective, the difference between physics and physiology has become rather definite between the nineteenth and twentieth centuries. It is true, on the other hand, that modern research in various fields of science is recovering a multidisciplinary approach reminiscent of the interests and methodologies characterizing the great natural philosophers of the Enlightenment (although often based on the collaboration among different scientists). Nowadays physiologists, particularly those applying strongly quantitative methods in biological research, are referred to as to "biophysicists," a term that underlines the renewed interaction and interchange between biology and physics.

Concluding on Volta, we should remark that, if it is true that in his last writings he tended to exalt the "physical" character of his research, this was not with the intent of asserting the primacy of physics over physiology. This was much more because he aimed at unifying all known forms of electricity. It is indeed not by chance that the most important controversy in which he was involved in after the invention of his battery, was not so much that concerning the physical or animal nature of Galvani's electricity. It was the problem of the correspondence between the artificial (and thus "physical") electricity of the electrostatic machines, and that similarly "artificial" (and "physical") electricity of his battery, the instrument that he had initially indicated, as we know well, as the *organe électrique artificiel*.

9

From Galvani to Hodgkin and Beyond

THE CENTRAL PROBLEM OF ELECTROPHYSIOLOGY IN THE LAST TWO CENTURIES

IN THIS CHAPTER we provide an outline of the electrophysiological research on nerve conduction and muscle excitability after Galvani and Volta. In the next chapter we will go into some depth on the analysis of modern understanding in this field.

As we have mentioned en passant in various parts of this book, Galvani and Volta were faced with an apparent dilemma that did not allow any possible agreement between their respective views. This was so at least within the limits of the scientific knowledge and conceptual schemes of their day. The Bologna doctor and the Pavia physicist each believed they had found the true cause of the electric disequilibrium responsible for nerve excitation and muscle contraction. For Galvani, this disequilibrium was intrinsic to the animal organism, whereas for Volta it originated from the metallic arc used to connect the excitable tissues of the frog preparation. Neither of them could recognize the validity of his adversary's conception. This was because by so doing each would have had to accept the existence of another cause for a phenomenon from the one which had already been considered a sufficient cause. Only the development of electrophysiology after Galvani and Volta would show that the dilemma was apparent rather than real. As a matter of fact, it was a false dilemma because a third possibility existed (*tertium datur*). There was indeed another explanation capable of accounting for the need for both the animal electricity invoked by Galvani and of the electromotive power of metals conceived by Volta.

The historical period initiated by Galvani with his research in 1780 would be ideally concluded with the studies carried out in England (at Cambridge University) by Alan Hodgkin and his collaborators in the period of 1934–1952. Hodgkin's studies would show that nerve signals are a genuinely electrical phenomenon, due to the existence within animal tissues of a form of electricity in a condition of disequilibrium, as Galvani had supposed at the end of the eighteenth century. At the same time, Hodgkin's research accounted for the need of another form of electrical disequilibrium, in order to put "animal electricity" in motion and produce the nerve signal. In the circumstances of the experiments with metallic arcs this disequilibrium was produced by Volta's "metallic" electricity.

In some way, the problem that was at the heart of the debate between Galvani and Volta would be the central problem of electrophysiology for the last two centuries. Indeed that period, various generations of scientists (initially mainly Italian, afterward German and eventually English) trying to unveil the mechanism of nerve conduction, would be faced by this problem. In this chapter we will not try to expose in a systematic way the history of electrophysiology from Galvani to the present day. We will limit ourselves to retracing the struggles of these scientists in their aim to understand the elusive nature of the involvement of electricity in neuromuscular physiology. In this way we will go through the paths of that complex and fascinating investigation itinerary referred to as the "Galvani-to-Hodgkin path." By doing this, we will be in a better position for presenting the modern physiological understanding of nerve conduction process. This book has indeed been written with the idea that there should not be any drastic separation between science and its history, or in the words of Goethe, that great poet who was strongly interested in science, "that history of science is science itself."[1]

9.1 MEASURING ANIMAL ELECTRICITY

After the invention of the Voltaic battery, there was extraordinary progress in the physical investigations of electricity that dominated the scientific panorama of the first half of the nineteenth century. In the beginning we find the successful electrochemical experiments of William Nicholson and Anthony Carlisle and particularly their dissociation of water, carried out in the same year as publication of Volta's invention. Afterward, besides pursuing further the electrochemical dissociation of various salts, leading to the discovery of many new chemical elements, the focus would shift to the chemical processes responsible for the production of electricity in the battery. These studies were pursued mainly by Humphry Davy and Michael Faraday, Davy's pupil and successor at Royal Institution in London, where an enormous battery of over 2,000 elements was constructed. They culminated in Faraday's fundamental laws of electrochemical phenomena. New horizons were afterward

---

[1]Goethe's statement appears in the foreword of the didactic part of his *Zur Farbenlehre* (*On Theory of Colours*, 1810).

opened by the discovery of magnetic processes associated to electric currents made by Hans Christian Oersted, and by the subsequent demonstration, given by Faraday, of the reciprocity of electric and magnetic effects. These achievements would play a decisive role for both technological progress (industrial production of electricity, invention of electric engines) and for both experimental and theoretic developments in modern physics.

In the field of physiology there was, however, no substantial new achievement for about three decades after Galvani. The influence of the apparent triumph of Volta over Galvani is well illustrated by the interpretation given by the Italian physicist Leopoldo Nobili of his experiments, carried out in 1828, in which he had succeeded for the first time in measuring animal electricity of a frog by using a physical device. Nobili had considerably improved the sensitivity of the electromagnetic galvanometer by reducing the influence of the earth's magnetism on the magnetic needle by means of his *astatic* device. He prepared a frog in Galvani's manner and put the skinned legs with the intact muscles in a vessel full of salt water while the portion of the frog with the cut muscle surface and lumbar nerves was dipped in a second similar recipient. In accordance with Galvani's results, a contraction was produced when a contact was established between the two vessels by means of moist cotton. A current flow was measured when the cotton was removed and the circuit closed through the galvanometer. Nobili called this current a "frog current" (*corrente di rana*) or "proper current" (*corrente propria*) (see Moruzzi, 1964; Mazzolini, 1985).

In a somewhat paradoxical way, the Italian scientist attributed scant significance to his discovery. He ascribed the measured current to the thermoelectric effects produced by the unequal cooling of nerve and muscle tissues by water evaporation. He did not recognize in them the manifestation of a genuinely biological process. This conclusion demonstrated the poor standing of animal electricity theory after Volta and his battery. It is appropriate to remark here that Nobili was then interested in the study of thermoelectric phenomena, and thus particularly "prepared" to perceive in the investigated phenomena the agency of thermal processes (another instance of the importance of mental structures in the path of scientific discoveries).

The first real progress in electrophysiology after Galvani was due to the work of another Italian physicist, Carlo Matteucci, who was born in Forlì, near the Adriatic coast, and was appointed professor at the University of Pisa in 1840. In 1838 Matteucci repeated Nobili's experiment and correctly interpreted the measurement obtained with the astatic galvanometer as evidence of animal electricity as conceived by Galvani (Finger & Piccolino, 2011; Moruzzi, 1964; Piccolino, 2011; Piccolino & Wade, 2013). The first success with the instrumental recording of animal electricity was thus the achievement of a physicist. This shows how difficult it was to separate, even in that period, physics from physiology in this field of studies. As a matter of fact, the investigation of animal electricity was then one of the favorite research fields of many apparently genuine physicists and chemists (like Becquerel and Gay-Lussac in France and Faraday in England).

In his studies Matteucci showed that that the current could be studied even in a preparation made up exclusively of muscle tissues, that is, without nerves. This result undermined Nobili's interpretation of the electricity as produced by unequal evaporation from the two tissues. In his experiments on muscle preparations, Matteucci noticed the deflection of the galvanometric needle was produced only if the muscle was cut, and one electrode was placed in the intact surface and the other on the cut side.

Matteucci succeeded in providing an unequivocal demonstration of the biological origin of muscle current with an ingenuous experiment, which somehow brought again within the biological domain the image of the stacked-disks battery that Volta had derived from the fish electric organ. Matteucci cut many frog thighs and placed them in series, in such a way the intact surface of a thigh touched the cut surface of the next one. The deflection of the galvanometric needle was found to increase progressively with the increase in the number of the thighs of the battery. This excluded the possibility that the current measured could be due to the electrochemical effects of the contact between the galvanometer electrodes and the muscle tissues (Matteucci, 1843, 1844). Matteucci used the expedient of the thighs battery in order to increase the force of the biological electricity in various experiments. A battery of frog thighs was more effective in inducing the contraction of a galvanoscopic frog at both the closure and opening of the circuit (Fig. 9.1). The biological battery could also be used to produce electrochemical effects. The exclusion of the involvement in these experiments of the electric power of metals was of great importance for someone like Matteucci who aimed at supporting Galvani's views against Volta's classical objections.

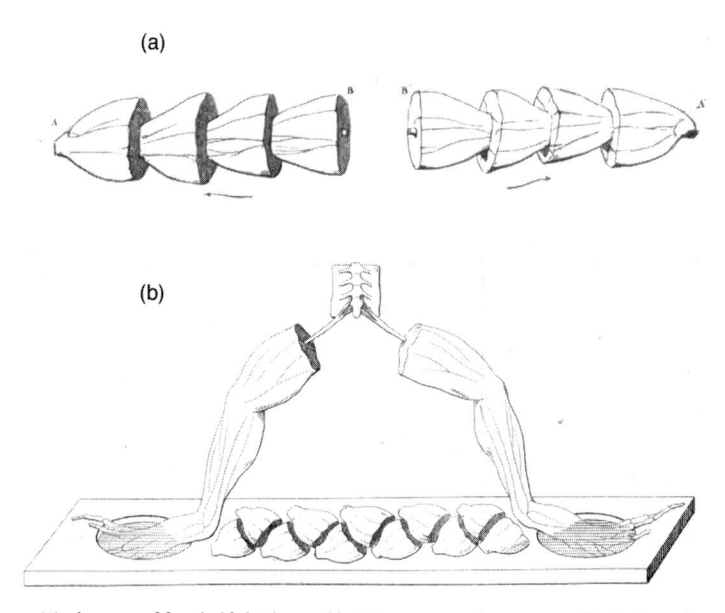

FIGURE 9.1. The battery of frog half-thighs used by Matteucci to demonstrate the biological origin of the electricity measured in muscle (*a*), and, moreover, for strengthening its effect in order to induce contractions in the prepared frogs (*b*) (from Matteucci, 1844, 1846).

Another important support for the biological origin of the current measured between the intact and cut muscle surfaces came from Matteucci's observation that the current disappeared when the muscle fell in a state of prolonged, spasmodic contraction or tetanus (Matteucci, 1844). This could happen both spontaneously or as a consequence of experimental manipulations (such as the application of *nux vomica*, a pharmacological preparation rich in strychnine).

Matteucci's experiments were thus providing objective and instrumental evidence for the existence of animal electricity. The existence of a "muscle current" seemed, moreover, to confirm Galvani's supposition about the presence of an electrical disequilibrium in muscle fibers, which could be considered as a biological equivalent of the Leyden jar. In fact, the current measured by Matteucci between the intact and cut surfaces of the muscle is an expression of the difference in the electric potential that exists between the interior and the exterior parts of muscle fibers (as well as in the nerve fibers). The lesion of the muscle created a path of reduced resistance allowing the electrode placed on the cut surface to sense, at least partially, the intracellular potential. As we shall see, this would only appear in a clear form much later, with the elaboration of the "membrane theory" of bioelectric potentials by Julius Bernstein in the period of 1902–1912.

Matteucci's observation that muscle current disappeared during tetanus stimulated the German scientist Emil du Bois-Reymond to commence investigations in this field. Another fundamental discovery was made by Matteucci with his experiments on the "induced twitch," first described in 1842. The nerves of a frog ("galvanoscopic") preparation were placed on the intact muscle surface of another preparation. When a contraction is induced in this last frog by using an electric or mechanical stimulus, a contraction suddenly arose in the galvanoscopic preparation. It is interesting to note here how the experimental arrangement used by Matteucci in this experiment seems to have been inspired by the experiments in which Galvani used his preparations to detect the electricity of the torpedo (see Figs. 9.2 and 9.3).

In the induced-twitch experiment, Matteucci excluded the possibility that the contraction induced in the galvanoscopic frog could be due to an electrical artifact. He initially considered it as the expression of an electric influence of the first contracting muscle of one preparation on the motor nerve of the galvanoscopic preparation (Matteucci, 1844, p. 134).

The great Italian physiologist, Giuseppe Moruzzi, remarked that with the induced-twitch experiment Matteucci held in his hands the keys to modern electrophysiology: the electrical influence exerted by an active muscle of a preparation on the nerve of the other preparation is the expression of the electrical signal propagating along the surface of excitable fibers (both muscle and nerve: Moruzzi, 1964; see also Piccolino, 2011). This signal represents the typical spread of excitation in muscle fibers and information along nerves. Matteucci was not able, however, to perceive the importance of his discovery, and, in a paradoxical way, eventually he abandoned the correct explanation given at the beginning, by negating the involvement of a propagating electric current in the phenomenon.

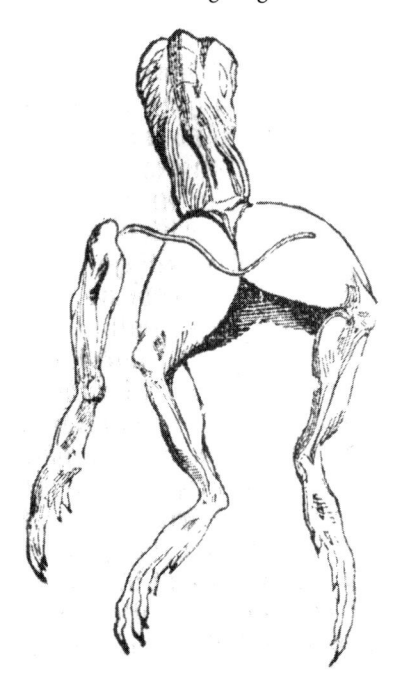

FIGURE 9.2. The phenomenon of "induced twitch" observed by Matteucci: When a contraction is produced in the prepared frog (by any suitable stimulus), also the "galvanoscopic" leg (whose nerve lies on the thigh of the first preparation) often contracts (from Matteucci, 1847).

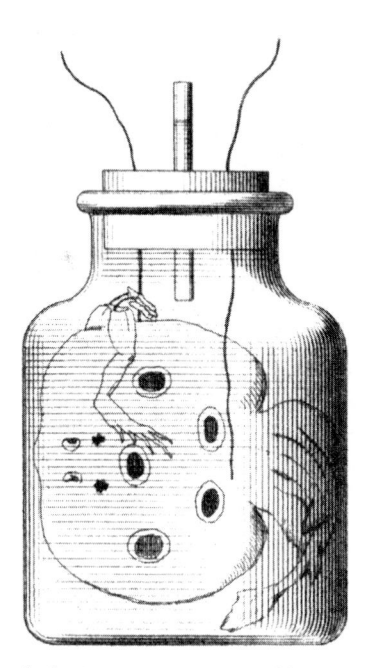

FIGURE 9.3. An experiment in which Matteucci uses a prepared frog in order to reveal the shock of a torpedo closed in an airtight bottle. The experiment aims at ascertaining whether the shock production is associated to a change of volume of the electric organs (from Matteucci, 1844).

After Matteucci, the study of the electric phenomena involved in the excitation of nerve and muscle was continued mainly by du Bois-Reymond, who initially confirmed the demonstration given by the Italian scientist of the existence of a muscle current. Moreover, he confirmed Matteucci's observation that the current measured between the intact and the cut surface of the muscle disappeared during tetanic contraction. In addition, du Bois-Reymond showed that a similar current was present in nerves, too, and that this current also disappeared during prolonged nerve excitation. In 1848 the electric phenomenon associated with the excitation of nerve or muscle was referred to by du Bois-Reymond as *negative Schwankung* (meaning negative "variation" or "oscillation": du Bois-Reymond, 1848–1884). It represents the first evidence of the event that would later be called "action current" or "action potential." The term "negative" indicated two correlated aspects of this phenomenon. One was because during the excitation the membrane surface becomes negative compared to a distant, inactive region; the other was because the potential difference between the intact and cut surface decreases (or disappears altogether) when the excitation wave invades the intact nerve segment (see Finger & Piccolino, 2011).

Despite the importance of his discoveries, it must be said that du Bois-Reymond exerted a somewhat unfavorable influence on the understanding of the role of cell membranes in the genesis of bioelectric potentials (see also Chapter 1). This role had been foreseen by Galvani. As already mentioned, the Bologna doctor located the electric disequilibrium responsible for nervous conduction and muscle contraction at the surface of the separation between the interior and the exterior of the muscle fiber. This was a really ingenuous intuition particularly because, in Galvani's time, cell theory had yet to be formulated, and the idea of the existence of a membrane separating the interior and the exterior of excitable fibers was an inspired but vague conjecture.

Du Bois-Reymond elaborated a model of electrical conduction in nerves and muscles that took no account of the existence and the importance of the fiber membrane in the phenomena of electric excitability. This was partially because he was influenced by the magnetic polarization theories of the period. These were developed to account for the observation that magnetic properties of a body persist even when it is broken in various parts. It was also because of a series of artifactual results in du Bois-Reymond's research that convinced him of the existence of potential differences between different regions of the intact surface of the excitable fibers and even in intact animals (including humans; see Finger & Piccolino, 2011).The model that he developed to account for the electrical properties of excitable fibers was based on the assumption that the fiber was made up of an ordered arrangement of microscopic particles ("electric molecules") bearing one positive charge in their equatorial region and two negative charges in their polar regions. The propagated electrical signal would then be a consequence of an alteration in the ordered state of these particles present in the resting fibers (see Fig. 9.4).

Another fundamental event of nineteenth-century electrophysiological research was the measurement of the speed of nerve transmission obtained by another German

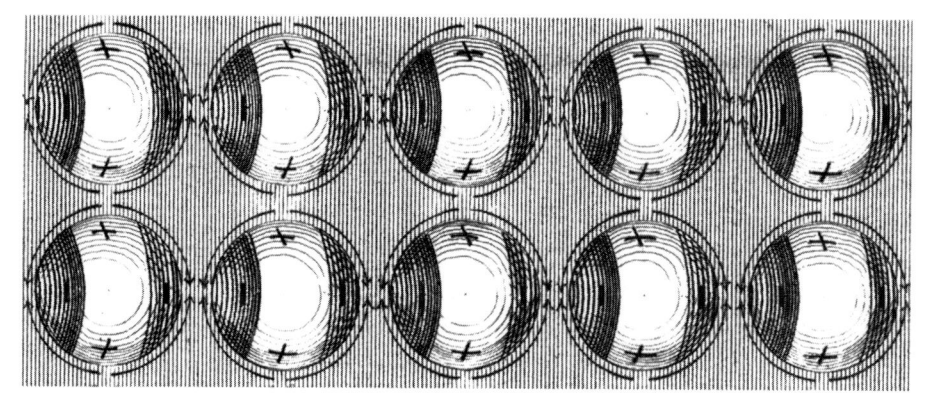

FIGURE 9.4. The model of "electric molecules" developed by du Bois-Reymond in order to account for the electrical phenomena in nerve and muscle fibers (from du Bois-Reymond, 1848–1884, vol. 1).

scientist, Hermann von Helmholtz, in the period 1850–1852. This event did not just represent a technical achievement because the simple fact of being able to measure nerve conduction speed apparently contradicted the scientific notions of the time. The great Berlin physiologist Johannes Müller, the teacher of du Bois-Reymond and Helmholtz, was convinced that the nerve principle should be something similar to light and thus capable of propagation along the fiber at extremely high speeds. In his opinion the distances needed to measure such events were not available in nerve physiology.

Notwithstanding Müller's scepticism, Helmholtz succeeded in measuring the nervous signal speed by means of a relatively simple experiment. In 1850 he employed a method derived from the ballistic procedures used in artillery to measure the speed of cannon balls or bullets (Helmholtz, 1850a-c). In a neuromuscular frog preparation he first measured the time between the application of an electric shock to the nerve and the ensuing contraction of the muscle. This was made with a device in which the movement of the frog's leg automatically acted on the time-measuring device. After the first measure, Helmholtz displaced the point of application of the electric shock nearer to the insertion in the muscle and noticed that the interval decreased. He ascribed the difference between the two measurements to the time required by the nerve signal to propagate between the two stimulated points. In this way he calculated a conduction speed of a few tens of meters per second (instead of thousandths of kilometers as implied in Müller's criticism).

In 1851 Helmholtz used the "Kymographion" or "Myographion" (a recently invented device based on a rotating smoked drum) in order to record graphically muscle contraction and thus calculate the speed of nerve conduction. The results obtained in this corresponded to that provided by the ballistic method and they were illustrated in an article published in 1852 (Fig. 9.5: Helmholtz, 1851, 1852).

Helmholtz's experiments represented a true milestone in the history of science. For the first time they enabled an instrumental and precise measurement of an event related with

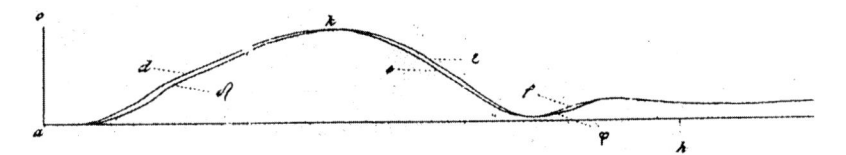

FIGURE 9.5. A graphical plot of the experiments carried out by Helmholtz with his *Kymographion* in order to measure nerve conduction speed. The two tracings monitor the contraction of the frog leg induced by the electrical stimulation of the nerve at two different distances from the insertion of the nerve into the leg muscles (muscle contraction in the ordinate against time in the abscissa) (from Helmholtz, 1852).

the most elevated and elusive processes of animal physiology, an expression of nervous phenomena, classically considered akin to immaterial entities like psyche or soul. This was a real leap forward for science and for the imagery of the era. Despite the progress in anatomical and physiological understanding of nervous organization, it was still difficult for the scholars of the age to recognize that the propagation of signals having to do with the sensation or the act of will did not occur instantaneously but required a finite, measurable time.

Soon after measuring the speed of nerve conduction, Helmholtz communicated the news of his achievement to his father, August Ferdinand, a professor of literature at the lyceum at Potsdam. In the response that the father wrote a few days later, we can see the difficulty that a man of the nineteenth century had in accepting the philosophical implications of these experiments:

> As regards your work, the results at first appeared to me *surprising*, since I regard the idea and its bodily expression not as successive, but a single living act, that only becomes bodily and mental on reflection: and I could as little reconcile myself to your view, as I could admit that a star that had disappeared in *Abraham's* time should still be visible. (Koenigsberger, 1906, p. 67)

It is interesting to note that, in his communication to the Paris Académie des Sciences, Helmholtz indicated the time required for nerve conduction as *temps perdu*, a phrase which would reappear with another meaning in nineteenth-century literature with the masterpiece of Marcel Proust (Helmholtz, 1850c; see Piccolino, 2003c).

There are various possible reasons why Helmholtz decided to try the experiments considered impossible by Müller. One derived from the awareness of a strange anomaly in the astronomical measurement of the era. To determine the position of a star in the vault of the heavens, astronomers needed to observe the star with a telescope and simultaneously take the precise time into account. For the time measurement they listened to the beats of a chronometer while they waited with attention for the passage of the star. For the entire operation they depended thus on a double input, visual and acoustic. It often happened that the position of the celestial bodies determined in this way by various astronomers were substantially different. In some occasions this had regrettable consequences. One such instance took place in 1796 at the Greenwich observatory, where the Astronomer Royal, Nevil Maskelyne, dismissed his assistant, David Kinnebrook, for his gross errors (on the average Kinnebrook "saw" a star to pass on the sky about half a second after Maskelyne: see Mollon & Perkins, 1996; Piccolino, 2008).

The differences between various observers were the rule rather than the exception. This was why astronomers (or other scientists concerned with the theory of instrumental errors) started to be concerned with the possible causes of the problem. Eventually, thanks mainly to the work of the great German astronomer and mathematician Friedrich Wilhelm Bessel, it appeared that the differences, far from depending simply on the skill of a particular observer, were systematic. Moreover, they could be partially anticipated, in as far as the errors made by a given astronomer tended to be constant. In some cases they were even bigger than those implied in the affair of the Greenwich astronomers. This had been indeed the case with the measurements performed together by Bessel and other two important German astronomers, Henrik Johan Walbeck and Friedrich Wilhelm Argelander. To make comparable the observations made by different astronomers, it was decided to calculate a so-called personal reaction for each one, meaning the average difference of his measurement with respect to a defined standard. Eventually, the possibility was considered that the individual differences might depend on physiological factors. They could reflect the different times needed by each individual astronomer to perceive the image of the celestial bodies or the chronometer beats, or by the time needed for their mental elaborations or motor reactions. This implied the idea that nervous and mental processes, long considered as belonging to the timeless domain of psychic-metaphysical events, would require a defined time to be effected.

It must be said that it was probably easier for astronomers than for other scientists to conceive the possibility of a "lost time" in the perception of an object. They were indeed accustomed to the idea that the image of a star perceived at a given time could reflect the conditions of the celestial body thousands of years before (as mentioned by Helmholtz's father in his letter). It is, moreover, interesting to notice that Helmholtz carried out his experiments in Königsberg, at the Albertina University, where he had been teaching since 1849, after having being called to the chair of physiology when he was just 29 years old. In Königsberg there was an important observatory that had been directed, until his death in 1846, by the same Bessel interested in the theory of astronomical errors. Upon his arrival to Königsberg, Helmholtz got acquainted with Bessel's successor as director of the Observatory, August Ludwig Busch, who was also concerned with the theory of astronomical and instrumental errors. The conversations with Bush (and possibly also with another important scientist of the Albertina, the mineralogist and physicist Franz Ernst Neumann) probably contributed to stimulating the young professor's interest in the personal reaction and its neurophysiological counterpart (Cahan, 1993; Meulders, 2001; Piccolino, 2008).

Another possible reason that induced Helmholtz to attempt the experiment depended on considerations connected with the model of bioelectricity developed by du

Bois-Reymond. As already mentioned, this model was based on an ordered arrangement of "electric molecules," undergoing a propagated disturbance during the conduction of the electric signal. It was likely that this molecular disturbance would have required a defined time for propagation, and this time might be long enough to be measured with physiological techniques.

Whatever the matter might have been, and despite the historical importance of Helmholtz's achievement, the low conduction speed measured in his experiments seemed to cast doubts on the electrical nature of the nerve signal. This was because it appeared to be much slower than the flow of current along an electric cable (or other physical conductors). The idea started to be formulated that nerve conduction was possibly a chemical rather than a physical event.

To re-establish a confidence in the analogy between nerve signals and electrical phenomena associated with nerve conduction (i.e., the *negative Schwankung*), it became imperative to demonstrate that two phenomena propagated at the same rate.

Initially Du Bois-Reymond did not succeed in his attempts to measure the propagation speed of the *negative Schwankung* because his electrometers were not capable of recording such short durations. The task would be eventually undertaken by Julius Bernstein, a pupil of Helmholtz and du Bois-Reymond. In 1868 Bernstein succeeded in this technically demanding experiment, by using an especially developed device called a "differential rheotome" (the word *rheotome* meaning "current slicer": see Fig. 9.6).[2]

Besides obtaining the first measurement of an electric signal in a nerve, with his experiments Bernstein could provide the first accurate recording of the time course of nervous excitation. He showed that this consisted of a transient phenomenon of about 1 millisecond duration. In some of his experiments he noticed that the current measured during the excitation phase surpassed the level of the resting current existing between the intact and an injured portion of the nerve. This implied that the excitatory event responsible for nervous conduction was not simply the destruction of the resting electrical potential. After Bernstein the phenomenon (later termed "overshoot") was practically forgotten and it reappeared in the main path of electrophysiology with a famous experiment carried out in 1939 by Hodgkin and Huxley on the giant nerve fiber of the squid (see Section 9.3).

After Helmholtz and Bernstein, it became evident that nerve signals, despite their electrical nature, did not follow the simple laws of the physical propagation of a current along

---

[2] Bernstein, 1868 (see Schuetze, 1983). The apparatus enabled the measurement of a very short event (less than a millisecond) despite the great inertia of the galvanometric needle. It was based on the action of a rotating wheel that opened a temporal window for a very short duration, making recordings possible through a *timing, sampling, and holding* procedure. The sampled interval was changed by varying mechanically the delay between the stimulus and the time recording window. Having determined, through a long and patient process, the time course of the electrical response induced on a given portion of the nerve by the stimulation applied at a certain distance, the position of the stimulated point was changed. In a way similar to Helmholtz's more simple experiments monitoring muscle contraction, it was possible to measure the time needed by the "negative oscillation" to propagate between the two different stimulated points.

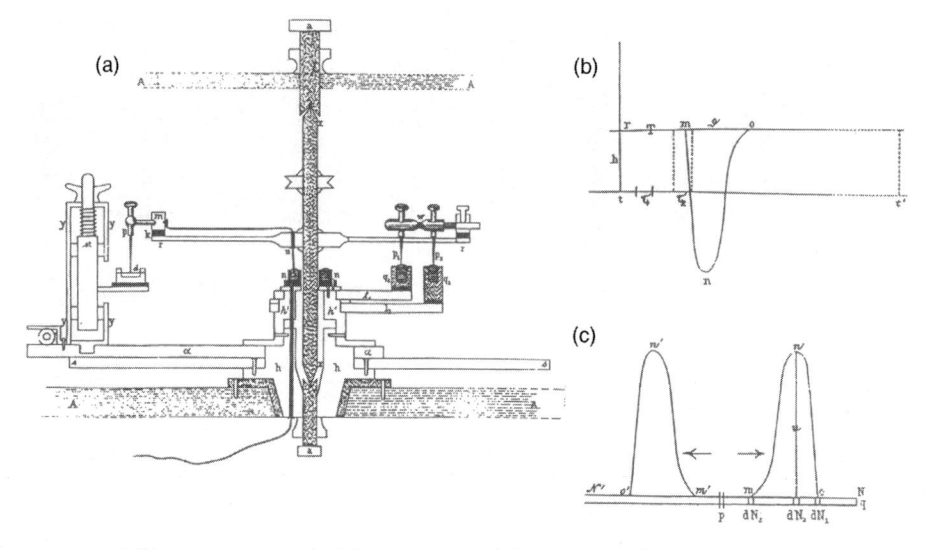

FIGURE 9.6. The experiment in which Bernstein succeeded in measuring the time course of the "action current" associated with the nerve signal: (*a*) the "differential reothome" used by the German scientist and (*b* and *c*) two of his tracings obtained with this method (from Bernstein, 1868, 1871).

a cable, and it involved a more specific and elaborated mechanism. The progressive elucidation of this mechanism, which will appear in its full complexity with the electrophysiological research carried out between the nineteenth and twentieth centuries, would provide a key to new understanding the true reason of the difficulty encountered by Galvani and Volta in reaching a reasonable compromise between their opposed views.

9.2 NERVOUS CONDUCTION: PROPAGATED ELECTRIC SIGNAL AND THE FIRING OF A TRAIN OF GUNPOWDER

The subsequent important progress in the physiology of nerve and muscle excitability was still due to the endeavor of German scientists. One of them was Ludimar Hermann; he was a student of du Bois-Reymond but eventually engaged in a bitter controversy with his teacher. Hermann directed attention to a fundamental aspect of electrical phenomena implied in nerve conduction: The propagation of the signal along a nerve fiber consisted of a negativity of the external fiber surface, and, at the same time, in order to stimulate the nerve it was necessary for the external surface of the fiber to become negative (Hermann, 1899; see Lenoir, 1986 and Piccolino & Finger, 2011). It could therefore be surmised that, in physiological conditions, the negativity of the external fiber surface brought about by the arrival of the *negative Schwankung* (called by Hermann "action current") would act as a stimulus for the nerve segment ahead. In this way a circle could be closed capable of accounting for the propagation of the nerve signal (let us notice here en passant that a similar explanation held true for Matteucci's "induced-twitch" experiment).

Hermann (who had elaborated a model of the generation of bioelectric potentials completely different from that of du Bois-Reymond) assumed that the nerve conduction could be due to a local excitation of the resting zone, induced by a current flow from the active zone. To account for this current diffusion, he proposed his "local-circuit" theory, which somehow corresponds to the model of current flow along physical cables formulated some years before by Lord Kelvin (Kelvin, 1855). This theory was based on the idea that nerve fibers are made up in their central part by a conductive core, separated from the external medium by a partially insulating membrane. Any electrical disturbance originating at a point of the fiber could influence nearby regions by means of the local-circuit currents passing through the internal core, the insulating membrane, and the external fluid (Fig. 9.7).

Another fundamental contribution to the physiology of nerve and muscle fibers was due to the work of the same Julius Bernstein, who—as seen earlier—had already succeeded in another landmark achievement in this field (i.e., the measurement of the propagation speed of the *negative Schwankung*). Bernstein applied the theory of electric potentials to the excitable fibers; it had its origins in the theory of ionic diffusion developed by Walther Nernst (Nernst, 1888; see De Palma & Pareti, 2011, and Finger & Piccolino, 2011). He assumed that in the resting state, excitable fibers would generate a diffusion potential across their membrane due to its selective membrane permeability for potassium ions. This theory did take into account the higher internal concentration of these ions compared to the extracellular medium. According to Bernstein, because of this specific permeability, an internal potential negativity was produced when potassium ions with their positive charge moved outward, with no possible compensating movement of other ions (to which the membrane was supposed to be impermeable; see Fig. 9.8). Much more will be said in the next chapter on Bernstein's theory and on the electrochemical conception of bioelectric potentials, which represents the basis of modern electrophysiology.

To account for electric excitation of muscle and nerve fibers, Bernstein assumed that the excitation consisted of the sudden disappearance of electrical polarization present in the resting fiber (i.e., in a state of "depolarization"). This would result in a sudden and generalized increase of membrane permeability to all ions. Because of the importance of

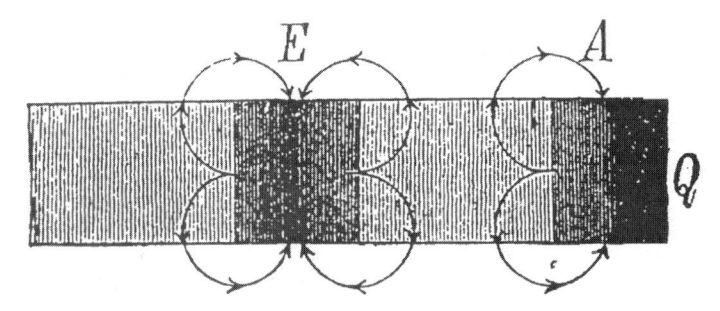

FIGURE 9.7. The local currents flowing between the active zone (E) or the injured region (A) and the neighboring areas of the muscle (from Hermann, 1879).

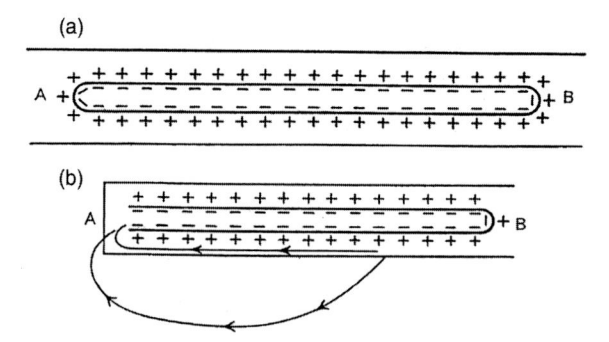

FIGURE 9.8. Bernstein's scheme of the theory of membrane polarization (*Membrantheorie*): (*a*) a muscle fiber with an excess of negative charges inside with respect to the external surface of the membrane; (*b*) a lesion has removed the local barrier to the passage of ions and, by consequence, the zone in A has become negative with respect to the external surface of the fiber, this resulting in a local flow of positive charges from B to A. According to Bernstein, an event similar to a breakdown of the membrane is produced during the excitation process, with A representing then the active zone (from Bernstein, 1912).

the membrane in these processes, Bernstein's theory was referred to as *membrane theory* of excitation. One of the implications, largely confirmed by subsequent studies, was that the electrical resistance of the membrane would decrease during excitation (Bernstein, 1902, 1912; see Hille, 2001).

With Bernstein, electrophysiological investigation began focusing again on the fiber membrane, abandoning the "electric molecules" model of du Bois-Reymond.

This still ill-defined surface of separation between the internal and external medium started acquiring a more specific chemico-physical connotation. Apparently one of the reasons leading Bernstein to develop this particular view of the excitation process was the similarity between the electrical negativity of the action current and that of the injured fiber. It is likely that the injury made the surface negative because it exposed the internal milieu of the fiber toward the external electrode. It seemed thus safe to assume that something similar occurred during the *negative Schwankung*, and that nerve (or muscle) excitation consisted of a reversible "break" in the membrane allowing for the generalized flow of ions resulting in the destruction of the resting potential.

After Bernstein, the progress of the knowledge in this field was mainly due to the work of English physiologists.[3] A fundamental acquisition of the English school concerned the "all-or-none" character of the electrical excitation in nerves and muscles. In primordial form, this character had already emerged with the studies on cardiac excitability performed by Felice Fontana in the eighteenth century (see Chapter 3.1). In 1871, an American physiologist, Henry Pickering Bowditch, who was working in Ludwig's laboratory at Leipzig, made some important observations by studying then isolated apex of a frog heart. He noticed that if an electrical stimulus was just strong enough to produce the

---

[3] Readers interested to the historical development of English physiology in the eighteenth century can consult, among others, Geison (1978).

contraction, a further augmentation of its intensity did not result in any further increase in the strength of contraction. On the other hand, if the intensity of the stimulus was gradually reduced, the contraction did not progressively diminish but remained constant until it disappeared suddenly when the stimulus intensity was less than a given level (referred to as its "threshold" by Bowditch).

For years, however, this "all-or-none" law seemed to belong exclusively to the cardiac muscle. This was because the contraction elicited in skeletal muscles, either by direct stimulation or by stimulation of the motor nerve, increased in a relatively progressive way upon augmenting the stimulus intensity. In 1902, the English physiologist Francis Gotch formulated the hypothesis that this behavior was due to an increase in the number of fibers excited, rather than to the increase of the contractile response in any single fiber. After Gotch, Keith Lucas (who worked in Cambridge) succeeded in demonstrating that, besides the heart fibers, other excitable tissues followed the all-or-none law. In 1905, by an accurate dissection, Lucas reduced the number of active muscle fibers in the frog dorso-cutaneous muscle to about 20 and stimulated the contraction with a direct electrical stimulus applied to the muscle. By increasing the intensity of the stimulus, he found out that the ensuing contraction did increase in a discontinuous way, in small discrete steps, the number of which largely corresponded to the number of active fibers.

In 1909 Lucas performed a similar experiment using an analogous preparation but induced the contraction through a stimulus applied to the motor nerve. Again it appeared that the response augmented in a discontinuous way, along discrete steps, the number of which corresponded in this case to the number of active nerve fibers (Fig. 9.9). Lucas could not complete these studies because he died in 1916 as a consequence of an air accident during World War I. Part of his research was published in 1917 in a volume entitled *The Conduction of Nervous Impulse*. This volume was edited by his student, Edgar Douglas Adrian, a name that would become famous in the history of electrophysiology (see Fig. 9.10).

A more direct demonstration of the "all-or-none" character of nervous signal emerged from the first recording of the electric activity of single nerve fibers obtained by Adrian

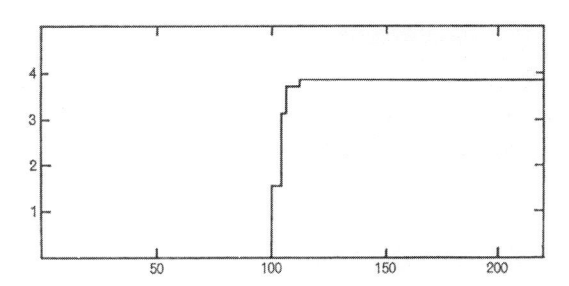

FIGURE 9.9. Lucas's experiment illustrating the "all-or-none" behavior of the contractile response in the muscle fiber. The abscissa represents the intensity of the electrical stimulus applied to the motor nerve and the ordinate the mechanical response of the muscle *cutaneus dorsi* of the frog. Notice the sudden transitions in the ordinate when the stimulus intensity is increased within a critical range of intensities. Every transition corresponds to the contraction of a single muscle fiber (from Lucas, 1909).

FIGURE 9.10. Edgar Douglas Adrian (1889–1977) at the epoch of his first electrophysiological studies (by courtesy of the Library of the Trinity College, Cambridge).

in 1926. In his first study, using the "capillary electrometer" (invented in the previous century by Gabriel Lippmann) and a valve tube to amplify the recorded potentials, Adrian succeeded in recording the electrical oscillations produced by single sensory fibers of both frogs and cats in response to physiological stimuli (Fig. 9.10).

In a second study, made in collaboration with the Swedish physiologist Yngve Zotterman and published in 1926, Adrian investigated the electrical activity of the sterno-cutaneous nerve of the frog, on the assumption that it contained receptors capable of signaling muscle tension. The muscle was carefully sectioned in order to minimize the number of active sensory fibers. Under these conditions the electrical response was found to consist of a train of impulsive deflections, with a constant amplitude and duration. These represented the activity of a single nerve fiber. Increasing stimulus intensity resulted in an increase of the frequency of the impulses, with no change in amplitude and duration (Fig. 9.11). Similar responses were confirmed in a variety of nerve (and muscle) fibers. It appeared thus that a "pulse-code-modulation," based on a variation of impulse frequency, was the way by which nerve fibers usually code information at both sensory and motor levels (Adrian, 1928, 1946; Fig. 9.12).

Another property of the nerve signal had been detected by Lucas and Adrian in a previous experiment. Small segments of the nerve of a frog neuromuscular preparation were more or less completely inactivated through exposure to alcoholic vapors (or by other means capable of interfering with the conduction process). A long exposure time was needed to block nerve conduction when the treated segment was short. A shorter time was required for the block when the treated segment was longer. A particularly interesting thing happened when two short segments of the same nerve were treated. If they were

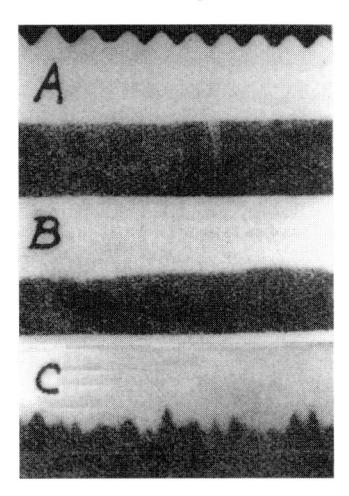

FIGURE 9.11. The first recording of the electric activity of a single nerve fiber obtained by Adrian in a sensory nerve using a sensitive amplifier (capillary electrometer). The fiber innervates mechanical receptors situated in the muscle, which measure the stretching of the muscle. The upper trace in (*A*) is a temporal calibration made with a diapason vibrating at 200 hertz. In (*A*) and (*B*), there are no responses because either the nerve is injured (*A*) or no mechanical stimulus is applied to the muscle. In (*C*) the muscle is being stretched by the application of a suitable weight, thus leading to the appearance of electrical deflections, which correspond to the impulses discharged by single nerve fibers (from Adrian, 1926).

separated by a relatively long untreated portion, it was found that the time needed for the block corresponded closely to that necessary in the case of treatment applied to a single short segment (Fig. 9.13). This behavior was interpreted as evidence that, in the case of partial block, the signal emerging from the first treated segment was capable of regenerating to its full amplitude, before diffusing to the second treated segment downstream.[4]

The words written by Lucas with reference to Adrian's experiments penetrate a fundamental feature of the explanation of nerve signal propagation and thus deserve an ample quotation:

> A disturbance, such as the nervous impulse, which progresses in space must derive the energy of its progression from some source; and we can divide such changes as we know into two classes according to the source from which the energy is derived. One class will consist of those changes which are dependent on the energy supplied to them at their start. An example of this kind is a sound wave or any strain in an elastic medium which depends for its progression on the energy of the blow by which it was initiated [...]. A second class of progressive disturbance

---

[4] Despite its great importance for the progress of physiology, this experiment and its conclusions was strongly criticized by Gen-Ichi Kato, professor of physiology at Keio University of Tokyo. The Adrian-Kato debate had some adventurous aspects, with Kato and collaborators trying to bring, via the Trans-Siberian railways, more than 100 Japanese giant toads from Tokyo to Stockholm for an experimental demonstration (see Piccolino, 2003b).

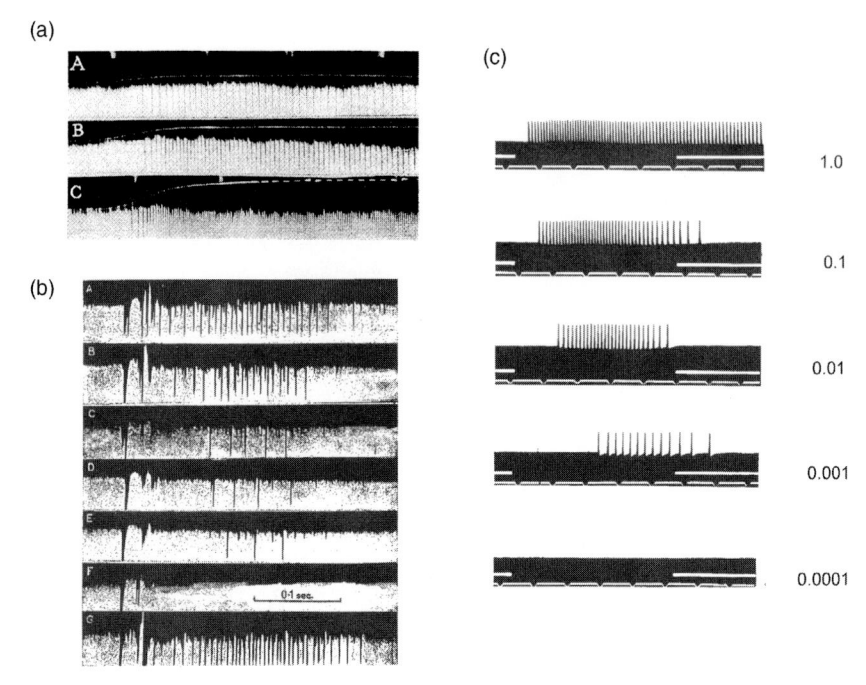

FIGURE 9.12. The typical impulsive electrical responses used by the nerve system for the transmission at distance of the information, obtained in an ancient recording based on the use of extracellular electrodes. In (*A*) and (*B*) the impulses correspond to downward-going deflection of the dark trace, whereas in (*B*) they correspond to the upward deflections. (*A*) Responses induced in the vagus nerve of the cat by lung insufflations with increasing volumes of air (A to C). (*B*) Discharge of nervous impulses coming from the cerebral cortex of the cat and recorded in the fibers of the pyramidal tract, in response to various stimuli, and in different experimental conditions. (*C*) Responses induced in a single fiber of the optical nerve of an arthropod (*Limulus polyphemus*) in responses to light stimuli of intensity progressively decreasing from top to bottom. Notice in all the tracings the possible variation of the impulse frequency while the amplitude remains almost unchanged (from Adrian, 1934 [*A*]; Adrian & Moruzzi, 1939 [*B*]; and Hartline, 1934 [*C*]).

is one which depends for its progression on the energy supplied locally by the disturbance itself. An example of this type is the firing of a train of gunpowder, where the liberation of energy by the chemical change of firing at one point raises the temperature sufficiently to cause the same change at the next point. Suppose that the gunpowder is damp in part of the train; in this part the heat liberated will be partly used in evaporating water, and the temperature rise will be less, so that the progress of the chemical change may even be interrupted; but if the firing does just succeed in passing the damp part, the progress of the change in the dry part beyond will be just the same as though the whole train had been dry. The recovery of the nervous impulse after its reduction in a narcotised tract of nerve suggests that the disturbance transmitted may be of the second type, depending for its progression on the local supply of energy from a source distributed along the nerve fiber. (Lucas, 1917, pp. 23–24)

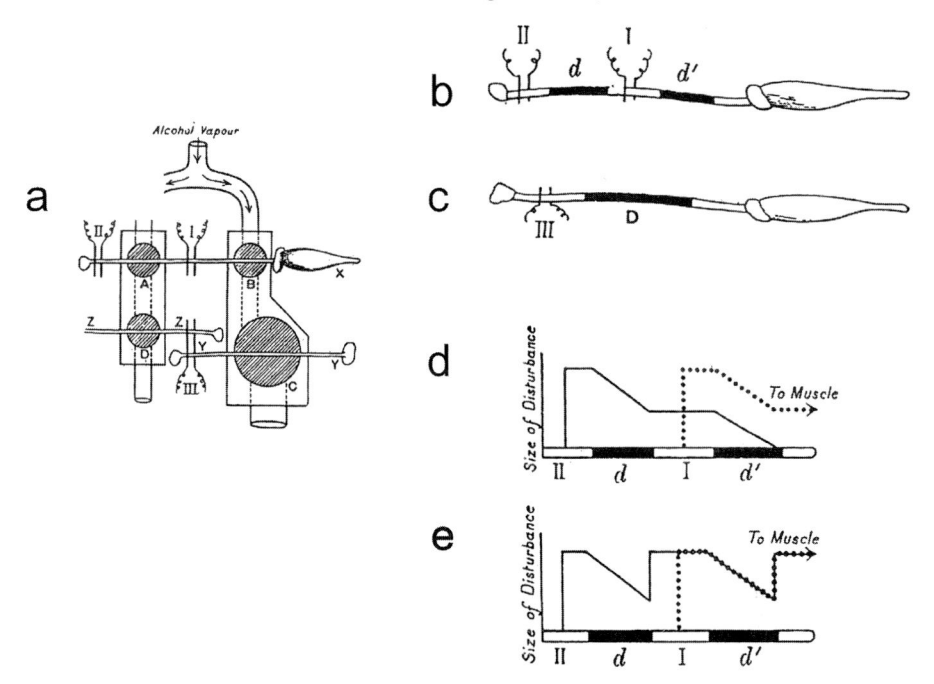

FIGURE 9.13. Adrian's experiment of nerve conduction blockade. (*a*) The apparatus used to apply alcohol vapors (and other narcotics) to segments of the frog nerve. The conduction is monitored by the muscle response produced by electric stimulation in various regions of the nerve as indicated. The apparatus allows for the simultaneous application of alcohol vapors to different segments of the nerve. (*b–c*) Experimental arrangements with the various stimulation electrodes (I, II, and III) situated in different segments of the node. In this and the following diagrams (*d–e*) the zones undergoing the treatment with alcohol are represented in black. (*d*) Scheme illustrating the interpretation of the experiment based on the block applied to two segments separated by an untreated zone, assuming that the signal does not regenerate after passing through the first treated segment (a possibility that Adrian rejected). (*e*) Scheme illustrating the interpretation of the same experiment within the framework of the hypothesis, entertained by Adrian, that the signal was capable of fully regenerating after passing through the first treated segment. (from Adrian, 1912).

One might recall here that the analogy between the explosion of gunpowder and the muscle contraction had been already evoked by Felice Fontana in his elaborations on the theory of irritability, and had probably influenced Galvani in his electric research (Chapter 3.1). It was also invoked by Helmholtz when he published in full form the results of his measurement of nerve conduction speed (Helmholtz, 1850b).

From the work of Lucas and Adrian it was clearly evident that the amplitude and time course of the propagated signal depended in an essential way on the local condition of the nerve in the region where it is recorded (or otherwise assessed). It did not depend, however, on the intensity of the stimulus or on the local conditions in the zone where it was generated (with the only proviso that the stimulus must be strong enough to evoke it and, moreover, that its propagation must not be totally blocked in between).

In Lucas's volume, and also in the initial research of Adrian, the nerve signal was indicated with the noncommittal phrase of "propagated disturbance." This expression served to distinguish it from the local effects induced by electric stimulation near the site of its application. These can be the only effects when the stimulus is below the threshold necessary for inducing the full, propagated signal. The phrase "propagated disturbance" also served to avoid any definite statement about the electrical nature of the nerve signal. There was undoubtedly substantial evidence that this signal was constantly associated with an electric event. This notwithstanding, it was unclear whether electricity represented the fundamental process of nerve conduction, or—on the contrary—it was simply one of its manifestations (as heat production, chemical changes, etc.), that is, an epiphenomenon of a more essential, underlying process.

It might seem surprising that the electrical nature of nerve signals was still a matter of debate, one century after Galvani, and decades after Matteucci and du Bois-Reymond. As a matter of fact, theories of nerve conduction based on chemical processes have challenged the electrical hypotheses for a long time. For instance, in 1946 David Nachmansohn proposed that a chemical substance, acetylcholine, could serve as a "messenger" for nerve signal propagation, and this hypothesis remained in the pathways of science until 1975 (see Nachmansohn & Neumann, 1975).

The unequivocal demonstration that nerve signals are genuinely electrical, as Galvani had supposed at the end of the eighteenth century, had to await the modern era of electrophysiology. This period started with the experiments by Hodgkin and collaborators in Cambridge, which would lead to a complete elucidation of the mechanisms underlying the generation and propagation of nerve signals. The origins of these modern developments can, however, be retraced to the studies of Lucas and Adrian.

Prior to considering some detailed aspects of Hodgkin's research, we must recognize that there were some logical foundations for difficulties in accepting the idea that nerve signals are electrical events. Since the experiments of Helmholtz, it seemed as though the conduction speed of nerve signals was too slow when compared with the flow of electricity along a physical cable. Moreover, as Helmholtz himself had remarked, temperature changes modified the conduction speed more than would be expected for a purely physical process. A strong dependence on temperature is generally considered as evidence of an underlying chemical mechanism. This resulted in the supposition that nerve impulses were basically chemical phenomena. This was so even though the appropriateness of reasoning based on temperature changes was questioned by some authors (see, for instance, Hill, 1921).

From the studies of Adrian, Lucas, and some of their predecessors, it appeared increasingly evident that electrical stimuli below threshold (i.e., weaker than that needed to induce the propagated disturbance) were nonetheless capable of augmenting the excitability in a restricted region near the site of stimulation. By this mechanism, a subthreshold stimulus could lead to the development of a fully propagated event when applied in combination with another subthreshold stimulus. Among other things, this explained why an ensemble of electrical stimuli in rapid succession (a "train" of electrical pulses) could be more effective than a single stimulus in exciting a nerve. These observations suggested that the local responses did not follow the all-or-none law which applied to the propagated response.

The different characteristics of the two response types were also evident from an analysis of the transient modifications in excitability following the application of stimuli of varying intensity. The full response induced by stimuli above threshold was followed by a temporary abolition or reduction of the excitability, a phenomenon called "refractoriness." This contrasted with the local increase of excitability brought about by subthreshold stimuli. Despite the obvious differences between the local and propagated events, there was a clear functional correlation between the two types of phenomena. First, the local response could develop into a full, propagated event once the stimulus was increased to exceed the threshold level. Moreover, if the progress of the full nerve signal was blocked by some treatment applied to a restricted portion of the nerve, a local increase of excitability was usually observed in the region beyond the block. This meant that an event similar to that occurring in the case of local, subthreshold stimulation was associated with the propagation of the fully developed nerve signal, an idea corresponding to the implication of Hermann's "local circuit" theory. In ordinary conditions (i.e., in the absence of the block), the advancing wavefront of the propagated disturbance would be capable of increasing the excitability of the segment downstream to such a point as to stimulate locally the generation of a new propagated disturbance.

9.3 THE INVOLVEMENT OF ANIMAL ELECTRICITY IN NERVE CONDUCTION DEMONSTRATED

The year 1934 can be considered the beginning of the most modern phase of electrophysiology. In this year a student of Trinity College Cambridge, Alan Hodgkin, started a series of studies on the effect of local nerve blockade in the frog neuromuscular preparation (basically the same preparation used by Galvani about 150 years earlier). As was the case for his teacher, Adrian, Hodgkin published the first results of his research in the *Journal of Physiology* at the age of 23 years (i.e., in his case in 1937, exactly two centuries after Galvani's birth). He stimulated the extremity of the sciatic nerve far from its insertion in the muscle and, in between, he produced a local block by cooling a short segment of the nerve. Under these conditions he analyzed the changes of excitability induced in the region downstream to the block (with respect to the direction of nerve signal propagation). Initially he used muscle contraction to monitor the effects of the stimuli and later he recorded the electric potential with an electronic amplifier. The initial purpose of this experiment was to ascertain whether the arrival of a stimulus near the blocked region might be accompanied by a local change in the membrane resistance, detectable after the block, as implied by Bernstein's membrane theory. Hodgkin placed the test electrodes downstream from the blocked region and noticed that subthreshold stimuli might be capable of evoking responses if they were applied in appropriate time relations with the arrival of a propagated impulse from the other side of the block. After rejecting the initial hypothesis based on a possible resistance change, he interpreted the results as evidence of a local increase in excitability. This explanation fit with Hermann's local-circuit theory. It was supported by the measurement of the potential changes observed downstream to the block (Fig. 9.14). These potentials had the characteristics of local responses propagated in a spatially declining way with a geometrical law similar to that implied by the propagation of the increase in excitability.

Hodgkin's observations strengthened the case for a tight relation between the propagation of the full-blown nerve signal and the local diffusion of electrical events in nerve. This relation was confirmed in 1939 by the results of studies performed in both crab and squid nerve in which he studied the change of conduction speed induced by modification of the electrical resistance in the medium surrounding the nerves. He could show that the speed decreased (and the nerve was eventually blocked) when the resistance was progressively decreased (as by immersing it in oil or by reducing drastically the volume of the fluid around the nerve); on the other hand, the speed increased (or was re-established) when the resistance was increased (as by using metallic conductors).

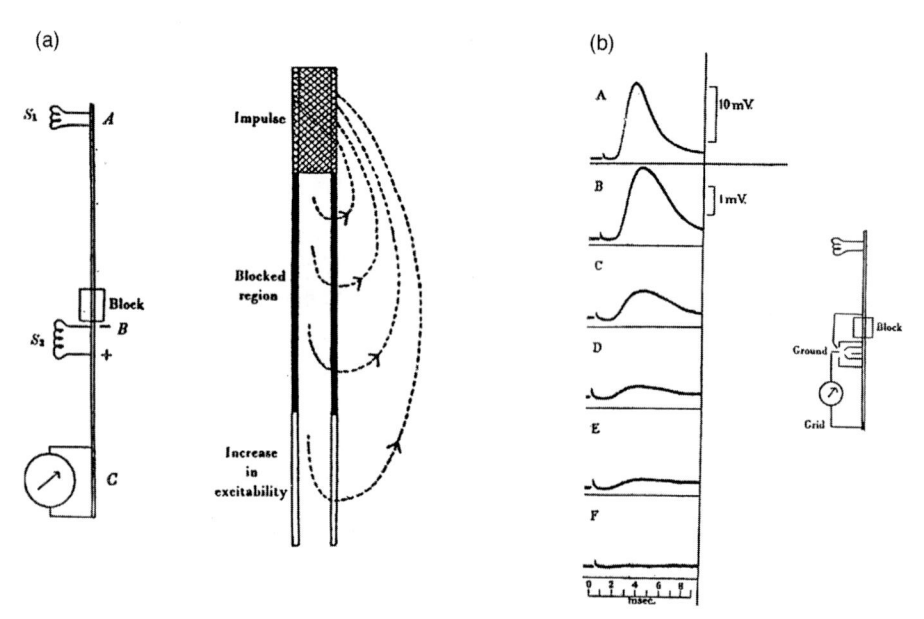

FIGURE 9.14. (*a*) The initial Hodgkin experiment of nerve conduction block by localized cooling in the frog nerve muscle preparation. The increase of excitability in the region distal to the block tested by the electrodes S_2 is observed only if the stimulus is applied with a time delay, with respect to the conditioning stimulus S_1, corresponding to the time of nerve conduction from the sites of stimulation (from Hodgkin, 1992). (*b*) Hodgkin's electric recording of the potential that propagated in a decremental way in a short nerve tract downhill after application of a cold block to a 3.5-mm nerve tract. In this panel, A is the extracellularly recorded action potential proximal to the block; B–F are the potentials distal to the block recorded at an amplification that is about five times greater than in A, at the following distances (from the block): (B) 1.4 mm; (C) 2.5 mm; (D) 4.1 mm; (E) 5.5 mm; (F) 8.3 mm (from Hodgkin, 1937).

These experiments in which Hodgkin used either conductive or insulating substances in order to modify nerve conduction could be considered as the first decisive evidence for the idea that electricity was fundamentally involved in nerve conduction, and that it was not simply an epiphenomenon of an underlying event. As we shall now see, Hodgkin was to provide much more substantial demonstrations of this idea with experiments that would fully confirm Galvani's intuition about the involvement of animal electricity in nerve and muscle excitability.

In a study carried out on the crab nerve and published in 1938, Hodgkin was able to demonstrate that local electrical responses, similar to those observed in blocked nerves, could be detected near the site of application of a subthreshold stimulus. With stimuli that produced a local positivity on nerve surface (i.e., of polarity opposite to those capable of stimulating), the response could be fully accounted for by a passive propagation along the fibers. With negative stimuli the responses were mirror images of these only when using weak stimulus intensities. Stronger stimuli produced responses characterized by a progressively longer queue. By increasing further the intensity these responses eventually developed suddenly in impulsive events of great amplitudes which invaded the entire nerve extension. These results confirmed what was already anticipated by the studies of Lucas and Adrian on the local excitability changes induced by local stimuli in frog nerves.

There was an important problem, however, that remained to be investigated. This concerned the relationship between the local, subthreshold potentials and the full-blown responses. It was also necessary to clarify how the local potential changes associated with the progress of the propagated impulse were involved in the conduction progress. These achievements required a series of progress in the experimental preparations and in electronic technology. These advances were fostered by the "discovery" made in 1936 by John Zachary Young of the existence in the squid of a giant nerve fiber (or axon), particularly suitable because of its size for electrophysiological investigations. Using this preparation in 1939, and by employing extracellular electrodes, Kenneth Cole and Howard Curtis in America were the first to provide a direct demonstration of the decrease in fiber membrane resistance associated with nerve conduction (Fig. 9.15). Moreover, in the same year Hodgkin and Andrew Fielding Huxley in England and Cole and Curtis in America succeeded in inserting an electrode inside the squid axon, thereby measuring directly the transmembrane potential in the resting state and during the excitation.

The existence of a membrane polarization implied in Bernstein's theory was confirmed by these studies (in the experiments of Hodgkin, Huxley, and Katz the interior of the fiber being about 50 millivolts negative with respect to the exterior; Fig. 9.16). It appeared unexpectedly that during excitation the action potential overshot the zero level by several millivolts, the interior thus becoming positive with respect to the exterior. This result was clearly in contrast to the prediction of Bernstein's theory, which stipulated a disappearance of the resting polarization during excitation (and not a potential reversal). It was, on the other hand, in agreement with the observation made by the same Bernstein (and confirmed afterward) that the amplitude of the action current could exceed the

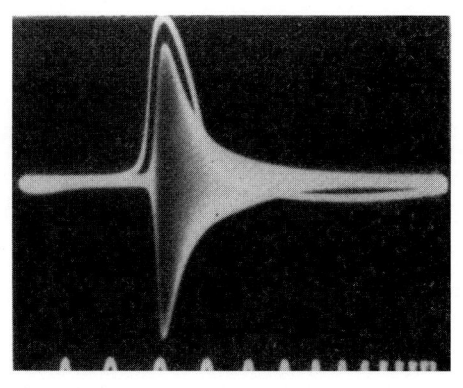

FIGURE 9.15. Evidence for membrane resistance decrease during the discharge of an action potential. Upper white line: action potential; white-dark band: measure of the membrane impedance obtained with the Wheatstone bridge method by applying a high-frequency (20 kHz) sinusoidal signal to two electrodes placed on the opposite site of a giant axon. From a measure of the impedance changes obtained at various frequencies (and proportional to the width of the band) the change of conductance was estimated to be approximately 40 times at the peak of the action potential relative to rest. Time marks: 1 ms apart (from Cole & Curtis, 1939) © 1939 Rockefeller University Press. J. Gen. Physiol. 22:649–670.

negativity of the resting injury current (Bernstein, 1868; Shaefer, 1936). Hodgkin and Huxley were obliged to postpone the interpretation of their experiments because of the events of World War II.

In 1949, Hodgkin and Bernard Katz were able to show that the action potential in the squid nerve decreased in amplitude by decreasing the extracellular sodium concentration. This suggested that nerve impulses were produced by a transient increase in the

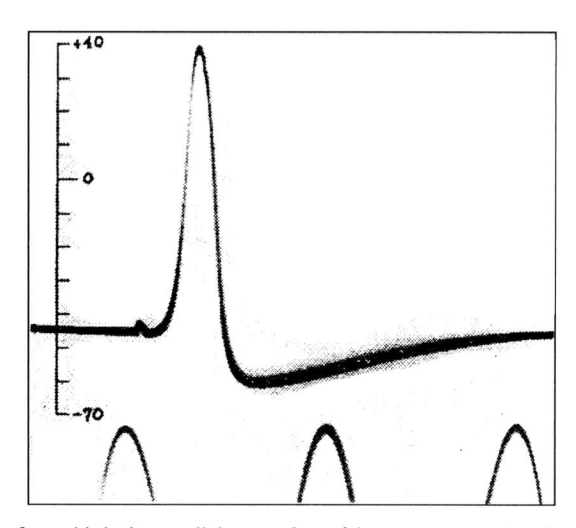

FIGURE 9.16. The first published intracellular recording of the action potential in the squid axon. Time course of the difference between the internal and external potential, in the resting state and during the discharge of an action potential. Time mark, 500 Hz. Notice the large positive overshoot of the membrane potential during the action potential, which contrasted with the expectation of the Bernstein's theory (from Hodgkin & Huxley, 1939).

sodium permeability of the membrane (Fig. 9.17). As we shall see in the next chapter, the membrane potential, which in the resting state depends heavily on potassium ions, became dominated by sodium during excitation. This is the consequence of a specific increase of membrane permeability to sodium induced by the electrical stimulus. During the action potential the membrane becomes much more permeable to sodium than to potassium. This represents an important revision to Bernstein's theory based on the assumption that, during the excitation process, the membrane would become unselectively permeable to all ions.

The new theory was based on the transition between two states of permeability of the membrane that could account for the electrical processes underlying nerve excitation. To verify the theory it was necessary to study the changes in membrane permeability and

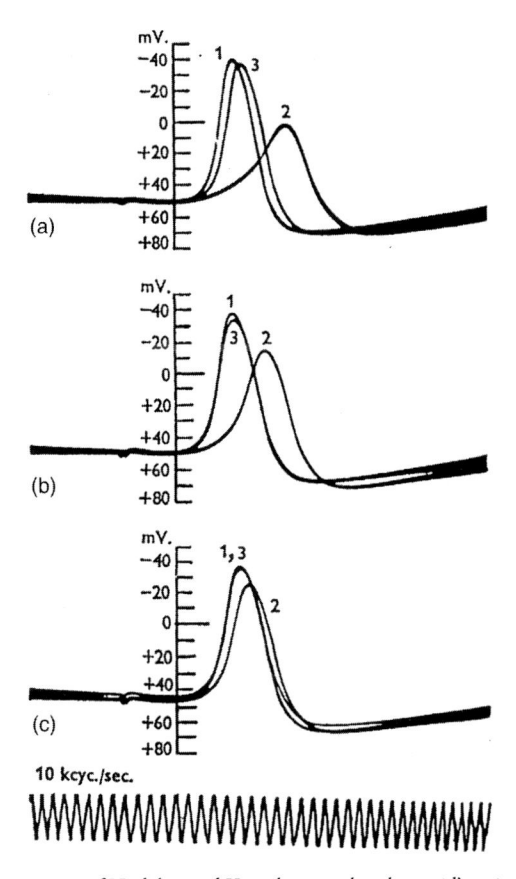

FIGURE 9.17. The experiment of Hodgkin and Katz showing that the squid's action potential decreases in amplitude if the extracellular concentration of sodium is reduced. This finding was interpreted as evidence suggesting that an influx of sodium is fundamental to the nerve event. In any panel, 1 is the control recording, 2 the tracing obtained in low sodium concentration, and 3 the tracing recorded after returning the preparation to the normal ionic conditions. In (a) the sodium concentration was reduced to 33% of the concentration in the normal perfusion liquid (seawater), in (b) to the 50%, and in (c) to 71% (from Hodgkin & Katz, 1949).

electrical processes accompanying the development of the action potential in precise and quantitative ways. A study of this type was difficult using conventional techniques. This was because the "regenerative" or "explosive" character of the action potential did not enable the study of the relation between current and voltage during the excitation process. These difficulties were similar to those encountered by a chemist wishing to analyze the dynamics of a chemical process underlying the explosion of a mass of gunpowder by simply triggering the process with fire.

To obtain accurate information on the electrical events underlying the genesis of the action potential, it was necessary to subdue in some way the explosive character of this phenomenon. It was, in other words, necessary to constrain the membrane so that it assumed and maintained various defined potential levels, despite its tendency to change the potential rapidly once the threshold was exceeded. The long-sought-after technique capable of allowing for such studies was the "voltage clamp," devised in 1949 by Cole and George Marmont. It was applied soon afterward, in a masterly way, to the giant axon by Hodgkin and Huxley, partially in collaboration with Katz. The results of their studies, published in a series of articles appearing in 1952, represent a milestone in the research on animal electricity (Hodgkin & Huxley, 1952a-d; Hodgkin, Huxley, & Katz, 1952).

By using the voltage-clamp technique, Hodgkin and Huxley provided the decisive, unequivocal demonstration that nerve signals are genuinely electrical events or, as Hodgkin expressed it in his Nobel Lecture delivered in 1963, "the action potential is not just an electrical sign of the impulse, but is the causal agent in propagation" (see Hodgkin, 1972). From Hodgkin and Huxley's studies, the implication of electricity in the generation of nerve signals would appear under a double aspect. On the one hand, the action potential is produced by the discharge of a particular form of electrical energy accumulated at the two sides of the nerve membrane; an energy that is produced as a consequence of physiological processes which clearly belong to the life domain (and are thus "animal" in Galvani's sense). On the other hand, a sudden change of the membrane potential is needed in order to initiate the complex events eventually resulting in the discharge of the energy accumulated by the membrane.

This emerged clearly from the comparison of the membrane current evoked by voltage pulses of opposite polarity applied to the squid giant axon with the voltage-clamp technique. The currents induced by "hyperpolarizing" stimuli (i.e., by stimuli that increase the internal negativity of the axon) were found to be of small amplitude and fully accountable for by simple charge movements caused by the applied stimuli. These currents were directed inward, meaning that positive charges entered the membrane toward the intracellular compartment, as could be expected as a consequence of the increased intracellular negativity. Their amplitude and time course corresponded to calculation based on the purely physical properties of the membrane, that is, its electric resistance and capacitance.

A different thing happened with depolarizing stimuli, that is, with stimuli that made the internal potential less negative (or eventually positive), particularly when these stimuli were of an amplitude which would be above threshold in the absence of the voltage

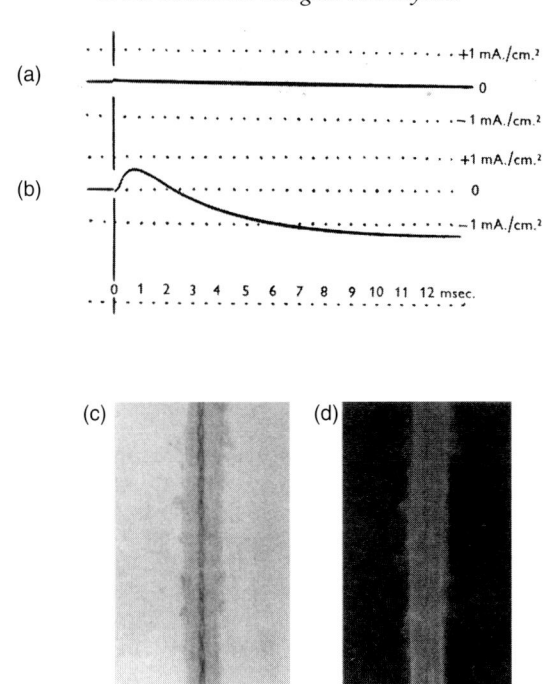

FIGURE 9.18. Voltage-clamp currents elicited in squid axon by a hyperpolarizing (*a*) and a depolarizing (*b*) voltage step. Contrary to the current conventions, in both tracings an upward deflection indicates a current entering into the cell, while a downward deflection indicates a current flow from the inside to the outside of the cell. In the upper panel the potential is displaced from the resting value of −65 to −130 mV, and this results in a very small inward flow of current (almost undetectable at the used amplification). In the lower panel the voltage is displaced from −65 to 0 mV, resulting in a large biphasic current with an early, inward-directed component carried by sodium ions that enter the cell, and in a late, outside-directed component corresponding to the outflow of potassium ions; (*c* and *d*) two photomicrographs of the squid axon penetrated with intracellular metallic electrodes for voltage-clamp experiments, obtained with transmitted light or dark-field illumination, respectively (from Hodgkin, Huxley, & Katz, 1952).

clamp. The initial phase of the responses was then dominated by an inward current. The direction of this current could not be accounted for by the simple movement of charges induced by the change of membrane potential. As a matter of fact, a movement toward the positivity of the intracellular potential should produce a movement of the positive charges from the intracellular to the extracellular compartments, that is, an outward current. The fact that in these conditions the current was inward indicated that it was not simply and directly produced by the modification of the electric field across the membrane caused by the stimulus. It was instead the consequence of modifications of the membrane properties induced by the depolarizing stimulus which brought about a movement of electric charge under the action of *a preexisting energy gradient*. This gradient originated from the metabolic activity of the cell. The membrane changes consisted of an increase in the permeability to sodium ions, allowing the entry of these ions under the action of their electrochemical gradient.

As we shall see in detail in the next chapter, sodium ions have a tendency to enter into the cell because they are more concentrated in the extracellular compartment compared to the cell interior (by about 10 times), and also because at rest the membrane is hyperpolarized, that is, the intracellular potential is negative (sodium ions carry a positive charge and have therefore a tendency to go toward negative potentials). In the resting state, however, they cannot enter the cell because of the relative membrane impermeability to their passage. To enable their entry, the membrane needs to become permeable to them. This is what happens when the intracellular potential moves toward a positive potential (i.e., when the membrane gets depolarized). Hodgkin and Huxley assumed that the depolarization changed the permeability to sodium ions by modifying the state of some membrane particles capable of regulating its capability to allow for the passage of these ions. This process was referred to by them as membrane "activation" for sodium. The influence of the transmembrane potential on the permeability appeared not to be instantaneous, requiring a definite—although short—time. In the terminology of biophysical studies on membrane excitability, sodium permeability would be indicated as "voltage dependent" and "time dependent."

A key feature in the control of membrane permeability to sodium emerged from the Hodgkin and Huxley studies. The entry of sodium due to the permeability change induced by depolarization makes the cell interior more depolarized because it involves an influx of positive charges. In turn, this depolarization makes the membrane more permeable to sodium by acting on the intramembrane particle regulating sodium permeability. This results in a further increase in permeability and consequently in a further depolarization. This process, to be dealt with in some detail in the next chapter, accounts for the all-or-none character of the nerve signal and for the threshold behavior of excitability to electrical stimuli. In this view the threshold would be the level at which the inward current carried by sodium ions exceeds the outward movement of positive charge induced by the depolarizing stimulus per se (Hodgkin, 1951). This "autoregenerative" cycle controlling the entry of sodium ions (called the Hodgkin cycle) is at the heart of membrane excitability. It accounts for the double electrical dependency of nerve signal generation: a signal that is generated by the release of a preexisting electrical energy (actually an electrochemical energy) and which, moreover, requires an electrical influence for the electrical energy to be put in action.

The electrical behavior of the nerve-fiber membrane during the generation and conduction of nerve signals is accurately described by the model developed by Hodgkin and Huxley ("Hodgkin-Huxley" or simply "HH-model") in all its fundamental aspects. It accounts for the all-or-none character of nerve impulses. It describes in a satisfactory way the time course of the local potential and of the propagated impulse and, moreover, of the change in membrane resistance and permeability to various ions during the excitation process. Furthermore, it explains the strong temperature dependency of this process (already noticed by Helmholtz) as consequence of the great influence that temperature exerts on the membrane mechanisms controlling the ionic currents involved in the generation of the action potential (these mechanisms were later referred to as membrane *gating*).

Many of the prediction of the HH-model have been confirmed by subsequent studies. The question left relatively unresolved by the studies of the two English physiologists concerned the mechanism of ionic permeation across the membrane. The hypotheses oscillated between two possibilities. One assumed permeation through microscopic devices similar to the pores of a sieve; the other was based on the presence of specific intramembrane particles capable of transporting the different ions. The research carried out after Hodgkin and Huxley had clarified the mechanisms of ionic permeation involved in membrane excitability with a resolution at a molecular level. This represented probably the most important achievement of contemporary electrophysiological studies. The success of these studies has made possible a new, extraordinary technique allowing for the recording of the elementary electric events of the membrane; this is the *patch-clamp* technique invented in 1976 by Erwin Neher and Bert Sakmann. These events are the minute currents that pass across a single molecule of the permeation mechanism present in the membrane. This technique has produced a true revolution in the progress of electrophysiology, which can probably be compared to that produced about two centuries ago by Galvani with his prepared frogs (or more recently to that connected to the invention of the voltage clamp).

The *patch clamp* consists of bringing a glass electrode with a relatively large size (about 1 micron tip diameter) in a very tight apposition to the external surface of the membrane. With tight apposition, called "seal," the current passing through the microscopic membrane patch is forced to flow through the electrodes, and through the amplifying and recording apparatus (Fig. 9.19).

Patch-clamp studies have definitely proved that membrane currents involved in the processes of electrical excitability are due to ions moving along their electrochemical gradients across molecular structures embedded in the membrane. As we shall see in the next chapter, these structures, called "ionic channels," are complex protein molecules characterized by the presence in their core of a aqueous pore allowing for the passage of ions, notwithstanding the hydrophobic characteristics of the membrane lipid composition. The most relevant event in recent studies of ionic channels has been the high-resolution X-ray analysis of the channel achieved by Roderick MacKinnon and collaborators in their study of a channel permeable to potassium ions (Doyle et al., 1998; Jang et al., 2003). This has led to an understanding of the three-dimensional molecular structure of the channel and of the permeation mechanism.

From the etymological viewpoint the term *channel*, being akin to canal, induces the image of "water pathway," capable of allowing for the transit of mobile elements. It also connotes ducts that impose a mechanical restriction to this transit. Channel and canal have indeed the same origin in the Greek term καννα (which in turn derives from the Akkadian *qanû*, also meaning "reed," "narrow tube"). Indeed, one of the limitations imposed on ion flow by these channels is mechanical, and this explains why their

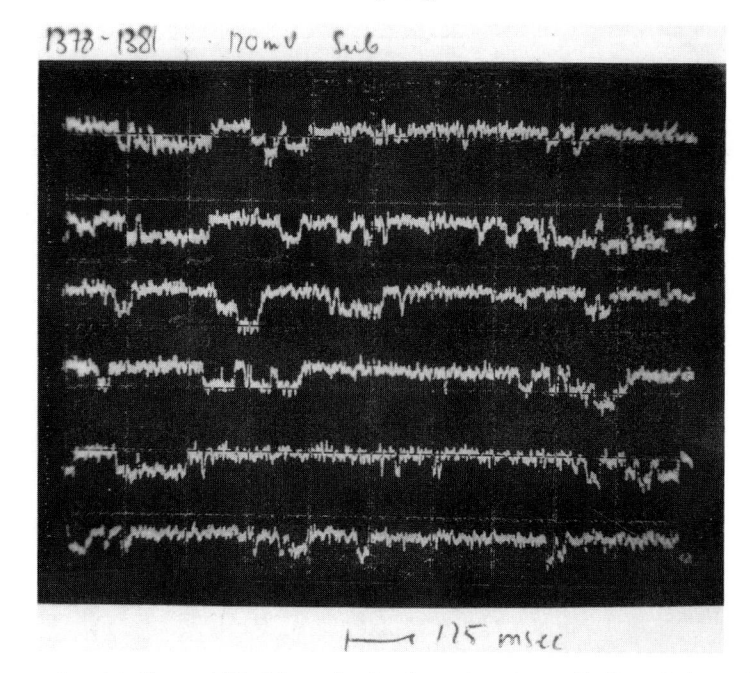

FIGURE 9.19. An originally unpublished figure showing the tracings obtained by Erwin Neher and Bert Sakmann on November 8, 1975, which were considered by them as their first clear evidence of single-channel currents (by courtesy of Erwin Neher).

dimensions are of the order of magnitude of the permeating ions. Because of this, ions bigger than a given dimension cannot pass across the membrane, rather like the pores of a sieve. On the other hand, an important aspect of the selectivity of the different channels for the various ions has to do with the need that ions have of losing, during their passage, the water accompanying them (see Sakmann & Neher, 1995).

With the notions acquired during this historical outline of electrophysiological progress after Galvani and Volta, we are now prepared to understand the biophysical processes responsible for the generation of electrical signals in nerves (and in muscles as well).

10

Neuromuscular Excitability

THE MODERN EXPLANATION

THE RAPID OUTLINE of the electrophysiological development since Galvani and Volta, presented in the previous chapter, simplifies the task that we are tackling now, that is, to illustrate the present understanding of nerve conduction and muscle excitability. These were the problems at the heart of the discussion of the two eighteenth-century scholars.

In our exposition of modern physiological knowledge, we will try—as far as is possible—to limit the use of specialized terms, and we will not assume that the reader has a specific familiarity with the scientific problems we are considering. We are aware that this will oblige us to deal with matters in a way which might seem to some excessively elementary. Indeed, our purpose is to address ourselves to readers provided with a general background but lacking a specialized and technical knowledge of the field. As discussed at the beginning of this volume, we would like to make science comprehensible to a large readership, with the aim of recovering that unit of culture that existed in olden times, before the separation into the "two cultures" of our present tradition. This was so at the time of Galileo, who called the book in which he was expounding his conception of the universe a "poem." This was still so at the time of Goethe (and thus of Galvani and Volta), when the great German poet declared that he considered his *Theory of Colours* and his discoveries in the field of comparative anatomy more important than his *Faust*.

10.1 CELL MEMBRANE AND IONS: A MACHINE GENERATING ELECTRIC POTENTIALS

The crucial element in the genesis of electric potentials in living organisms is represented by the plasma membrane (or cell membrane), the thin structure that encloses the internal (or intracellular) compartment of the cell, thus separating it from the extracellular part. In the cell membrane there is indeed the complex molecular machinery responsible for the electrical phenomena of living beings, and particularly for the processes underlying the excitation and conduction of electrical signals in nerve and muscle.

The membrane has a thickness of about 7 nanometers (one nanometer corresponds to one-billionth of a meter and is generally indicated as 10^{-9} meters). It is therefore invisible to the optical microscope, which normally does not resolve structures below 0.2 micrometers (one micrometer being one-millionth of a meter). The cell membrane is made up predominantly of lipids, that is, fatty materials. Lipids have generally a very low affinity for water (they are hydrophobic). The adaptive reason for this can easily be understood by taking into account that both intra- and extracellular compartments are basically two water milieus. If membranes were hydrophilic (i.e., had affinity for water), water and the various substances dissolved in it could pass easily through them. This would make it impossible to maintain the difference in the composition between the intracellular and the extracellular compartments, with the consequence that cell mechanisms and eventually life itself would become impossible. The development of life on our planet has become possible through the development of membranes. These were necessary in order to maintain the new evolutionary acquisitions that were being produced through mutational processes since the appearance of the primordial forms of life (self-replicating proteins or nucleic acids) in low-salinity seas billions of years ago. In this way the "advantageous" evolutionary achievements could form the basis for new developments capable of making the organisms more able to survive (and generally more complex).

The intracellular, membrane-delimited space contains various organelles, some of which (like the cell nucleus, mitochondria, the endoplasmic reticulum, and the Golgi system) are also enclosed by membranes with specific characteristics. The milieu in which these structures are located (the so-called citosol) is basically a water solution with a saline composition substantially different from the extracellular medium.

As indicated earlier, a fundamental function of the cell membrane is to "separate" and thus keep the differences between the internal and external media; this is essential for the maintenance of life and for evolution. On the other hand, the processes of life require reciprocal exchanges between the two compartments. This exchange must make possible the inward movement of nutrients, of oxygen, and of all other materials necessary for the energetic and structural processes; it must also allow for the elimination of the waste products of cell metabolism. Moreover, in many living beings, cell membranes must allow for the passage (in one or other direction) of the information necessary for the interaction of the cell or the organism.

There are various mechanisms allowing for such transport through membranes that were initially developed for preventing the permeation of water, solutes, and other substances. These mechanisms are based on structures of variable complexity, and they are generally selective, in the sense that they allow for the preferential or exclusive passage of specific substances.

We will not enter into the details whereby all these mechanisms operate. We will limit ourselves to considering two main types of membrane structures that underlie bioelectric phenomena by allowing the passage of ions (i.e., of particles bearing one or more positive or negative electrical charges) between the internal and external media. The most important ions for the genesis of membrane electrical phenomena are sodium (Na^+), potassium (K^+), and chloride (Cl^-); the first two are cations (they bear a positive charge as indicated by the + superscript), whereas the third is an anion (it bears a negative charge as indicated by the − notation). Another ion that plays an important role in the processes connected more or less directly to the electrical phenomena of the excitable cells is calcium (Ca^{2+}), a divalent cation (i.e., an ion bearing two positive charges). Despite its importance, we will not deal with the role of calcium ions in bioelectric phenomena due to the space limits on our discussion of this topic.

The first structure that we will examine is rather complex, on both the side of its structural organization and of its function. We should nevertheless describe it first because in some ways it precedes the other, both logically and operatively. It is called a sodium-potassium pump (Na^+-K^+ pump) or, alternatively (Na^+-K^+ ATPase). We will understand the meaning of these two expressions by describing the role that the structure plays in the cell membrane. Let us say first that the main difference in the ionic composition between the intracellular and the extracellular compartment concerns the distribution of Na^+ and K^+. The intracellular compartment is rich in K^+ and poor in Na^+, whereas the opposite is the case for the extracellular medium. These differences are due to the action of the Na^+-K^+ pump. As implied by one of the terms used to indicate it, this "pump" forces the movement of both sodium and potassium across the membrane. Sodium is pumped out, whereas potassium is pumped in. The Na^+-K^+ pump is a real molecular machine and, as it occurs in all machines, it needs energy for its functioning. The pump receives its energy via a molecule endowed with abundant chemical energy, adenosine triphosphate or ATP. The pump draws energy from ATP by breaking one of the two high-energy phosphate bonds, which results in the transformation of ATP in ADP, that is, adenosine diphosphate. This process accounts for the second name of the pump, that is, Na^+-K^+ ATPase (ATPase meaning the capability to break down ATP through and hydrolytic mechanism). In its normal function the pump transports three Na^+ ions to the outside while it transports two K^+ inside. Due to this process, there is plentiful sodium in the extracellular medium and the cell interior has an abundance of potassium, which accounts for the difference in the composition of these ions between the two compartments. The energy need of the pump is due to the fact that it must work in an "uphill" direction, forcing ions to go against their gradient, from the medium where they are less concentrated toward the medium where they are more concentrated.

A particularly good example of the pump's operations comes from experiments in which the cell permeability of red blood cells is increased artificially. Ions and other components of the intracellular and extracellular compartments can thus freely pass across the membrane with the result that the cell loses potassium and gains sodium with, eventually, the complete disappearance of any concentration difference. If, however, red blood cells are allowed to repair their membrane and receive adequate energy supplies (for instance, by adding ATP in the extracellular medium), then after a short delay the gradient in the concentrations of both sodium and potassium is re-established.[1]

The Na^+-K^+ pump is present practically in all cells of the organism. In nerve fibers it is particularly active, since a large part of the fiber energy (between 50% and 70%) is normally consumed by the pump to create and maintain the physiological ion concentrations at the two sides of the membrane. Given the importance of its role, one could expect that an alteration of pump function could significantly affect fundamental cell processes. This is indeed the case. In fact, some drugs act by modulating the activity of this pump, like some of those used in the treatment of heart failure (among them the derivatives of digitalis or strophanthus). It would be difficult to explain how an interference with the activity of Na^+-K^+ pump could produce beneficial effects on heart function. Interested readers are referred to some suitable treatises on physiology or biophysics (as, for instance, Hille, 2001). They would enjoy other intellectual pleasures in addition to discovering the reasons for such seemingly arcane processes. In particular, they would learn that a question posed centuries ago in the history of medicine (what is the action mechanism of digitalis and similar cardio-active drugs) has been answered by laboratory research conducted with a totally different aim: to ascertain how cells regulate their ionic composition.

Let us now confine ourselves to considering how the pump intervenes in the genesis of membrane electrical phenomena. As a matter of fact, the pump produces a small change in membrane potential by itself. This is because on every cycle the cell loses a positive charge and thus its internal potential becomes more negative (this is because three Na^+ ions go out while only two Na^+ ions go in). This effect is, however, modest, and generally it contributes only minimally to the genesis of the bioelectric potential. The membrane potential difference, which is in the order of about 70 millivolts (a millivolts being one-thousandth of a volt, the unit of measure of electric potential, thus indicated in honor of Alessandro Volta), is to a large extent only an indirect consequence of pump activity.

---

[1] The results of these experiments, first carried out by the Hungarian biochemist George Gardos, have had great theoretical as well as practical relevance. For instance, in the blood kept at low temperature the red cells tend progressively to lose potassium and to gain sodium. As a consequence, the blood becomes unsuitable for transfusion, because they would be broken down if injected in the circulation of the patient with serious consequences. To re-establish their normal ionic concentrations, it is necessary, prior to the transfusion, to incubate the blood at physiological temperature in the presence of energy-rich substrates.

To understand how this happens, we need to consider the second mechanism whereby ions can pass through the membrane, that is, that represented by the ionic channels, already alluded to in the previous chapter. These channels are conceptually and operatively simpler than the pump, in as far as they behave as aqueous pores of sizes comparable to those of the ions that they allow to pass through. The pores are lined with an internal hydrophilic coating, which enables the bidirectional passage of the ions, according to their gradients (i.e., according to their concentration difference and to the influence of the electric potential).

In cell membrane there are various types of ionic channels. They are usually classified on the basis of the ions they allow to permeate and, moreover, with relation to the mechanism controlling their opening. Recent biophysical studies, combined with advances in molecular biology techniques, have unveiled the great variety of ionic channels present in the membranes of nerve cells, as well in other types of cells in living organisms (both animal and vegetal). We will confine ourselves to considering exclusively two types of channels that are permeable to potassium ions and one type of channel that is permeable to sodium ions. One of the two channels permeable to K^+ is normally open in the resting membrane (the leak K^+ channel), whereas the other opens only when the membrane gets excited. Also the channel permeable to Na^+ opens up exclusively when the membrane is excited, and it does it following a very peculiar time course that is of outstanding importance for the kinetics of nerve signals.

Let us now consider the behavior of membranes in the resting state, a condition in which only a type of K^+ channel is open, the so-called leak channel. Because of the pump action, the cell has an excess of K^+ inside and an excess of Na^+ outside. As a consequence of their larger external concentration, Na^+ ions have a tendency to enter, but they cannot do so because in the resting state the Na^+ channels are closed. In the same condition K^+ can go out because the leak K^+ channel is open and K^+ spread outward through these channels. Let us now examine what happens by following, step by step, the outward passage of K^+. The outward K^+ results in a loss of positive charge from the intracellular to the extracellular medium, in addition to a very modest increase of extracellular K^+ concentration (and a parallel decrease of the intracellular concentration). Because of this loss, as the process goes on, the intracellular compartment becomes negatively charged with respect to the extracellular medium. In other words, the intracellular potential would become increasingly negative with respect to the extracellular medium. The change in potential across the membrane brought about by this charge movement would depend on the quantity of charge going outward and on a physical characteristic of the membrane corresponding to its electrical capacity according to the formula already elaborated by Volta for the physical capacitors:

$$T = \frac{Q}{C}$$

where T is the electrical tension, Q the charge, and C the capacity. The formula is now commonly written in a slightly different way:

$$V = \frac{Q}{C}$$

where the tension is substituted by the potential indicated in V (volts).

However, the progressive buildup of a difference of electrical potential between the interior and the exterior of the cell would tend to slow down, and eventually to block, the further outward passage of K^+ because of the positive charges carried by potassium ions. In other words, the creation of an inward negativity would produce an influence opposite to that consequent to the concentration gradient on the movement of K^+. Because of this, K^+ would tend to go out, while due to the intracellular negativity, K^+ would tend to go in. We can expect that eventually an equilibrium will be reached between the two opposite tendencies, with the result that no net flux of K^+ would occur across the membrane. This is just what happens when the difference in potential and the difference in the K^+ concentration would eventually be in a relation defined by a fundamental law of electrochemistry, the law of Nernst, so indicated from the name of the German scientist who developed the theory of electrochemical potentials in the nineteenth century:

$$V = \frac{RT}{F} \ln \frac{[K^+]_e}{[K^+]_i}.$$

As mentioned in the previous chapter, Bernstein was the first to formulate the hypothesis that the bioelectric potentials were generated as a consequence of the selective permeability of the resting membrane to the potassium ions which were at a higher concentration inside the cell in comparison to the extracellular compartment. He applied the laws of the electrochemical potential developed by Nernst to his "membrane theory" (Chapter 9.2).

Let us know examine the Nernst law, by defining, first, the various terms of the relation. V is the intracellular potential, R is the gas constant, and T is the absolute temperature (i.e., the temperature measure in degrees Kelvin with respect to absolute zero, i.e., about—273.16°C). The operator ln stands for natural logarithm (i.e., the logarithm on base 2.72). F is the Faraday constant, which corresponds to the charge of a mole of monovalent ions, meaning ions that bear, like K^+ and Na^+, a single charge. In the Nernst law $[K^+]e$ and $[K^+]i$ represent, respectively, the external and the internal concentration of K^+.

Without going into too much detail, we will try now to grasp, at least at an intuitive level, the reason why the relation between the intracellular potential and the transmembrane potassium concentration has this form and involves these parameters.

Nernst's law establishes a relation between an electrical potential (which is a measure of the electrical forces acting on charged particles) and a chemical potential (the potential which derives from the diffusion of particles moving under the action of thermal energy). Thermal energy depends on the temperature, thus accounting for the presence of T in the equation. It can easily be understood that if there were no thermal energy (which theoretically occurs at absolute zero), there would be no chemical energy. In our particular situation, potassium ions would have no tendency to spread out, and consequently there would be no potential difference across the membrane. Gas constant R is a measure of the energy gained by a gas (or by a solution) when the temperature is increased by one degree Kelvin. This constant is referred to as a standard quantity of a gas (or solute), that is, a mole, which means to a number of particles corresponding to the Avogadro number (a very big number, 6.022×10^{23}, something like 600,000 billions of billions of particles).

Now let us try to understand the meaning of F, Faraday's constant. This is the measure of an electrical unit, and precisely the measure of the electric charge carried by a mole of particles each bearing a single charge (for example, a mole of monovalent ions like K^+ and Na^+). It results from multiplying the elementary charge e by the Avogadro number, leading to the figure of 9.648×10^4 Coulombs per mole (i.e., about 100,000 Coulombs per mole). This is a huge electrical charge, such that if distributed over the Earth's surface it would increase the Earth potential by about 140,000 volts.

As mentioned the Nernst law establishes a relation between an electric potential and a chemical potential. The larger the chemical potential (and at a given temperature this would depend on the ion gradient concentration), the larger the electric potential would be, according to the particular logarithmic relation defined by the Nernst formula. At equilibrium, the chemical force that acts on a particle (or on a defined number of particles such as those contained in a mole) would be balanced by the electrical force. This latter force depends on electrical potential difference and on the charge of the particle (or of a mole of particles). For a given electrical potential, the force acting on a particle (or a mole of particles) would depend on the charge of the particles. This accounts for why there is the Faraday constant in the denominator of the Nernst formula.

It is less intuitive to explain why the relation between the electrical potential and the ratio of transmembrane concentration has a logarithmic form. We limit ourselves to saying that, because of the properties of the logarithmic function, the electric potential augments in an additive manner when the concentration ratio increases in a multiplicative way. To give an example, at physiological temperatures the potential difference would increase by about 60 mV for a 10-fold augmentation of the ratio of monovalent ions. We should thus expect an intracellular potential of about –60 mV for a 10-fold ratio of potassium concentration, of –120 mV for a 100-fold ratio, and of 180 mV for a 1,000-fold ratio.

The measurement of the resting intracellular potential by various techniques shows that the internal side of the membrane is negative and the value of the potential corresponds rather closely to that predicted by Nernst's law. The potential measured in most nerve cells is around –70 mV, while that calculated on the basis of Nernst's law is between –75 and –80 mV. The correspondence is fairly good, even though the difference between measured and calculated potential is systematic and always in the sense that the predicted potential is more negative than the measured one. We will see later the reasons for this difference.

We will neglect this difference for a moment and assume a perfect correspondence between the experimental and the theoretical values, that is, those predicted by Nernst's law. This being the case, potassium ions would be in a condition of equilibrium. In other words, there would be no net movement of potassium across the membrane because the electrical forces acting on these ions would be exactly counteracted by chemical forces.

Viewed in an historical perspective, the Nernst equilibrium might serve to elucidate a problem that appeared somewhat mysterious to the eighteenth-century physiologists. How is it possible that an electrical disequilibrium could exist inside animal bodies, if body liquids are electrically conductive and thus capable of dissipating any disequilibrium that would possibly arise?

In point of fact, the electrical disequilibrium existing at the two sites of the membrane is necessary to keep a more global equilibrium, due to the unequal distribution of potassium ions at the two sides of the membrane. If, by an accidental and temporary perturbation the electrical difference were to disappear (or to change in some other way), the electrical potential would soon be re-established to counteract the chemical difference. This would happen because—as previously described—the tendency of potassium ion to go outward would not be hindered by electrical forces until the conditions of the Nernst equilibrium were created.

Until now we have rapidly outlined the functioning of the complex "machine" responsible for the generation of resting membrane potential, based on the interplay of the action of the pump and ionic movement through membrane channels. We are now in a better position to understand how, as a consequence of a sudden change of the membrane properties, the electrical potential of excitable fibers would be modified with the development of the action potential.

Before doing this, let us make some historical remarks. The existence of electrical differences at the two sites of the cell membrane, due to the Nernst equilibrium (as first suggested by Bernstein), seems to confirm the ingenuity of Galvani's hypothesis of animal electricity, that is, of an electricity present in a state of imbalance between the interior and the exterior of excitable fibers. On the other hand, the electrochemical potential as conceived by Nernst is the potential of a particular Voltaic battery, called concentration battery (or concentration cell). It corresponds to the battery that Volta hypothesized when he attempted to produce an electrical effect by the contact of two liquid conductors, in the absence of any metal. In a somewhat ironic way, Galvani's supposition of an electrical machine responsible for animal electricity would be found to be based on a functioning principle issued as an historical consequence of Volta's research on the electromotive action of dissimilar conductors.

The consideration of the electrochemical mechanisms underlying the resting potential of excitable fibers is a prerequisite for understanding the processes of the generation of electrical signals that encode the flow of nerve information in animal organisms. In the resting state, the membrane is in a condition of quasi-equilibrium for the distribution of potassium ions. In the same conditions, the situation of sodium ions is quite different, because the tendency of sodium to enter into the cell because of the concentration gradient is not counteracted by a suitable electrical difference. On the contrary, the negative potential of the intracellular compartment accentuates the tendency of the positively charged sodium ions to flow inward. Despite this, sodium cannot enter into the cell to any substantial degree because the membrane is almost completely impermeable to their passage.

The electrochemical disequilibrium for sodium ions existing at the two sides of the membrane represents the fundamental form of energy allowing for the generation and propagation of the nerve signal. This energy is distributed along the entire length of the nerve fibers, and a similar condition exists also in muscle fibers. This is necessary in order to allow for the propagation of the electric signal despite the physical difficulty opposed to this process by the longitudinal electrical resistance that can be extremely high in some nerve fibers. We remark here en passant that, even though the propagation of the electrical impulse is basically similar in nerve and muscle fibers, there is indeed no direct passage of the electrical excitation from nerve to muscle. In this particular case (as it also normally occurs at the contact between different nerve cells), the passage of the signal is due to a chemical substance (or "transmitter"), accumulated at the terminations of the nerve fiber which are in close contact with the muscle surface. The arrival of electric impulses at the nerve termination brings about the release of this transmitter, which spreads out toward specialized "receptors" present on the muscle surface. As a consequence, a localized change of the muscle membrane potential is produced, which eventually results in the generation of a full-blown electrical impulse in the muscle (and in the activation of the contraction mechanism).

As far as the generation of the electrical impulse in nerve (and muscle) fibers is concerned, the onset of the process requires a perturbation capable of "breaking" the condition of almost absolute membrane impermeability to sodium ions that characterizes the fiber at rest. The nature of the triggering process resulting in an increase in membrane sodium permeability has to do with a crucial characteristic of a type of channel present in the membrane. These channels, already considered in the previous chapter (and indicated as Na^+ channels), are selectively permeable to sodium ions. However, at the resting potential of the cells (i.e., when the membrane is hyperpolarized by about -70 mV) these channels are closed and sodium cannot pass. A crucial feature of the Na^+ channels is that the "gate" controlling the opening-closing state starts to open once the membrane begins to be depolarized (i.e., when the intracellular potential moves toward a positive direction). Let us now consider what happens if that occurs. Sodium ions start entering into the fiber; following their electrochemical potential, this produces a further depolarization of the membrane (Na^+ carry a positive charge). Consequently, as the channel gate opens, more and more ions enter, which results in the explosive cycle considered in the previous chapter (Hodgkin cycle).

The bulk entry of sodium caused by the full opening of a Na$^+$ channel does not, however, remain for long. One of the reasons is that, as the internal potential becomes less negative (and eventually positive), the force driving sodium ions inside decreases. Sodium entry ceases when membrane potential becomes positive enough such as to correspond to the electrochemical potential for Na$^+$ (as predicted by Nernst's law applied to the asymmetrical concentrations of this ion, i.e., about + 60 mV, the internal side of the membrane being positive with respect to the external side). At this point the chemical energy due to the higher sodium concentration in the extracellular medium (causing an inward movement of Na$^+$) is compensated by the energy due to positive intracellular potential inside (which favors sodium exit).

In fact, during the excitation the membrane does not reach the exact Nernst equilibrium potential for sodium but only approaches it. One of the reasons for this is that, during the excitation process, the leak conductance for potassium present at rest remains active, thus partially counteracting the consequence of sodium channels opening.

It must be said that in principle the Nernst equation is only valid for a membrane permeable to a single ion, and it could not properly be applied to the condition of excitation in which there is more than one channel open. It could not be applied even at the resting state because in this condition, in addition to the main permeability for potassium, there is a very small permeability for sodium ions. Since, however, this resting sodium permeability is much smaller than the permeability to potassium, it can be neglected with no great error in the estimate of the resting potential.

More generally, when a membrane is permeable to several ions, another more complex equation must be used to calculate the membrane potential instead of the Nernst equation. This is known as the Goldman-Hodgkin-Katz, from the names of the scholars who developed it (Goldman, 1943; Hodgkin & Katz, 1949). We are not going into the details of this more complex equation, and we will limit ourselves to saying that a membrane permeable to several ions would attain a potential which depends on the ion concentrations at the two sides of the membrane and on their relative permeability. For instance, in a membrane permeable to sodium and potassium, the potential would depend on the concentrations of the two ions and would be more influenced by the ion to which it would be more permeable. In the case that the permeability to one ion greatly exceeds the permeability to the other, the Goldman-Hodgkin-Katz equation would reduce to Nernst's law for the predominant ions. The membrane potential would thereby correspond to that predicted by the Nernst equation for this ion. In the other cases it would be intermediate, to a variable degree, between the values of the Nernst equation for the two ions. The resting membrane is about 100 times more permeable to potassium than to sodium, and therefore its potential is reasonably near to the Nernst equilibrium potential for potassium. The small resting sodium permeability accounts for the small deviations from the Nernst value for potassium, a deviation that is always toward positivity (this is because sodium equilibrium potential is about +60 mV).

In a comparable way, since during the excitation the membrane is permeable to both ions (although with reversed permeability ratio, being now about 10 times more permeable to sodium), the potential attained would be less positive that the Nernst equilibrium potential for sodium (it would about +50 mV).

The action potential is a rapid phenomenon, which usually develops in less than 1 millisecond, as first shown by Bernstein with his ingenious technique of the "differential rheotome" (see Chapter 9.1). There are important reasons for that, which have to do with the capacity of nervous systems to encode and transmit information. To understand the need for short signals, let us try to figure how nerve fibers could convey information, if the impulse were of long duration (i.e., if the condition corresponding to the bulk entry of sodium, with a potential of about + 50 mV, were to last for seconds or minutes, instead of for less than a millisecond).

Let us make a practical example in this respect, by considering the behavior of our visual system in a normal living condition, that is, the rather common situation of a pedestrian who is going to cross a road. Being a prudent person, this pedestrian looks first carefully to the one and then the other side in order to be sure that no vehicle is approaching. We assume that the information meaning "no vehicle coming" is communicated to his brain through one or more fibers in the optic nerve which change their potential from −70 to +50 mV to signify precisely that (the neurophysiologic process is indeed much more complicated but on purely theoretically grounds we can make this simplifying assumption). The pedestrian decides then to cross the road. Suppose that the particular state of the fibers (+50 mV) is maintained for a long time (say 10 seconds), during which time a car (a Ferrari) suddenly approaches at a very fast speed. Even though our pedestrian continues looking from left to right while he crosses the road, his brain remains under the influence of the fibers saying, with their +50 mV potential, "no car coming." This would mean that no signal would be transmitted from the motor centers of the brain to the muscles in a way appropriate to trigger a suitable stop-and-escape reaction, with the worst possible consequence for our poor pedestrian.

We could do better than give an example dealing with a car or a pedestrian in order to indicate the necessity of brief nerve signals. This is because brief impulses have developed during evolution, millions of years before the appearance of humans and more or less fast cars. However, the situation we have envisioned helps to understand the need for short electrical impulses in nerve fibers. Within the interval of 1 millisecond or less the membrane should come back to the resting state (−70 mV), ready to remain in this state or to signal a new impulse, depending on the possibly of fast temporal evolution of objects in the visual world (or for communicating to muscles motor commands that can vary in a similarly fast way).

There are two main mechanisms responsible for the termination of the excitatory phase of electrical impulse in the fiber membrane. The first, and more important one, depends on an intrinsic characteristic of the sodium channels responsible for the impulsive electrical signal. On one side, these channels open rapidly when the membrane moves toward a positive potential (i.e., is depolarized). On the other side, a depolarization starts another, slower process, which results in a closure of the channels (this process operates on a different gate than that responsible for the rapid opening induced by the depolarization). By this second process, the sodium channel starts closing when the potential approaches the maximum depolarized level. As a consequence, the potential would start coming toward a negative condition corresponding to the Nernst equilibrium potential for potassium, because of the prevailing potassium conductance of the leak channels.

By itself this mechanism would be sufficient to bring back the membrane potential to the resting state. It would require, however, a relatively long time, since the current passing through the leak channels is of modest intensity. The need for brief nerve signals has eventually made recourse to another mechanism capable of accelerating the falling phase of the return to the resting state (repolarization) after the peak of the action potential. This is based on other channels, which—as with the leak channels—are permeable to potassium, but capable of allowing a more substantial passage of these ions. These channels are normally closed in the resting state and they are opened by depolarization, allowing for a massive efflux of potassium, which tends to bring the membrane potential back toward negativity. There are indeed various types of potassium channels having these features. A common characteristic of these channels is that, compared to the sodium channels, they open at a relatively slow rate when the membrane is depolarized. Because of this, the membrane first depolarizes to a largely positive potential (due to the bulk entry of sodium) and afterward returns rapidly to the negative potential (when potassium efflux dominates).

10.2 THE ELECTRIC MECHANISM OF NERVE CONDUCTION AND MUSCLE EXCITATION

The nerve impulse (or action potential) is—as already mentioned—the mechanism for encoding and transmitting information that the nervous system has developed along the evolutionary process. It is characterized by a stereotyped amplitude and time course. Indeed, none of these parameters depends on the amplitude of the stimulus or of the physiological cause that had produced the impulse. As to the dependency on the stimulus intensity, nerve impulses obey the "all-or-none" law (Chapter 9.2), being produced in its full amplitude when the stimulus exceeds a certain threshold. In typical nerve cells the threshold value corresponds usually to a depolarization of about 20–30 mV relative to the resting potential.

Another fundamental feature of the stimulus with relation to its effectiveness in inducing the discharge of a nervous impulse is a fast time course. A fast membrane depolarization, capable of bringing the potential suddenly from the resting value of −70 mV to about −40 mV is usually effective in producing the impulse. However, if the same potential of −40 mV is attained in a gradual way, the stimulus might turn out to be unsuccessful. This is the consequence of a property of the membrane called "accommodation," which depends largely on the kinetic characteristics of the processes that regulate the gating of the voltage-dependent ionic channels of the membrane. An electrical stimulus with a slow onset phase is ineffective for two reasons. First, it produces the inactivation of the sodium channels while it activates them; secondly, it activates the delayed potassium channel responsible for the membrane repolarization. This is to say, with a slow depolarizing stimulus there is no change to exploit effectively the faster rate of opening of the sodium channels, relative to the slow characteristics of inactivation of these same channels, and of the slowness in the opening of the delayed potassium channels.

Electrophysiologists are well aware of the need for fast electrical stimuli in order to excite effectively nerve and muscle fibers, and this is why they normally use electrical pulses with extremely rapid variation rate (in the order of a microsecond, that is, a millionth of a second). Seen from an historical perspective, this temporal feature of the excitation process accounts for the particular effectiveness in Galvani's experiments of rapid electric stimuli (as spark discharges) in exciting the neuromuscular frog preparation. Galvani was impressed by the phenomenon, and this is why he opened *De viribus*, with the chance observation of the frog convulsion brought about as a consequence of the sparking of a distant electrical machine. In the circumstance of this crucial experiment, the sudden discharge of the machine produced, by a combination of electromagnetic and electrostatic effects, an extremely fast change in the membrane potential of the nerve and muscle fibers traversed by the lines of force of the electric inductive influence. As mentioned in Chapter 4, during his experiments Galvani noticed that a sparking electrical stimulus from a relatively distant machine was much more effective than that resulting from a direct connection of the preparation with the prime conductor of the machine. Being a sagacious observer, Galvani was struck by the "novelty of the matter" and "by the wonder" and started studying the phenomenon. As it happens, with the direct connection to the machine much more electrical fluid passed from the machine to the preparation, and nevertheless, this way of application of the stimulus turned out to be less effective than the spark at distance (Chapter 4.2 and 4.3).

Within the physiological conceptions of his time, Galvani was strongly influenced by the Hallerian theory of irritability. Within this framework, he was prepared to accept the idea that the stimulus was not directly producing the effect (i.e., muscle contraction) but was instead triggering a principle intrinsic to the animal preparation which was the effective cause of the physiological effect. By developing this conceptual paradigm both logically and experimentally, Galvani eventually arrived at the idea of an electricity present in a condition of disequilibrium within the animal tissues (see Chapters 4 and 5). He conceived thus the idea of an animal electricity responsible for nerve excitation and muscle contraction, thereby laying down the foundations of modern electrophysiology. On his side, Volta had become aware of the need for using fast electrical stimuli in order to obtain some physiological reactions (for instance, muscle contraction and visual sensations). He had approached the conception of a property similar to the "accommodation" of modern physiologists with his notion of the *repos d'équilibre* (Chapter 8.1).

Galvani noticed another feature of the relation between the stimulus and physiological response; it was the particular relation existing between the intensity of the electrical stimulus and that of the muscular contraction. He noticed that the strength of the contraction did not augment in a progressive way by increasing the intensity of the stimulus, but it increased rapidly for certain augmentations of the stimulus intensity, and then eventually, and rather abruptly, it stopped increasing and remained constant (it "saturated" according to the modern terminology). With stimuli of decreasing intensity it could happen that, after a phase of relatively smooth decrease, the contraction decreased suddenly or ceased altogether in an abrupt way (Chapter 4.3).

These aspects noticed by Galvani in his experiments are correlated with the highly "nonlinear" character of the stimulus–response relationship in the electrical excitability of nerve and muscle fibers. As discussed in the previous chapter, this relationship appeared particularly evident in the research of Lucas and Adrian. We now know that the lack of a simple proportionality in the response of nerve and muscle fibers to electrical stimulation is the expression of the autoregenerative or explosive character of the processes underlying the generation of the action potential in any individual fiber.

There is an important aspect of the electrical behavior of the nerve (and muscle) that we need now to discuss in some detail, after having mentioned it en passant previously. This is the reason why excitable fibers developed a complex mechanism of generation and conduction of electrical signals, and why they did not simply base electrical conduction on a simple, physical propagation of current, as in a metallic cable. Why do we need such a complex machine based on pumps that require energy, a variety of ionic channels with complex features and with specific kinetic properties (and in some excitable cells the number and variety of channels greatly exceed the few channel types considered here)? In addition to being complex, this system is also metabolically expensive, since—as mentioned earlier—a nerve cell spends a great part of its energy fueling processes connected with the electrical excitability of the membrane. Why is there not a simple and economic process?

A simple process was indeed in the mind of Volta when he assumed that the electric disequilibrium produced by the contact of two different metals was the sufficient cause for the current flow along nerve fibers. Because of his simplistic conception of the mechanism of electric signaling in excitable fibers, Volta was confident of his idea that no assumption of an intrinsic disequilibrium of electricity inside animal tissue was required. As a matter of fact, this is still the idea that many modern physicists have of the process of nerve conduction. Physicists are indeed accustomed to the idea that an electrical signal can propagate over great distances along a metallic cable with no important attenuation. Since animal tissues are electrical conductors, and since the length of nerves is relatively short, it is at first view difficult to speculate why nature has not made recourse to a simple "cable-like" electrical conduction in nerve fibers. It seems as if nature, usually considered as simple and economic in its operations, has preferred a definitely prodigal and "baroque" system in order to propagate the electrical signals in the nervous system. To understand this apparent extravagance of nature, we need to make some physical considerations that—as we shall see—would demonstrate the impossibility for nature to conduct electricity along nerve fibers in a simple, cable-like manner.

Against a physicist like Volta (and against his followers at a time separated by more than two centuries), we need to develop some of those quantitative arguments which generally reflect the force of physical reasoning. In this respect we could quote a passage from Volta, sometimes proudly cited by more or less modern physicists, in order to indicate a certain superiority of physics over other scientific disciplines. The passage appears in the first *Memoir on Animal Electricity*:

> What good can one expect, especially in physics, if things are not reduced to degrees and measures? How can one evaluate the causes, if not only quality, but also the quantity and intensity of effects are not determined? (VEN, I, p. 27)

If we make recourse to quantitative arguments for the electrical conduction in nerve fibers, we need to notice first that bodily fluids, although capable of conducting electricity, are much less conductive than the classical "good conductors," that is, metals. This point was well put in evidence by Cavendish and afterward recognized by Volta himself (see Chapter 8.4). The specific electric resistance (or resistivity) of a biological liquid similar to the axoplasm (i.e., the cytoplasm contained inside a nerve fiber) is indeed about 10^8 times bigger than the resistivity of a metal like copper or silver. This means that a wire made of axoplasm would conduct about 100 million times poorer than a metallic wire. In other words, if a copper wire of a certain diameter would allow conduction for a distance of 1,000 km, then a "wire" of axoplasm of the same diameter would be similarly effective only for the distance of 1 cm.

Things become even more complicated for nerve fibers when we consider that in order to convey a great quantity of information in a particular nerve, we need to make fibers very thin. This is because as the diameter of the fibers is increasingly reduced more and more fibers with a nerve of a given size can be accommodated in a given space. Physical considerations show that the resistance of a wire having circular section increases when the diameter of the fiber decreases, with a proportion that depends on the second power of diameter. That is, if the diameter gets 10 times smaller, the resistance would be 100 times greater. By keeping in mind this relation let us now compare a big copper cable of the diameter, say, 1 cm, with another cable, also made of copper, but extremely thin, of the diameter of, say, 1 micron. A simple calculation can show that the resistance of the thin cable would be 100 million times larger than that of the thick cable.

If, on the other hand, the thin cable, instead of being made of copper, is made of axoplasm, then the resistance would be 10^{16} larger than that of the big copper cable. The physical difficulty in trying to force an electric signal in such a thin cable of, say, 1 meter length, would be comparable to that encountered in attempting to transmit an electrical signal for a distance of 10^{16} meters along the big copper cable. This is an immense distance, 10,000 million million kilometers. To portray how long such a distance is in a vivid way, we can repeat the efficacious example given by Hodgkin, who calculated that this distance exceeds by far the distance between the Earth and Saturn, the most distant of the classical planets (Hodgkin, 1964; see Chapter 9.1).

To transmit effectively an electric signal for such huge distances, we would need to use very high electrical potentials, in order to face the unavoidable energy losses and current leaks. These would certainly cause the breakup of the conductive materials. In the specific circumstances of nerve fibers, the problem of current leaks would be particularly serious, because the fibers are surrounded by an electrically conductive liquid medium. As in marine cables, we could attempt to address this problem by wrapping our axoplasm wire with an electrically insulating sheet. This is, however, feasible only within certain limits. To keep the overall size of the fiber small, we could not use insulating sheets that are too thick, and this would necessarily limit their insulating power.

The problem of transmitting through an electric cable system up to distances greater that those supported by the physical characteristics of the cable is not, however, insurmountable. From a formal point of view it is similar to that encountered with systems of optical communication, when people wished to transmit over very long distances. These systems have been in use for centuries. For example, between the eighteenth and nineteenth centuries, Lloyds of London was informed of the maritime traffic at the port of Leghorn by an optical telegraph system. This system was based on a series of optically transmitting towers, each one of which transmitted a bright optical signal to the next station in the line, when a visible signal arrived from the previous station. The system was laborious and time consuming but permitted communication over distances much longer that those covered by simple, direct, luminous signaling.

Let us turn now to the electrical system and suppose that an electrical engineer needs to transmit over a distance of, say, 10,000 km through an electric cable telegraph by using metallic cables capable of supporting transmission only up to a distance of 1,000 km. The engineer could solve this task in a way formally similar to that considered earlier for optical telegraphy. He would interpose a series of electrical stations along the transmission line, at a distance of 1,000 km from each other. When a large electrical signal generated from the first station arrives attenuated but still detectable, the receiving station will enhance the signal and send it on to the next station, and so on to the last station. For this solution to be successful, each station must have some means of boosting the signal before sending it on to the next station, and for this it needs to have a local source of electrical energy. This system could be automated using an appropriate device (an electric or electronic relay) capable of triggering the production of large electrical signals upon the arrival of smaller incoming ones, through the use of local energy.

With such arrangements, however, what really circulates from the first to the last station is not the energy of the original signal; it is only the information associated with it- that is, the command leading to its generation. Using eighteenth-century wording, one could say that nothing of the electric "fluid" present at the first station would reach the last one, although the arrival of the signal at the end of the transmission line is dependent upon the electrical actions occurring at the initial station. The mechanism allowing for the transmission of electrical signals along nerve fibers, particularly in the large fibers of mammals, is, in fact, similar to that in the model outlined earlier. Like the electrical cables used in underwater transmissions, these fibers need to be enveloped by a highly insulating material, because of the conductive and shunting nature of the fluids surrounding them. The biological coating serving this role is the myelin sheath, which is made up of many lipids (the oily, potentially insulating substance envisioned by Galvani in his experiments, see Chapter 5). The insulating coating, however, is not continuous in these nerve fibers. There are interruptions that are much shorter in length than the coated segments, where the fiber membrane is exposed to the extracellular space. In 1878, the French histologist Louis-Antoine Ranvier gave the first good description of the bead-like appearance of covered nerve fibers (Ranvier, 1878: see Fig. 10.1). Today, we know how glial cells that wrap around the axon form the myelin sheath, and we use the eponym "nodes of Ranvier" to denote the gaps in the axon's coating where there is no myelin. At a functional level, the nodes of Ranvier are the reception-transmission stations in the process of conveying an electrical signal along a long axon. The myelinated segments, in contrast, correspond to the insulated cables for the transmission of the signal from station to station, until the end is reached. Every node is endowed with local energy, due to the asymmetric distribution of ions at the membrane sites, along with the complex machinery needed for the production of relatively large-amplitude electric impulses (about 100 mV, or 0.1 V).

An impulse is automatically activated once a small electrical signal (less than about 10 mV) reaches the node. Under physiological conditions, the activating signal comes

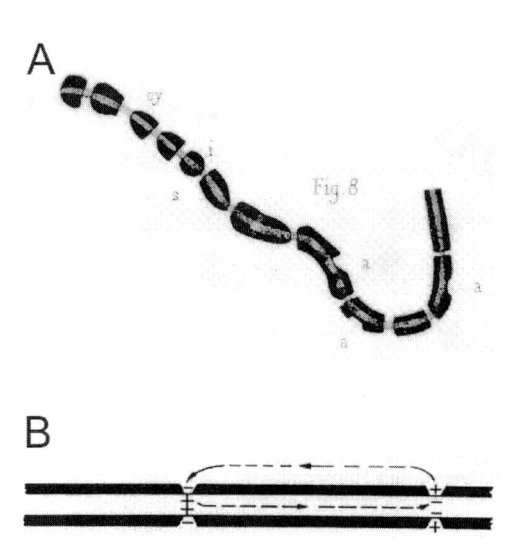

FIGURE 10.1. (*A*) An image of a frog nerve fiber with its overlying nodes of Ranvier (from Ranvier, 1878). (*B*) A scheme illustrating how the current involved in nervous conduction jumps from one node to another by saltatory conduction (from Hodgkin, 1964).

from the adjacent, previously activated node via the transmission through the internodal segment of the fiber. The impulse is transmitted to the following node in the sequence in a partly attenuated way and regenerated there. The process then continues until the fiber terminates, the repeated "leaping" of the event from node to node inspiring the phrase "saltatory" (meaning leaping, hopping, or jumping) conduction (Fig. 10.1). Under experimental conditions, the activation of a node (with the resulting local discharge of an impulse) can be brought about by an external electrical stimulus capable of producing a small but adequate modification of the nodal membrane potential.

Here is the solution adopted by nature in its evolutionary course to solve the apparently insurmountable difficulty of transmitting an electrical signal over long distances, without attenuation, along very thin nerve fibers made up of poorly conductive matters. It is an astute solution, and the complex underlying processes were certainly far from the reach of the conception of eighteenth-century science.

We can now come back to Galvani and Volta, and to their "dilemma." In many of the experiments that they conducted, the external electrical stimulus was provided by the application of a metallic arc. The electrical potential generated at the contact of two metals is generally of the order of 1 V, and, although the current flow is largely dispersed in the passage across the muscle and nerve tissues, this is still sufficient to trigger activation at the nodes of Ranvier. Also, the contact of biological liquids with a metal or metals (and under some conditions a direct contact between living tissues without any intermediate body) can produce a sufficient electrical stimulus.

In this context we need to note that muscle fibers are also capable of generating electrical impulses in a way that is fundamentally similar to nerves, even though muscles are generally much less sensitive to external stimuli. But muscle fibers do not have an insulating myelin coat with nodes of Ranvier, so the propagation and transmission of the electrical impulse take place in a continuous way. These are not, however, just characteristics of muscle fibers. The smallest nerve fibers of vertebrates, and all nerve fibers in invertebrates, lack myelin and conduct the electrical impulse in a continuous, nonsaltatory way.

Importantly, the relationships between nerves and muscles in both vertebrates and invertebrates are normally nothing like the tight electrical relationship between muscle and nerve that Galvani envisioned with his Leyden jar model. As mentioned in the previous chapter, nerve and muscle fibers communicate only via chemical signals. Moreover, both have a local reserve of energy necessary for the conduction of their own electrical signals.

So now we can appreciate why the Galvani–Volta dilemma was not a true dilemma, and why both Galvani's animal electricity and Volta's metallic electricity were required to bring about nerve stimulation and consequent muscle contractions. Under their experimental conditions, the metallic electricity would have activated only the local processes in the nodes, which might eventually result in a full impulse. The energy of the full electrical signal, however, was not derived from this external source but depended on the local energy accumulated in the membrane of the node. In the absence of the intrinsic electricity at the node, the metallic electricity would have been transmitted for only a short distance (of the order of microns or perhaps a few millimeters, depending on the fiber type and diameter). Thus, it would not have reached the end of the fiber, so as to bring about even the smallest muscle contraction, because of the extremely high electric resistance of the nerve fibers involved.

The metallic electricity of Volta was only the stimulus, the "excitatory cause," not the "effective cause" of the electric charge movement responsible for the nervous signal. It acted by triggering the process that released the electricity accumulated between the internal and external compartment of the fiber, electricity that was thus fully "animal," in Galvani's sense. The necessity of the triggering action of the external metallic electricity was due to the fact that the intrinsic animal electricity could not flow, because of the impermeability of the membrane to current flow in the resting state.

In physiological conditions, there is nothing like metallic electricity to trigger the release of the intrinsic electricity accumulated at the two sites of the cell membrane. As is evident from the considerations expressed earlier, the triggering stimulus acting on a given node of Ranvier is the small current arriving from the previous node. This is due to the local transmission of the current associated with the discharge of the full impulse in the previous node. It is a current of rapid onset and rapid termination, which is normally the case for electrical events associated to the discharge of electrical impulses in nerve cells. This accounts for the fact that nature has evolved a system whereby the stimulation of the impulse discharge is brought about preferentially by a sudden and impulsive stimulus. On this basis we can understand why in Galvani's experiments an impulsive stimulus, of the type produced by a spark discharge of an electrical machine, was so effective in stimulating the frog preparation.

Surely the complex mechanisms underlying the discharge of electrical signals in excitable fibers could not be understood within the limits of science at the time of Galvani and Volta. Nonetheless, within the framework of Hallerian theory of irritability, and particularly following the development of Haller's theory by Felice Fontana, Galvani was ready to grasp the idea that the external stimulus was only stimulating an internal process, based on a local energy. Eventually he could identify this process with the functioning of an electrical "animal machine" and the local energy as his "animal electricity." Using modern language to make explicit the principles of the Hallerian theory, we could say that what is communicated in the passage from stimulus to response is not energy, but information, a command to which the physiological system responds with the action for which it is programmed. Galvani was ready to interpret his experiments in this perspective, and this led to his fundamental conception of the existence of animal electricity. In this respect Galvani was more modern than Volta, who was anchored to scientific conceptions more connected with the idea that physical processes involve exclusively the transmission of forces or energies.

An unprejudiced view of the case of Galvani and Volta, far from leading us to discussions about the superiority of one sector of a science compared to another, should stimulate considerations on the global character of the scientific endeavor. We should also bear in mind the complex and unpredictable pathways leading to scientific discoveries, sometimes based on the interplay of apparently unrelated disciplines. On the one hand, who could have foreseen that Volta's battery would be somehow derived from Galvani's frogs? And on the other hand, who could have predicted that the contact of two metals would be so important in unveiling the electricity concealed within the nerves and muscles of animals, and thus prompt the emergence of an entirely new field of physiological science—one of fundamental importance for understanding the functioning of the nervous system?

Rather than retaining rigid separations between the various sectors of sciences, modern scholars should keep an open and attentive attitude, aimed at grasping the richness of the complex systems and of the "machines" created by both inanimate and animated nature. This is the attitude entertained by the great "natural philosophers" of the eighteenth century. To repeat one of the arguments presented throughout this book, the case of electric fishes helped both Galvani and Volta to recognize that electricity could be developed in both animal bodies and in artificial devices according to principles apparently heterodox with respect to the physical laws of the era. If the scientists who followed had paid more attention to these same fishes, they could have discovered the existence of the mechanisms of electroreception and electrolocation (already implicit in Walsh's observations) much earlier. Moreover, they could have appreciated much earlier the artificial systems of radar detection and other forms of electric location.[2]

As we have already remarked, Volta, a scientist who has been considered an emblem of the superiority of physics over physiology, was ready to recognize the superiority of nature over art. With the optimism of a scientist of the Enlightenment, and looking at the achievements of animated nature, he expressed the wish that one was not far from the possibility "that art could also arrive there."

---

[2] On the theme of electric fishes, one could ask whether modern electrophysiology has clarified the physiological mechanism underlying the production of electric shocks. This is certainly the case. In 1953, just 1 year after Hodgkin and Huxley had been unveiling the mechanisms of impulse generation in nerve fibers, Richard Keynes and His Martins-Ferreira discovered the process whereby an intense shock is produced by the electric organs of the electric eel. However, we have not pursued the detail of these and following studies on electric fishes here because this topic has been extensively covered in a recent book (Finger & Piccolino, 2011).

> Is mastery of old Volta's ideas
> about galvanism necessary
> to run or repair a dynamo?
>
> —BLOCH (1949/1992), p. 30

11
Concluding Remarks

WE MUST ADMIT it—this book of ours is unusual and at times it might not have been easy to read. Some might have been surprised to find in these pages expressions that seem to belong to quite different fields of knowledge: "Hallerian irritability" and "nervous force" side by side with "patch-clamp"; "magic squares" and conducting arcs together with "voltage-dependent membrane channels," "ionic pumps," and the Nernst formula of electrochemical potential. Surely, this creates some difficulty for the readers and also for the authors, as we shall see later on. The choice has not been casual, however, nor have the authors' intentions been to astonish readers with a great exhibition of items of knowledge coming from different fields, which would, in the end, only confuse them. Indeed, as we have tried to show in more than one place in the book, it is only by widening the conceptual as well as the temporal horizons beyond the traditional distinction between the "two cultures," that it is possible, in our view, to understand the most significant, and most neglected, aspects of Galvani's research, also with respect to his controversy with Volta.

As an example, let us consider the description of the electrical phenomena of the membrane, to which we have devoted almost two chapters, based on the use of technical terms and concepts like ionic channels, pumps, or electrochemical potentials, which are at first sight very distant from the notions of the physiology of the end of the eighteenth century. Although the use of these terms may expose us to the charge of employing unduly terms and concepts that would not make their appearance until later, it is indeed by referring to these concepts that it is possible, in our view, to highlight one of the significant features of Galvani's research and of his theory of animal electricity. For the Bolognese scholar, "animal" electricity could not be assimilated to "common" electricity, not because they differed in some essential quality but because the former implied the presence of a specific "machine" in the organism. More precisely, it implied the presence of a complex mechanism—structural as well as functional—capable of producing an electrical disequilibrium, of keeping it in a specific place inside the animal tissues (i.e., the muscular fibres), and of allowing its being used in fundamental physiological processes like nervous conduction and muscular contraction. The core element of modern electrophysiology, beyond the more or less technical terms and concepts it makes use of, consists of the fact that the membrane of nerve and of muscle fibers is actually a "machine," which produces and utilizes the electricity necessary to encode and transmit information to the excitable tissues. Despite a very high electrical resistance of the internal matter of nerve fibers, which constitutes an apparently insurmountable physical obstacle for "simple" electrical conduction, the membrane allows the electrical signals to be propagated in long and thin fibers and it is exactly in this sense that it can be considered a proper electrical animal machine, although in modern terms the membrane should be better designated as a "molecular machine."

The modern explanation of the electrical phenomena of the membrane sheds some light also on minor details of Galvani's and Volta's electrophysiological work. For example, when Galvani observed some contractions following a direct nerve-muscle or a nerve-nerve contact, he also noted that these contractions occurred particularly when the contact was established by means of the extremity of the cut nerve. Again, when Volta was faced with the experimental confutation of his observations that bimetal arcs were inefficient in provoking contractions of the viscera, he claimed that it was necessary to apply at least one of the metals directly to the muscle. These examples, if considered in terms of what we know now, show the great observational skill and experimental acumen of both Italian researchers.

A particularly relevant case in which the intersection of historical and modern perspectives can contribute to enrich the reconstruction of Galvani's and Volta's work is the understanding of the crucial scientific problem which was at play in their research, and which marked an irreconcilable aspect of the controversy between the two men of science. The experiments with metal arcs implied the effects of "two" types of electricity, namely animal electricity described by Galvani and metallic electricity described by Volta. Both of them were necessary for the production of contractions, in most of the experimental dispositions adopted by the two scientists, and that was indeed due to the functioning peculiarities of the electrical "machine" which is at the basis of neuromuscular physiology. Animal electricity, even when in a state of disequilibrium and thus ready to flow on its own energy, still requires a distinct (electrical) stimulus, of internal or external origin according to the circumstances, which, by activating it, can trigger the nerve signal responsible for the contraction. In physiological conditions, the stimulus responsible for activating the electricity in a state of disequilibrium in the excitable tissues derives from the intrinsic electricity itself, according to a complex self-regenerating mechanism, which we have already described in detail. In the case of the experiments with metal arcs, it was the contact between different metals which provided the external electricity acting as a stimulus on the nerve fiber, thus liberating the energy of the electrical disequilibrium intrinsic to the animal. Accordingly, to claim that in the experiments with metal arcs the frog's leg moved as a consequence of a passive effect of the electricity produced by the metals, as many still think (and some still write), and consequently that Volta was right and Galvani was wrong, is one of those stereotypes that can be overcome only by referring to modern electrophysiological notions. On these bases, any statement or question on how right or wrong should be distributed between the two scientists—an issue which has heavily conditioned many of the scholars who have dealt with the controversy between Volta and Galvani—loses credibility.

The need for the "two" types of electricity for the generation of muscular contractions in the experiments with metal arcs allows us to suggest an explanation for the very singular fact that, over two centuries ago, Galvani's animal electricity and Volta's metallic electricity met on the table where some "prepared" frogs were placed, and were thus discovered in a tight temporal and causal relationship. Usually, the contact between different metals produces voltage lower than 1 volt but, as we have already observed, most physical electroscopes available to the "electricians" of the second half of the eighteenth century could detect only potentials on the order of tens or hundreds of volts. By contrast, the "prepared frog" used by Galvani constituted a kind of extraordinary biological electroscope that was able to detect the weak electricity present in metals, thanks to its electrical and mechanical amplifiers described in the previous chapter. In fact, one could object that other physical apparatuses able to detect a very weak electricity did exist, such as the "doublers" or "multipliers" of electricity (especially the one designed by William Nicholson) as well as Volta's *condensatore*, and that these instruments were actually used by Volta when, in 1796, he demonstrated the existence of "his" metallic electricity without referring to a biological detector—the frog. So, without Galvani's frogs, the discovery of metallic electricity would have been probably only delayed by a few years or, maybe, by a few decades at most. However, beyond the fact that Volta's interest in metallic electricity was stimulated by Galvani's experiments, there are good reasons for believing that Volta's path to the battery would have been harder without Galvani and his frogs. As Giuliano Pancaldi has very effectively pointed out, one of the problems arising from the use of physical instruments capable of revealing weak electricity (like Nicholson's "doublers," for example) lay in the difficulty of clearly distinguishing the electricity to be measured from the residual electricity, that is, the electricity that remained in the instrument even without any contact with external generators (Pancaldi, 1990, 2003). The problem could be solved, as Volta did in 1796, by resorting to complex procedures capable of excluding the influence of the residual electricity present in the instrument. However, we cannot tell for sure what would have happened if Volta had not been certain from the outset about the existence of metallic electricity, a certainty that he had acquired from the experiments where metal arcs were applied to Galvani's frogs.

As we have already mentioned, many of the "electricians" who at the time were interested in the study of "weak electricity" did point out that under certain conditions, either natural or experimental, small "degrees" of electricity were produced. In his last scientific memoir, Volta himself acknowledged that in 1789 Nicholson, on the basis of experiments carried out with his doubler, had advanced the hypothesis that various metals could be a source of electricity. However, this observation was not developed further. This might have perhaps happened, we may be tempted to argue, because Nicholson had not resorted to Galvani's frogs. But there is still more. Indeed, even if Nicholson (or Volta) had been able to note that metals were capable of generating weak electricity, using only their physical instruments, they could hardly have derived as much from this observation as Volta did from the experiments based on Galvani's frogs. As a matter of fact, the event that marks a turning point in the history of electrical science is not so much the demonstration that the contact between different metals can generate an electrical disequilibrium, but, rather, that in some definite circumstances the contact between different metals can generate a continuous current. Volta derived this fundamental result from the observation that frogs contracted not only when the circuit was established through the metal arc but also when this circuit was opened, and he confirmed it with great clarity in the experiment with the tongue; this was a physiological experiment that constituted Volta's original development of the electrophysiological experimentation started by Galvani. Since contemporary electrometers were not capable of distinguishing between static and dynamic electricity, reliable evidence of the potentially continuous aspect of the current generated by the contact between different metals would have been very difficult to obtain in a purely physical context.

If this reconstruction is largely hypothetical, based more on possibilities than on historical facts, what the history tells for sure is that the two types of electricity (the "animal" and the "metallic" one) were discovered together, and that at the outset all started with a research aimed at investigating the physiological involvement of electricity in nerve and muscle function. Our story also reveals, moreover, how the discovery of animal electricity and electric current did not emerge out of a linear development, marked by "logical" and chronological steps; these could have been the invention of instruments to measure electricity, the construction of apparatuses to produce an electric current, and finally the measurement of the electricity involved in animal functions. Instead, what happened was due to the stubbornness of a Bolognese physician who did not abandon the idea of explaining the muscular contractions he had observed by means of an electricity intrinsic to the animal, despite some important experimental and logical difficulties he had encountered. Not only that: This doctor thought he could describe it all by resorting to the physical model of the "animal Leyden jar," which not only provided a plausible explanation for the mechanism responsible for the contraction but also allowed him to overcome the theoretical difficulties about the existence of an electrical disequilibrium inside bodies constructed from conducting materials. It is a paradox that as a consequence of this, although metallic electricity is conceptually less problematical than animal electricity—and somehow "preliminary" with respect to it—the latter electricity was discovered before the former.

If the pathways of the discoveries we are proposing may seem particularly tortuous, we should consider that it is hardly possible for totally new disciplines, like Galvani's electrophysiology or Volta's electrodynamics, to develop in a completely linear way. The first electrometer sensitive enough to measure animal electricity, namely the astatic galvanometer built by Leopoldo Nobili, exploited the electromagnetic effect produced by the current flow generated by a battery that passed through a metal winding. Such an instrument, based on a complex device aimed at reducing the effects of the Earth's magnetic field, thus required both the invention of the battery and an awareness of the phenomena of electromagnetic induction. But, in turn, the invention of the battery required the availability of instruments able to detect the weak electricity present in metals and to show the continuous nature of the current flow generated by metals. Crucially, such knowledge was made available by Galvani's experiments with frogs and by other procedures based on an electrical amplification due to the existence of an unbalanced animal electricity. Research pathways such as those of Galvani and Volta could not be linear due to the complexity and the mutual interconnection of the phenomena involved.

In this regard we may note that Galvani, in his experimental setting, was not able to separate the "source" of animal electricity from its detector, while such a problem did not affect Volta's research on metallic electricity. This marked a fundamental advantage on Volta's behalf. Galvani's difficulty can be exemplified by the comparison between the apparently similar experiments where two experimenters established a "chain" with each other to detect the electricity of the torpedoes (as in the case of Walsh and his assistant) or of the frogs (as in the case of Galvani and Raymond Rialp). In the case of the torpedo, a "topological" distinction between what generated the discharge (the fish) and who detected the discharge by perceiving its effects (the experimenters) was clearly evident. On the contrary, in Galvani's research the frog acted both as a generator and as a detector of animal electricity, while the two experimenters were involved in the circuit only as conducting elements (see Fig. 5.5 in Chapter 5). The distinction between source and detector was instead very clear in Volta's interpretation of the same experiment: The bimetal arc was the source of the electrical movement while the frog—that was like an extremely sensible electroscope—worked as a detector. This distinction was afterward confirmed in an even stronger way by substituting the biological detector with a physical instrument, such as Nicholson's "doubler" or Volta's *condensatore*. Hence, the possibility of eliminating the animal preparation, which, though allowing an efficient detection of metallic electricity, still remained biological in nature, thus introducing in the scheme some complex and potentially ambiguous elements.

We have already referred to some paradoxical aspects that apparently characterized Galvani's and Volta's investigative pathways. One particularly interesting fact is that the Bolognese "physician" adopted a physical model—the Leyden jar—while the Como "physicist" adopted a model derived from the biological world, such as the electric organ of the torpedo or of the eel of Surinam (or Guiana). However, there are some other remarkable facts that should be mentioned. For example, even if Volta eventually discarded Galvani's animal electricity, he was still interested in giving electricity a role in neuromuscular physiology. Indeed, he did not have problems in supposing that "an immaterial agent, such as the Will, exerts a real force and physical action," so that, by acting on its site (the "cerebrum"), it can provide the electrical fluid with the "little impulse" necessary to activate the latter and stimulate the "nervous force," which was the true agent of muscular contraction. In this way the "physicist" Volta made very different entities interact with each other and somehow acknowledged the failure of the attempt to give a physical interpretation of the mechanism responsible for muscular contractions. Instead, he recovered one of the theories developed in the physiological realm some decades earlier, which did not aim at explaining but at simply describing the phenomena involved in animal motion.

Taking a different attitude, Galvani preferred not to publish his research in 1782, as he realized he was not able to give a plausible explanation of the mechanism of muscular contraction but only a description of certain properties that referred to ill-defined concepts such as "nervous force" or "nerve-muscular force." The Bolognese doctor only published his results in *De viribus*, when he thought he had developed a coherent physical explanation of the role of electricity in neuromuscular physiology. This explanation—based on the model of the animal Leyden jar—implied an assimilation, though not complete, of the "nervous force" to the "electric forces," which were entities that were well established, if not completely known, in eighteenth-century natural philosophy. It is exactly this idea that *De viribus electricitatis in motu musculari* wanted to convey from its very title. It refers to the relation between electricity and physiological phenomena like Galvani's paper of 1782 (*On the Nervous Force and Its Relation with Electricity*), but in a much more peremptory and definite way. In this regard we may note that the modern English translation of *De viribus* published in 1953 by Margaret Glover Foley carries a misleading title—*On the Effects of Electricity on Muscular Motion*—that does not take into account the central idea of Galvani's theory, namely that electricity had a role in animal motion, and not simply that electricity caused some effects on this animal function (see Galvani, 1953a).[1]

In previous chapters we have repeatedly discussed how Galvani, at the beginning of *De viribus*, described the initial stages of his research by relating it to an apparently accidental event. The themes of "chance" (which plays a crucial role in the experimental path) and "fortune" (which helps the scientist in his study of natural phenomena) are a recurring

---

[1] Another English edition of *De viribus*, with the translation made by Robert Montraville Green (an anatomist) and with an introduction written by Giulio Cesare Pupilli (a physiologist), bearing exactly the same title (*On the Effects...*) was published in the same year (see Galvani, 1953b). It is interesting that none of the reviews of the two editions appeared on *Isis* (respectively in 1955 for the Glover Foley's translation and in 1956 for the Montraville Green's version) addresses specifically the problem of the mistranslation of the title. This occurs despite the punctual criticisms raised by both reviewers of the English rendering of Galvani's text (Heathcote, 1955; O'Malley, 1956).

element in Galvani's published memoir and appear associated to some moments that, in retrospect, Galvani classified as fundamental for the development of his research. Indeed, chance seems to have accomplished a central role not only in the so-called first experiment—in which contractions at a distance were observed—but also in the first observation of muscular contractions with metals (the so-called second experiment). As we have already seen, this accidental element has been constantly underlined by both Galvani's contemporaries and scholars who have come after, so that chance has become one of the main factors in Galvani's research, and fortune one of the most widespread stereotypes that accompanies his image still to this day.

We do not deny that an accidental component may have played a role in Galvani's research, as it often happens more generally in scientific activity of the past and present times. However, we think that a reconsideration of the relationship between Galvani and chance is needed, both with regard to the role of chance in stimulating and directing Galvani's research and to the way in which Galvani used this term in the published report of his investigative results. We have already reconstructed Galvani's scientific education, his professional and scientific interests, his experimental practice, and the problems that were considered as fundamental in the debate concerning life sciences in the second half of the eighteenth century. All these elements concur to explain why some frogs prepared in a specific manner were placed on his laboratory bench, next to some electrical instruments. Only a scholar like Galvani, who was trying to understand the relations between electricity and muscular motion using those specific animals and those particular instruments and who had in mind certain explicative paradigms developed in the physiology of the time (especially Hallerian irritability) could have had the "fortune" to observe, "by chance," phenomena like the contractions at a distance or the motions of frog's legs in experimental settings involving metals. Therefore, it should now be impossible to claim—as it has been in the past—that Galvani's discovery depended on some weird circumstances such as the preparation of a *bouillon* of frogs that a caring husband used to give to his beloved sick wife, or a situation of some frogs drying on a metal railing. Similarly, it now appears difficult to claim that Galvani's research was a "grope in the dark" until the light of chance came to illuminate the right road, as Emil du Boys-Reymond suggested in the nineteenth century, and many others after him have done (see Chapter 1).

These considerations led us to consider another significant aspect of the relation between Galvani and chance, namely the fact that in presenting the initial stages of his experimental activity, the Bolognese scholar highlighted the role played by chance. In fact, while the reference to chance is largely present in *De viribus*—in which Galvani described the experimental route he followed in developing his theory of animal electricity—it is completely absent in his following works, though in both his *Trattato sull'arco conduttore* and *Memorie sulla elettricità animale* there are descriptions of experiments which were as crucial as those reported in *De viribus*. What happened then to chance and fortune after the publication of *De viribus*? One might think that

Galvani—both lucky and fully prepared to support his good fortune—accidentally ran into some phenomena when his research was in its initial stages and consequently was more open to unpredictable results; then he took a more definite research path in which chance was less relevant than the problems aroused by his theory and by the experiments and arguments of the other protagonists of the debate on animal electricity, especially Volta. As we have seen in our reconstruction of Galvani's research after *De viribus* and of his confrontation with Volta, this explanation is certainly correct but is not the entire story. There is another element that should be taken into consideration in discussing Galvani's use of chance and that relates to the rhetorical dimension of scientific discourse.

When, at the beginning of *De viribus*, Galvani writes that his aim is to "present a brief and accurate account of the discoveries in the same order of circumstance that chance and fortune in part brought to me, and diligence and attentiveness in part revealed" (GDV, p. 363; translated in Galvani, 1953a, p. 45), he wants to clarify that he is going to describe a series of experiments actually performed and of observations actually made. As we have already seen, the accuracy of experimental reports was a central issue for natural philosophers since the origin of modern science and was still quite open in Galvani's time. The choice of presenting research results in a narrative form, the publication of detailed and realistic illustrations, the reference to authoritative witnesses, and the use of "virtual witnessing" were all means to convince the reader that what he was reading had really happened in the way it was described. In this regard, to present an observation as the result of some, at least partially, accidental circumstances may thus be considered another way to record that an event—in Galvani's case, the observation of contractions at a distance or the experiment with the frog placed on the house's terrace railing—had really happened as it was reported.

Therefore, it was not by *chance* that Galvani strengthened the relevance of *chance* in *De viribus* and avoided referring to it in his following works. Indeed *De viribus* was Galvani's first important work in this area and he was not well known in the "Republic of Letters." The "professor from Bologna," or the "Bolognese academician," as Galvani was identified in some reviews of *De viribus*, was completely aware that the first step in making his theory well received was to be accepted by his colleagues as trustworthy, especially as his theory dealt with a very controversial topic like the role of electricity in life processes. A way to gain such credibility depended on convincing other scholars that his report of experiments was accurate and that the observations he described had actually happened. Galvani revealed himself as very able in such a work of persuasion, and he strengthened as much as he could the "reality" of the content of his work. Accordingly, it seems reasonable to suppose that when he highlighted the relevance of casual elements in making the newest and astonishing observations reported in *De viribus* possible, Galvani was simply accomplishing such a persuasive aim by using a very efficient tool. The immediate success of *De viribus* and the great interest aroused by the experiments described in it clearly demonstrate that Galvani was a very efficient scientific communicator, an ability that historians have in general neglected to accord to him.

Among the incorrect appreciations of Galvani's style (and work) that have heavily conditioned the image of the Bologna scholar is that expounded in 1953 by Ierome Bernard Cohen, an important science historian from a physical background, in the introduction to an English edition of *De viribus*. This is a reference work for the English readers interested in Galvani because it is almost universally present in any important university library of the Anglo-American world. Cohen endorses uncritically du Bois-Reymond's stereotype of Galvani as an uncertain pioneer in an unexplored field and that of the superiority of physics on physiology in the research on animal electricity. Moreover, he warns the readers not to expect a clear and logical ensemble of data and conclusions in his writings. It is, however, at the style of Galvani's prose that Cohen levels particularly cutting remarks as, for instance, when he writes: "His sentences are so extremely long that the original thought is apt to be lost altogether—by the reader, and perhaps by the author too" (Cohen, 1953, p. 40).

In doing so, the American historian (who, by the way, in his introduction shows to be very poorly acquainted with the electrophysiological knowledge of his own time) is, in an evident way, applying to Galvani's style (the rich and elaborated Latin of a savant of the eighteenth century) the critical schemes used nowadays for the modern scientific publications usually written in English. This is a serious mistake for a historian who seems to ignore that the short and linear sentences suitable to a language poor of grammatical forms as is modern English (particularly that used in specialized scientific texts) cannot be applied to the Latin of the cultural communication of the *Ancien regime*. Because of the richness of grammatical and syntactic forms, Latin is indeed more inclined to a prose characterized by complex structural architecture, and certainly this was the prose expected by the "philosophers of nature" of Galvani's time, a cultivated readership of vast intellectual interest. Modern scientific texts are, on the contrary, generally addressed to people of surprisingly poor cultural dimensions outside the narrow field of their specific interests.

The harsh remarks of Cohen on the quality of Galvani's prose in *De viribus* are contradicted by recent studies, as, for instance, those of Maria Luisa Altieri-Biagi, a linguist and philologist certainly well acquainted with the Latin and Italian prose of the eighteenth century.[2]

The success of *De viribus* in Galvani's times testifies that not only the readers were capable of grasping the thoughts of the author but also that with this publication Galvani was able to gain a solid credibility in his field of scientific research and to become one of its main protagonists. Once this result was achieved, the rhetoric tools used in *De viribus*

---

[2] It is interesting to notice that in the other English translation also published in 1953 (Galvani, 1953b), there is no harsh remark on Galvani's prose like Cohen's. On the contrary, in the Preface written by Sidney Licht there is an allusion to the "eloquently simple Galvani original" (p. VI); moreover, in the introduction written by Giulio Cesare Pupilli, it appears the following statement: "All of Galvani's writings show exhaustive thoroughness and literary excellence" (p. XVI).

to gain consensus among other scientific researchers were no longer necessary. Accordingly, the style in which Galvani wrote his following works, such as *Trattato dell'arco conduttore* and *Memorie sulla elettricità animale*, was radically different from the one of *De viribus*, and chance disappeared from the experimental reports.

The important role played by the rhetorical and communicative dimension in Galvani's science shows that the Bolognese scholar did not have a narrow view of scientific culture, limited to the scientific milieu of his city, and he was not reluctant to confront the learned community. Contrary to a stereotype in the collective imagery of Galvani, he was very attentive to the cultural dynamics of his time and quite able to take advantage of them. However, to claim that the success gained by *De viribus*—and, more generally, the impact of Galvani's research on the science of his time—could be explained uniquely in terms of successful communicative strategies and stylistic fineries, would be too simplistic. The fact is that the experiments described by Galvani in his *De viribus* were successfully repeated in all scientific centres of the time, and his results were almost universally confirmed by scholars from different parts of the world, such as Florence, Pavia, Paris, London, Germany, and Northern Europe. In fact, the debate concerning Galvani's theory of animal electricity did not relate to the replication of his experiments, which had been a central issue in other fields of scientific research such as Haller's theory of sensibility and irritability, but mainly the interpretations of the experiments and observations carried out by Galvani and the other protagonists of the new field. Apart from the communicative strategies adopted by Galvani in order to persuade his readers about the reliability of what he reported, the credibility he was able to gain in the Republic of Letters was due to the fact that his experiments—especially those with metal arcs and coatings—were successfully repeated by others.

Galvani was perfectly aware that the repeatability of his experiments played a central role in determining the success of his work, as is demonstrated by the presence in his writings of very detailed descriptions of the experimental arrangements, the continuous reference to the cautions necessary in performing and repeating the experiments, and the insistence on the great number of experimental trials before establishing a result. Moreover, in *De viribus*, Galvani clearly distinguished between the description of the experiments, which covered the first three parts of the work, and their theoretical interpretation, which was developed in the final part entitled *Conjecturae et consectaria* (*Conjectures and consequences*). Such a distinction reflected a general attitude deeply grounded in Galvani's scientific practice, as is demonstrated by the fact that also in private documents (such as his laboratory notes) he distinguished between the report of the experimental procedure and considerations about the objectives of the research or the consequences of the results, which were classified as "reflections," "warnings," or "corollaries."

There are several reasons why we chose to consider the reconstruction of Galvani's experimental methodology and scientific practice, as well as of other scholars like Alessandro Volta, as a central issue of our book. Although Galvani has always been considered a great experimenter, nobody has so far analyzed this particular aspect of his scientific activity in full detail. This can only be done by examining all his published and unpublished works, by accurately following his investigative pathway, and by placing his research in the scientific context of his time. We do not want to recover here all the elements that characterize Galvani's approach to natural phenomena, and in particular to the study of living beings. As we have repeatedly emphasized, it is an approach that unites a great systematic effort with a very open-minded attitude toward the variation of the experimental conditions and the exploration of the suggestions and problems coming from previous results, especially when unexpected. Generally speaking, a similar way of reasoning, and in particular the ability to adapt the aims and methods of a research to new and often unexpected developments, characterizes several fields of research especially during periods of fundamental transformations that give rise to new interpretative paradigms or new scientific fields—as happened in Galvani's case with animal electricity. Indeed, Galvani's investigation can be considered as an important stage in the process that transformed eighteenth-century natural philosophy into modern experimental science.

Galvani's scientific approach was based on the interaction of elements derived from the great medical-anatomical tradition inspired by Galilean science and developed in Bologna mainly by Marcello Malpighi, and from the experimental philosophy—of Boylean and Newtonian origin—adopted by the Institute of Sciences in the early part of the eighteenth century. Galvani brought the frogs, which had been already used by Malpighi in some crucial stages of his anatomical research, into the electrician's laboratory, one of the main "theatres" where an experimental philosophy in rapid progress was staged. It is exactly the encounter of the animal preparation—the frog—with the physical instruments and apparatuses of the "room for experiments," and the ways in which Galvani made these elements interact with each other, that characterizes the novelty and originality of his research. Galvani explicitly admitted in his writings that such an encounter was not easily conceivable and, consequently, that in several respects it was problematic. For example, in *Memorie sulla elettricità animale* he addressed Spallanzani, whom he considered "a person quite skilled in experimenting" with the following words:

> Let me advise [you] that in order to prove the truth of the fact it is necessary to try similar experiments several times and with the greatest diligence and accuracy, as they are so delicate and so many are the circumstances that they require both on the part of the animal and on the part of the apparatus, that the experimenter easily lacks the opportunity to see what he is seeking. (GEA, p. 85)

The difficulties "on the part of the apparatus" related both to the organization of the different experimental elements on the working bench (for instance, the way to establish the contact between the conducting arc and animal parts) and to the conditions of the experiment (for instance, the condition of the metals or of the other conducting materials, the presence or absence of humidity in the muscles and nerves, the condition of the plate where these elements were located, etc.). In many regards these difficulties were similar to those typical of the experimentation carried out in the eighteenth-century *cabinet de physique* by natural philosophers, and especially by those who studied electrical phenomena. Other relevant problems depended "on the part of the animal" and contributed to making the experimental results variable and often doubtful. From the beginning of his electrophysiological research, Galvani had been aware of the great variability of animal responses even though all "physical" circumstances remained constant. However, he elaborated a series of strategies that allowed him to solve this experimental problem and, in some cases, he even managed to use such variability as an argument for his theory. One of these strategies consisted of developing an animal preparation that was both appropriate for his research and reproducible in a relatively easy way. As discussed in Chapter 9, this preparation became a standard element in the research path opened by Galvani, and it formed a fundamental aspect of electrophysiological science until the research carried out in the first half of the twentieth century by Alan Hodgkin.

Another relevant strategy adopted by Galvani consisted of attempting a "classification" of the specific variability obtained with the animal preparation. For example, in the *Trattato dell'uso e dell'attività dell'arco conduttore nelle contrazioni dei muscoli* he established a relation between the different degrees of the "natural muscular force of the animal preparation" and the experimental results. Sometimes he used such variability to obtain particular effects, as when, after remarking that the frog's leg was more excitable by electricity than its heart, he observed that a "very languid" torpedo can successfully stimulate the former but not the latter. Although he tried to overcome the experimental variability of the animal preparation by building such classifications, he still maintained a fundamentally cautious attitude. In the *Trattato*, for instance, after considering three degrees of the frog's strength (or vitality), he claimed that "there is no fixed temporal duration of these three different degrees of force, nor are they to be found in every animal"; as a consequence, it was necessary to "repeat the experiments more and more times, in many different animals with varying strength, in order to be sure of the results" (GAC, pp. 9–10).

If these remarks are insufficient to provide an exhaustive answer to the question of Galvani's contribution to the development of the scientific method in modern life sciences, they may contribute to the historical study of the transition from the natural philosopher of the eighteenth century to the experimental scientist of the nineteenth. This issue, however, is beyond the scope of our book (see, for instance, Shapin, 2008); instead, in concluding we would like to return briefly to the relationship between different fields of science and culture, an issue we dealt with in different parts of our book and which somehow characterizes it as the result of a joint project. In fact, during the long period of our collaboration we realized the great difficulty of working together, due to the different cultural fields we belonged to. Indeed, while one of us comes from a "hard" experimental science, the other comes from philosophical studies, and in particular from the history of science. As already noted in the *Preface* to the Italian edition of this book, such a situation implies a kind of bilateral "transformation"; thus, the hard scientist has taken on some characteristics of the historian, and vice versa, and we both discovered the advantage of studying a very complex and interesting episode of eighteenth-century science by integrating perspectives and methodologies from different fields.

The problem of different cultural approaches has re-emerged both in the planning phase of the book—for example, when we had to decide on its editorial style, its public and its conceptual organization—and, maybe even more acutely, during the process of writing. In particular, we had many difficulties with the use of some terms that are commonly used in our ordinary language without being apparently very problematic but that carry very different meanings for people belonging to the opposite sides of the so-called two cultures. Terms like "truth," "objectivity," and "validity" are often used by scientists even if they are interested in the "cultural" aspects of their discipline and aware that scientific knowledge is subjected to personal and social dynamics similar to those that characterize other, more fluid branches of knowledge. This use depends on the deeply held belief that their science is founded on the paradigm of reality and objectivity (though they know well that these are very difficult, if not impossible, to reach). Scientists tend to project this kind of utopian truth from the situation of their laboratory research to the historical work and the reconstruction of past science. Even when dealing with the science of different historical periods, they are often faced with problems like determining what hypothesis is "true," which result is "valid," or which result describes "how things really are." Notably, scientists are inclined to reason in this way even if they are ready to reconsider the complexity of the problems tackled as soon as they note that scientists in the past adopted concepts and notions that—although expressed in terms still in use—are very different from those used nowadays (such as the notions of "force" or "fluid" that are recurrent in Galvani's and Volta's case). Scientists are often inclined to reason in this way even when they note how difficult the attribution of objective—and unconditioned—"truth" to results and scientific hypotheses could be, or even though they clearly derive, from what they have experienced "in the field," that no available solution is able to completely solve a given problem. Analogously, noting that some apparently weak—and somehow "false"—conceptions have revealed themselves as relevant in promoting the scientific debate and scientific progress, is not a good reason to abandon such a tendency. Even in these cases, then, scientists are tempted by terms like "truth," or "objectivity," and by the possibility of referring—at least unconsciously—to the epistemological structures implied by them.

By contrast, historians of science tend to contextualize every single term and to treat with suspicion every reference to "factual" or "objective" elements that is considered out of an historical context. Moreover, when analyzing a scientific work, they avoid more or less consciously any reference to the existence of a hard core of "scientific truth" that lies beneath the structure—conceptual, as well as methodological and argumentative—on which they focus their attention. Even when they are convinced of the existence of natural laws, and although persuaded of the stability of these laws along centuries, historians usually consider natural phenomena as historically—and sometimes socially—constructed.

When scholars from different fields study the same historical topic, they bring not only different skills or different ways of analyzing data, but they also nourish different interests, so that they focus their attention on different aspects of the general problem they are dealing with. Usually, scientists are more interested in reconstructing the so-called history of scientific thought, or in understanding the way in which a past scientist has solved a difficult problem—either experimental or conceptual—without the availability of adequate instruments (like those present in modern laboratories). In a sense, even if they consider history fascinating, scientists tends to make scientific activity atemporal and, consequently, unhistorical. By contrast, even if historians too are interested in the conceptual and methodological aspects of science, they are often more attracted by the social and cultural dimensions of scientific activity and consider it their duty to reconstruct the contexts in which scientific knowledge is produced, the elements which stimulate and promote (or contrast) it, the way in which it is communicated and does (or does not) establish itself, and the identity of those people and institutions that participate in the scientific enterprise.

Historians are often more able than the modern scientists to understand the meaning and consequences of the appearance of a new instrument or experimental procedure in a *cabinet de physique* of the past, which though seemingly casual and rather irrelevant, might be derived by somewhat related research and might reveal themselves very important in the further development of the experimental pathway. Historians are usually more ready to catch the proper value of the metaphorical images that often recur in different but related experimental researches, and they are particularly interested in highlighting personal aspects of the scientific activity—like anxieties, hesitations, difficulties, or enthusiasms—that emerge from the study of private records of the laboratory practice but are absent in the published works. For them, these aspects define the human dimension of scientific knowledge and make scientific activity a fascinating form of human creation.

On the other hand, particularly in some sectors of modern historiography of sciences, the interest of historians in the sociological dimensions of the scientific enterprise, and for other aspects of its context, has become excessively dominant with respect to the more specific, and properly "scientific," features of science. This attitude can lead sometimes to a total miscomprehension of the personality of a scientist and of a particular phase of science history. Some modern texts of science history are, in an evident way, based on a neglect of the properly scientific aspects of the scientific enterprise. Together with the tendency of some science philosophers to problematize science in its core values, it may also have another unfortunate consequence, that of depriving the scientific enterprise of its ethical dimension, an essential characteristic of the intellectual revolution promoted by Galileo and his followers in the early modern age. It is not by chance that also nowadays science is the target of the attacks launched by fundamentalist systems of cultural power (as many "churches" around the world) in their attempt to undermine the revolutionary and incoercible intellectual force of the scientific endeavor (in our times the case concerns particularly Darwin and the theory of evolution, while four centuries ago it implied Galileo and the Copernican system).

Any unilateral vision of science and of its history is limiting and potentially misleading. The plurality of interests, as well as of approaches and methods, that characterizes historians and scientists in their approaches to the history of science, far from representing a sort of obstacle, should be considered a form of enhancement for the historical research. To this end, however, it is very important that they try to interact effectively with each other and to develop a common language that makes communication possible. As Marc Bloch once remarked in a different context, when approaching a specific historical problem, specialists from different fields might be able to find a common ground. He writes: "We simply ask both to bear in mind that historical research will tolerate no autarchy. Isolated, each will understand only by halves, even within his own field of study." And soon afterward he adds: "For the only true history, which can advance only through mutual aid, is universal history" (Bloch, 1949/1992, p. 39). Bloch's remarks, included, as it was, in a book that can be considered a sort of intellectual testament written in a dramatic period of the history of our civilization, can be taken as a suggestion to claim that it would be necessary to work for the recovery of a unity of knowledge beyond disciplinary distinctions and the separation of the two cultures. Only by reasoning in this way can the knowledge that has accumulated during past and present history be shared and made the basis of the conservation of collective memory, which is the real pillar of civilization and something we cannot deny without great danger.

In the specific field of the study of science and its history—the field to which our book belongs—it is necessary to overcome the distinction between the science "in its making" (or science tout court, as modern science and especially experimental science is often supposed to be) and the history of science. In fact, the former risks shutting itself off more and more due to the use of a language which is so technical that even the editors of the most important scientific journals are facing problems with a readership fragmented in many very small communities. In this regard, it is worth noting that some journals like those of the *Nature* group are beginning to publish historical articles written in a comprehensible language. On the other hand, the history of science risks losing interest in the eyes of the lay public if it appears engaged solely with the science of the past and interested only in relatively minor aspects of the scientific enterprise which might appear marginal to contemporary readership.

A modern engineer might correctly object against the need that a "mastery of old Volta's ideas about galvanism" be "necessary to run or repair a dynamo," as considered by Bloch in the passage appearing as an opening quotation of this chapter. On the other hand, all those engaged in science as a fundamental tool of acquisition and communication of human knowledge are entitled to look at the matter with a different perspective and to appreciate the importance of the historical path hidden beneath an instrument of common use in everyday life.

By considering science and its history as a unitary discourse, it is indeed possible to retrieve the cultural value of science, to show the fascination that scientific results can produce, and to highlight new and surprising aspects of scientific notions that seem apparently obvious. As an example, we may say that the reason and the charm of the complexity of membrane phenomena, responsible for the generation of nerve impulses, clearly emerge from the reconstruction of the problems and the difficulties that Galvani and Volta, as well as many other scientists after them, faced at the end of the eighteenth century in order to clarify the role of electricity in neuromuscular physiology.

An active collaboration between scientists and historians can thus favor the re-establishment of a unitary notion of scientific culture, but only on condition that they accept to interact fruitfully, overcoming consolidated mistrusts and trying to develop a common language. Crucially, such a language should not only allow an efficient interdisciplinary discussion but should also make possible the communication of science and its history to a lay public. This is what we have tried to achieve in our book.

BIBLIOGRAPHY

Abbri, F. (1984). *Le terre, l'acqua, le arie. La rivoluzione chimica del Settecento.* Bologna: Il Mulino.

Accademia delle scienze dell'Istituto di Bologna. (1999). *Discorsi e scritti in onore di Luigi Galvani nel bicentenario della morte, 1798–1998.* Bologna: Forni.

Adams, G. (1785). *An essay on electricity* (2nd ed.). London: Logographic Press.

Adanson, M. (1757). *Histoire naturelle du Sénégal.* Paris: J-B. Bauche.

Adelmann, H. B. (1966). *Marcello Malpighi and the evolution of embryology* (Vols. 1–5). Ithaca, NY: Cornell University Press.

Adelmann, H. B. (1978). A supplement to the correspondence of Marcello Malpighi. *Journal of the History of Medicine and Allied Sciences, 33,* 53–73.

Adrian, E. D. (1912). On the conduction of subnormal disturbances in normal nerve. *Journal of Physiology, 45,* 389–412.

Adrian, E. D. (1926). The impulses produced by sensory nerve endings. Part I. *Journal of Physiology, 61,* 49–72.

Adrian, E. D. (1928). *The basis of sensation.* London: Christophers.

Adrian, E. D. (1933). The all-or-nothing reaction. *Ergebnisse der Physiologie, 35,* 744–755.

Adrian, E. D. (1934). Afferent impulses in the vagus and their effect on respiration. *Journal of Physiology, 79,* 332–358.

Adrian, E. D. (1946). *The physical background of perception.* Oxford, England: Clarendon Press.

Adrian, E. D., & Lucas, K. (1912). On the summation of the propagated disturbance in nerve and muscle. *Journal of Physiology, 44,* 68–124.

Adrian, E. D., & Moruzzi, G. (1939). Impulses in pyramidal tract. *Journal of Physiology, 97,* 153–199.

Adrian, E. D., & Zotterman, Y. (1926). Impulses from a single sensory end-organ. *Journal of Physiology, 61,* 151–184.

Aepinus, F. U. T. (1756). Mémoire concernant quelques nouvelles expériences électriques remarquables. *Histoire de l'Académie Royale des Sciences et Belles-Lettres, 12*, 101–121.

Aepinus, F. U. T. (1762). *Recueil de differents mémoires sur la tourmaline*. St. Petersburg: Imprimerie de l'Académie des Sciences.

Aldini, G. (1794). *De animali electricitate dissertationes duae*. Bononiae: Typographia Instituti Scientiarum.

Aldini, G. (1804). *Essai théorique et expérimentale sur le galvanisme* (Vols. 1–2). Paris: Fournier et Fils.

Algarotti, F. (1737). *Il Newtonianismo per le dame, ovvero, dialoghi sopra la luce e i colori*. Napoli: s.n.

Alibert, J-L. (1801). Éloge historique de Louis Galvani. *Mémoires de la Société Médicale d'Émulation, séante a l'École de Médicine de Paris pour l'an VIII* (pp. I–CLXVI). Paris: Chez Richard, Caille et Ravier.

Altieri Biagi, M. L. (1998). La lingua della scienza nel secolo di Luigi Galvani. In M. Poli (Ed.), *Luigi Galvani (1737–1798)* (pp. 39–60). Bologna: Fondazione del Monte di Bologna e Ravenna.

Angelini, A. (Ed.). (1993). *Anatomie accademiche, Vol. 3. L'Istituto delle Scienze e l'Accademia*. Bologna: Il Mulino.

Arago, F-J. (1854). Éloge historique d'Alexandre Volta lu à la séance publique du 26 juillet 1831. In J-A. Barral & J. Baudry (Eds.), *Œuvres complètes* (Vol. 1, pp. 187–240). Paris: Gide et J. Baudry.

Arduini, F. (1988). *I luoghi del conoscere: I laboratori storici e i musei dell'Università di Bologna*. Milan: Silvana.

Armaroli, M. (1981). *Le Cere anatomiche bolognesi del Settecento: [exhibition] Università degli studi di Bologna, Accademia delle scienze, September—november 1981*. Bologna: CLUEB.

Atzori, F. (2004). Terminologia elettrica galvaniana: dal Saggio sulla forza nervea alle Memorie sulla elettricità animale. In F. Frasnedi & R. Tesi (Eds.), *Lingue stili traduzioni. Studi di linguistica e stilistica italiana offerti a Maria Luisa Altieri Biagi* (pp. 323–338). Florence: Cesati.

Atzori, F. (2009). *Glossario dell'elettricismo settecentesco*. Florence: Accademia della Crusca.

Azzoguidi, G. (1775). *Institutiones medicae in usum auditorium suorum* (Vols. 1–2). Bononiae: Typis Joan. Bapt. Sassi.

Bacon, F. (1626). The new Atlantis (posthumous). In W. Rawley (Ed.), *Sylva sylvarum: Or a naturall history. In ten centuries*. London: William Lee. [Modern edition: *Advancement of learning and new atlantis*. London: Oxford University Press, 1974].

Bajon, M. (1774). Mémoire sur un poisson à commotion électrique, connu en Cayenne sous le nom d'Anguille tremblante. *Observations sur la Physique, 3*, 47–58.

Baldelli, F. (1984). Tentativi di regolamentazione e riforme dello Studio bolognese nel Settecento. *Il Carrobbio, 10*, 10–26.

Bancroft, E. (1769). *An essay on the natural history of Guiana*. London: T. Becket & P. A. de Hondt.

Barbensi, G. (1967). Introduzione. In G. Barbensi (Ed.), *Opere scelte* (pp. 11–46). Turin: UTET.

Baruzzi, A. (Ed.). (1989). *From Luigi Galvani to contemporary neurobiology*. Padua: Liviana.

Bastholm, E. (1950). *The history of muscle physiology*. Copenhagen: Munksgaard.

Beach, E. F. (1961). Beccari of Bologna, the discoverer of vegetable protein. *Journal of the History of Medicine, 16*, 354–373.

Beauchamp, K. (2001). *A history of telegraphy. Its history and technology.* Stevenage, England: The Institution of Electrical Engineers.

Beccari, I. B. (1955). *Prolegomena institutionum medicarum* (G. Alberti, Ed.). Bologna: Cappelli.

Beccaria, G. (1753). *Dell'elettricismo artificiale e naturale libri due.* Turin: Campana.

Beccaria, G. (1772). *Elettricismo artificiale.* Turin: V. Franzini.

Beccaria, G. (1793). *Dell'elettricismo. Opere del G. Beccaria delle Scuole Pie, con molte note nuovamente illustrate* (Vols. 1–2). Macerata: Antonio Cortesi.

Belli, S. (1994). *Le camere di fisica dell'Istituto delle Scienze di Bologna (1711–1758).* Unpublished Ph.D. Dissertation, University of Bari.

Belloni, L. (1958). Francesco Redi, biologo. In *Celebrazione dell'Accademia del Cimento nel tricentenario dalla fondazione* (pp. 53–70). Pisa: Domus Galilaeana.

Belloni, L. (1966). La neuroanatomia di Marcello Malpighi. *Physis, 8,* 253–266.

Belloni, L. (1975). Marcello Malpighi and the founding of anatomical microscopy. In M.L. Righini Bonelli & W. R. Shea (Eds.), *Reason, experiment and mysticism in the scientific revolution* (pp. 95–110). New York: Science History Publications.

Benguigui, I. (1984). *Théories électriques du XVIII siècle.* Geneva: Georg.

Bennett, M. V. L. (1961). Modes of operations of electric organs. *Annals of the New York Academy of Sciences, 94,* 458–509.

Bentivoglio, M. (1996). 1896–1996. The centennial of the axon. *Brain Research Bulletin, 41,* 319–325.

Beretta, M., & Grandin, K. (Eds.). (2001). *A galvanized network. Italian-Swedish scientific relations from Galvani to Nobel.* Stockholm: Royal Swedish Academy of Sciences.

Bernabeo, R. A. (Ed.). (1992). *Il Sant'Orsola di Bologna, 1592–1992.* Bologna: Nuova Alfa.

Bernabeo, R. A. (Ed.). (1999). *Luigi Galvani (1798–1998) fra biologia e medicina.* Bologna: Clueb.

Bernard, C. (1865). *Introduction à l'étude de la médecine expérimentale.* Paris: Baillière.

Bernardi, W. (1992). *I fluidi della vita. Alle origini della controversia sull'elettricità animale.* Florence: Olschki.

Bernardi, W. (2000). The controversy on animal electricity in eighteenth-century Italy. Galvani, Volta and others. In F. Bevilacqua, & L. Fregonese (Eds.), *Nuova Voltiana. Studies on Volta and his times* (Vol. 1, pp. 101–112). Milan: Hoepli.

Bernardi, W., & Manzini, P. (Eds.). (1999). *Il cerchio della vita. Materiali di ricerca del Centro studi Lazzaro Spallanzani di Scandiano sulla storia della scienza del Settecento.* Florence: Olschki.

Bernardi, W., & Stefani, M. (Eds.). (2000). *La sfida della modernità. Atti del Convegno internazionale di studi nel bicentenario della morte di Lazzaro Spallanzani.* Florence: Olschki.

Bernstein, J. (1868). Über den zeitlichen Verlauf der negativen Schwankung des Nervenstroms. *Archiv für die gesamte Physiologie des Menschen und der Thiere, 1,* 173–207.

Bernstein, J. (1871). *Untersuchungen über den Erregungsvorgang in Nerven and Muskelsysteme.* Heidelberg: Winter.

Bernstein, J. (1902). Untersuchungen zur Thermodynamik der bioelektrischen Ströme. Erster Theil. *Archiv für die gesamte Physiologie des Menschen und der Thiere, 92,* 521–562.

Bernstein, J. (1912). *Elektrobiologie, die Lehre von den elektrischen Vorgañgen im Organismus auf moderner Grundlage dargestellt.* Braunschweig: F. Vieweg.

Berry, A. J. (1960). *Henry Cavendish*. London: Hutchinson.

Bertholon, P. (1780). *De l'électricité du corps humain dans l'état de santé et de maladie*. Paris: Chez P. Fr. Didot le Jeune.

Bertholon, P. (1786). *De l'électricité du corps humain dans l'état de santé et de maladie* (2nd ed., Vols. 1–2). Paris: Croulbois.

Bertoloni Meli, D. (Ed.). (1997). *Marcello Malpighi, anatomist and physician*. Florence: Olschki.

Bertoloni Meli, D. (2011). *Mechanism, experiment, disease: Marcello Malpighi and seventeenth-century anatomy*. Baltimore, MD: The Johns Hopkins University Press.

Bertucci, P. (2007). *Viaggio nel paese delle meraviglie. Scienza e curiosità nell'Italia del Settecento*. Turin: Bollati Boringhieri.

Bertucci, P., & Pancaldi, G. (Eds.). (2001). *Electric bodies. Episodes in the history of medical electricity*. Bologna: CIS.

Bianchi, N. (1874). *Carlo Matteucci e l'Italia del suo tempo*. Turin: Bocca.

Bilancioni, G. (1923). Galvani come studioso dell'anatomia del naso e dell'orecchio. *Storia della Scienza, 4*, 331–346.

Black, J. (1777). *Experiments upon magnesia alba, quick-lime, and other alcaline substances, by Joseph Black; to which is annexed An essay on the cold produced by evaporating fluids, and some other means of producing cold, by William Cullen*. Edinburgh: W. Creech.

Blezza, F. (1983). *Galvani e Volta. La polemica sull'elettricità*. Brescia: La Scuola.

Bloch, M. (1949). *Apologie pour l'histoire ou métier de l'historien*. Paris: Colin. [English edition, P. Putnam, *The historian's craft*. Cambridge, MA: MIT Press, 1992].

Boerhaave, H. (1743–45). *Praelectiones academicae in proprias institutiones rei medicae. Edidit, et notas addidit Albertus Haller* (Vols. 1–7). Venetiis: Apud Simonem Occhi.

Bohr, N. (1937). Biology and atomic physics. In *Celebrazione del secondo centenario della nascita di Luigi Galvani* (pp. 68–78). Bologna: Luigi Parma.

Bolletti, G. G. (1751). *Dell'origine e de' progressi dell'Instituto delle Scienze di Bologna*. Bologna: Lelio dalla Volpe.

Bonfioli, A. (1792). *Raccolta di poetiche composizioni in lode dell'illustrissimo, ed eccellentissimo Signor Dottore Luigi Galvani*. Bologna: Stamperia di Gaspare de' Franceschi.

Bonino, G. G. (1824–25). *Biografia medica piemontese* (Vols. 1–2). Turin: Bianco.

Borelli, G. A. (1680–81). *De motu animalium* (Vols. 1–2). Rome: Bernabò.

Boring, E. G. (1957). *A history of experimental psychology*. New York: Appleton, Century & Crofts.

Bowditch, H. P. (1871). Eigenthümlichkeiten der Reizbarkeit, welche die Muskelfasern der Herzens zeigen. *Bericht der Sächsische Gesellschaft (Akademie) der Wissenschaften, 23*, 2–39.

Boyle, R. (1744). New experiments physico-mechanical, touching the spring of the air, and its effects. In T. Birch (Ed.), *The works of the Honourable Robert Boyle*. London: A Millar. [Original work published in 1660].

Brazier, M. A. B. (1959). The historical development of neurophysiology. In J. Field (Ed.), *Handbook of physiology* (Vol. 1, pp. 1–58). Washington, DC: American Philosophical Society.

Brazier, M. A. B. (1963). Felice Fontana. In L. Belloni (Ed.), *Essays on the history of Italian neurology* (pp. 107–116). Milan: Istituto di Storia Della Medicina.

Brazier, M. A. B. (1984). *A history of neurophysiology in the 17th and 18th centuries. From concept to experiment* (Vols. 1–2). New York: Raven Press.

Bresadola, M. (1997). La biblioteca di Luigi Galvani. *Annali di Storia delle Università Italiane, 1*, 167–197.

Bresadola, M. (1998). Medicine and science in the life of Luigi Galvani (1737–1798). *Brain Research Bulletin, 46*, 367–380.

Bresadola, M. (2003). At play with nature. Luigi Galvani's experimental approach to muscular physiology. In F. Holmes, J. Renn, & H. Rheinberger (Eds.), *Reworking the bench. Research notebooks in the history of science* (pp. 67–92). Dordrecht: Kluwer.

Bresadola, M. (2008a). Animal electricity at the end of the eighteenth century: the many facets of a great scientific controversy, *Journal of the History of the Neurosciences, 17*, 8–32.

Bresadola, M. (2008b). Galvani, Luigi. In *Enciclopedia of life sciences* (DOI: 10.1002/978047-0015902.a0002801). Chichester, England: John Wiley and Sons.

Bresadola, M. (2008c). Medicina e filosofia naturale: l'indagine sul vivente a Bologna tra Seicento e Settecento. In A. Prosperi (Ed.), *Storia di Bologna* (Vol. 3.2, pp. 375–436). Bologna: Bonomia University Press.

Bresadola, M. (2011a). *Luigi Galvani. Devozione, scienza e rivoluzione*. Bologna: Editrice Compositori.

Bresadola, M. (2011b). A physician and a man of science: Patients, physicians, and diseases in Marcello Malpighi's medical practice. *Bulletin of the History of Medicine, 85*, 193–221.

Bresadola, M. (2011c). Carlo Matteucci and the legacy of Luigi Galvani. *Archives Italiennes de Biologie, 149* Suppl., 3–9.

Bresadola, M., & Pancaldi, G. (Eds.). (1999). *Luigi Galvani International Workshop. Proceedings*. Bologna: CIS.

de Brosses, C. (1799). *Lettres historiques et critiques sur l'Italie*. Paris: Ponthieu. [Partial English translation in C. de Brosses, *Selections from the letters of de Brosses*. London: Kegan Paul, 1897].

Buchwald, J. Z. (Ed.). (1995). *Scientific practice. Theories and stories of doing physics*, Chicago, IL: University of Chicago Press.

Cahan, D. (Ed.). (1993). *Hermann von Helmholtz and the foundations of nineteenth-century science*. Berkeley, CA: University of California Press.

Caldani, L. M. A. (1756). Sull'insensitività, ed irritabilità di alcune parti degli animali. Lettera scritta al chiarissimo, e celebratissimo signore Alberto Haller. In G. B. Fabri (Ed.), *Sulla insensitività ed irritabilità halleriana. Opuscoli di vari autori raccolti da Giacinto Bartolomeo Fabri* (Vol. 1, pp. 269–336). Bologna: Girolamo Corciolani.

Caldani, L. M. A. (1757). Sur l'insensibilité et l'irritabilité de Mr. Haller. Seconde lettre de Mr. Marc Antoine Caldani. In *Mémoires sur la nature sensible et irritable des parties du corps animal* (Vol. 3, pp. 343–490). Lausanne: M-M. Bousquet.

Caldani, L. M. A. (1759). *Lettera terza del Signor Dottore Leopoldo Marc'Antonio Caldani sopra l'irritabilità ed insensibilità Halleriana*. Bologna: San Tommaso d'Aquino.

Caldani, L. M. A. (1778). *Institutiones physiologicae, editio altera retractatior*. Patavii: Typis Cominianis.

Cameron Walker, W. (1936). The detection and estimation of electric charges in the eighteenth century. *Annals of Science, 1*, 66–100.

Cameron Walker, W. (1937). Animal electricity before Galvani. *Annals of Science, 2*, 84–113.

Canguilhem, G. (1952). *La connaissance de la vie*. Paris: Vrin.

Canguilhem, G. (1970). La constitution de la physiologie comme science. In C. Kayser (Ed.), *Physiologie* (Vol. 1, pp. 11–50). Paris: Flammarion.

Canterzani, S. (1791). De animalibus electrico ictu percussis. *De Bononiensi Scientiarum et Artium Instituto atque Academia Commentarii, 7*, 41–44.

Cantor, G. N., & Hodge, M. J. S. (Eds.). (1981). *Conceptions of ether. Studies in the history of ether theories, 1740–1900*. Cambridge, England: Cambridge University Press.

Carradori, G. (1793). *Lettere sopra l'elettricità animale*. Florence: Luigi Carlieri.

Carradori, G. (1817). *Istoria del Galvanismo in Italia*. Florence: All'insegna dell'Ancora.

Castellani, C. (2001). *Un itinerario culturale. Lazzaro Spallanzani*. Florence: Olschki.

Cavallo, T. (1786). *A complete treatise on electricity, in theory and practice with original experiments* (3rd ed., Vols. 1–2). London: C. Dilly.

Cavazza, M. (1990). *Settecento inquieto. Alle origini dell'Istituto delle Scienze di Bologna*. Bologna: Il Mulino.

Cavazza, M. (1995). Laura Bassi e il suo gabinetto di fisica sperimentale. Realtà e mito. *Nuncius, 10*, 715–753.

Cavazza, M. (1996). L'Istituto delle Scienze di Bologna negli ultimi decenni del Settecento. In G. Barsanti, V. Becagli, & R. Pasta (Eds.), *La politica della scienza. Toscana e stati italiani nel tardo Settecento* (pp. 435–450). Florence: Olschki.

Cavazza, M. (1997a). La recezione della teoria halleriana dell'irritabilità nell'Accademia delle Scienze di Bologna. *Nuncius, 12*, 359–377.

Cavazza, M. (1997b). The uselessness of anatomy. Mini and Sbaraglia versus Malpighi. In B. Meli (Ed.), *Marcello Malpighi, anatomist and physician* (pp. 129–145). Florence: Olschki.

Cavazza, M. (1999). Laura Bassi maestra di Spallanzani. In W. Bernardi & P. Manzini (Eds.), *Il cerchio della vita. Materiali di ricerca del Centro studi Lazzaro Spallanzani di Scandiano sulla storia della scienza del Settecento* (pp. 185–201). Florence: Olshki.

Cavendish, H. (1776). An account of some attempts to imitate the effects of the torpedo by electricity. *Philosophical Transactions of the Royal Society of London, 66*, 196–225.

Cecchelli, M. (Ed.). (1982). *Benedetto XIV* (Vols. 1–2). Cento: Centro Studi Girolamo Baruffaldi.

Clark, W., Golinski, J., & Schaffer, S. (Eds.). (1999). *The sciences in enlightened Europe*. Chicago, IL: University of Chicago Press.

Clarke, E., & Jacyna, L. S. (1987). *Nineteenth-century origins of neuroscientific concepts*. Berkeley, CA: University of California Press.

Clarke, E., & O'Malley, C. D. (1968). Nerve function. In *The human brain and spinal cord. A historical study illustrated by writings from antiquity to the twentieth century* (pp. 139–259). Berkeley, CA: University of California Press.

Cohen, I. B., (1953). Introduction. In L. Galvani, *Commentary on the effect of electricity on muscular motion: A translation of Luigi Galvani's De viribus electricitatis in motu musculari commentaries* (pp. 9–42). Norwalk, CT: Burndy Library.

Cohen, I. B. (1966). *Franklin and Newton*. Cambridge, MA: Harvard University Press.

Cole, K. S. (1949). Dynamic electrical characteristics of the squid axon membrane. *Archives de Sciences Physiologiques, 3*, 253–258.

Cole, K. S., & Curtis, H. J. (1939). Electric impedance of the squid giant axon during activity. *Journal of General Physiology, 22*, 649–670.

Coleman, W., & Holmes, F. L. (Eds.). (1988). *The investigative enterprise. Experimental physiology in nineteenth-century medicine*. Berkeley, CA: University of California Press.

Comitato per le celebrazioni del VII centenario degli Ospedali di Bologna. (1960). *Sette secoli di vita ospitaliera in Bologna*. Bologna: Cappelli.

Conrad, L. I. (1995). *The Western medical tradition, 800 BC to AD 1800*. Cambridge, England: Cambridge University Press.

Contardi, S. (2002). *La Casa di Salomone a Firenze. L'Imperiale e Reale Museo di fisica e storia naturale (1775–1801)*. Florence: Olschki.

Cosmacini, G. (1997). *L'arte lunga. Storia della medicina dall'antichità a oggi*. Rome: Laterza.

Cranefield, P. F. (1957). The organic physics of 1847 and the biophysicists of today. *Journal of the History of Medicine and Allied Sciences, 12*, 407–423.

Cremante, R., & Tega, W. (Eds.). (1984). *Scienza e letteratura nella cultura italiana del Settecento*. Bologna: Il Mulino.

Croone, W. (1664). *De ratione motus musculorum*. London: J. Hayes. [Afterward, in T. Willis, *Cerebri anatome nervorumque descriptio et usus*. Amsterdam: Apud Casparum Commelinum, 1667].

Crosland, M. P. (1963). The development of chemistry in the eighteenth century. *Studies on Voltaire and the Eighteenth Century, 24*, 369–441.

Cunningham, A., & French, R. (Eds.). (1990). *The medical Enlightenment of the eighteenth century*. Cambridge, England: Cambridge University Press.

Curtis, H., & Barnes, N. S. (1985). *Invitation to biology* (4th ed.). New York: Worth.

Dacome, L. (2007). Women, wax and anatomy in the "century of things." *Renaissance Studies, 21*, 522–550.

Daintih, J., & Gjertsen, D. (1999). *A dictionary of scientists*. Oxford, England: Oxford University Press.

Dallari, U. (1888–1924). *I rotuli dei lettori legisti e artisti dello Studio bolognese dal 1384 al 1799* (Vols. 1–5). Bologna: R. Deputazione di Storia Patria.

Daston, L. (1991). The ideal and reality of the Republic of Letters in the Enlightenment. *Science in Context, 4*, 367–386.

Davy, H. (1829). An account of some experiments on the torpedo. *Philosophical Transactions of the Royal Society of London, 119*, 15–18.

Davy, H. (1834). Observations on the torpedo, with an account of some additional experiments on the electricity. *Philosophical Transactions of the Royal Society of London, 124*, 259–278.

Davy, J. (1832). An account of some experiments and observations on the torpedo (*Raia torpedo, Linn.*). *Philosophical Transactions of the Royal Society of London, 122*, 259–278.

Davy, J. (1834). Observations on the torpedo, with an account of some additional experiments on its electricity. *Philosophical Transactions of the Royal Society of London, 124*, 531–550.

Delbourgo, J. (2006). *A most amazing scene of wonders: Electricity and enlightenment in early America*. Cambridge, MA: Harvard University Press.

De Palma, A., & Pareti, G. (2011). Bernstein's long path to membrane theory: Radical change and conservation in nineteenth-century German electrophysiology. *Journal of the History of the Neurosciences, 20*, 306–337.

Dibner, B. (1952). *Galvani-Volta. A controversy that led to the discovery of electricity*. Norwalk, CT: Burndy Library.

Di Marco, M. (1971). Spiriti animali e meccanicismo filosofico in Descartes. *Physis, 13*, 21–70.

Dini, A. (1991). *Vita e organismo. Le origini della fisiologia sperimentale in Italia*. Florence: Olschki.

Doyle, D. A., Morais, C. J., Pfuetzner, R. A., Kuo, A., Gulbis, J. M., Cohen, S. L., & MacKinnon, R. (1998). The structure of the potassium channel. Molecular basis of K$^+$ conduction and selectivity. *Science, 280*, 69–77.

du Bois-Reymond, E. (1843). Vorläufiger Abriss einer Untersuchung über den sogenannter Froschstrom und über die elektromotorischen Fische. *Poggendorff's Annalen der Physik und Chemie, Series II, 28*, 1–30.

du Bois-Reymond, E. (1848–84). *Untersuchungen über thierische Elektricität* (Vols. 1–2). Berlin: Reimer.

du Bois-Reymond, E. (1852). *On animal electricity: being an abstract of the discoveries of Emil Du Bois-Reymond* (H. Bence Jones, Trans.). London: Churchill.

Duchesneau, F. (1982). *La physiologie des lumières. Empirisme, modèles et théories*. The Hague: Nijhoff.

Dulieu, L. (1961). L'abbé Bertholon. *Cahiers Lyonnais d'Histoire de la Médecine, 6*(2), 3–25.

Eamon, W. (1990). *From the secrets of nature to public knowledge*. Cambridge, England: Cambridge University Press.

Eandi, G. A. (1792). Ragguaglio delle sperienze del Sig. Luigi Galvani accademico bolognese estratto da una lettera diretta al Sig. Conte Prospero Balbo. *Giornale Fisico-Medico, 2*, 94–109.

Emch-Dériaz, A. (1992). *Tissot. Physician of the Enlightenment*. New York: Lang.

Fabri, G. B. (Ed.). (1757). *Sulla insensitività ed irritabilità halleriana. Opuscoli di vari autori raccolti da Giacinto Bartolomeo Fabri* (Vols. 1–2). Bologna: Per Girolamo Corciolani.

Fabri, G. B. (Ed.). (1759). *Sulla insensitività ed irritabilità halleriana. Supplimento agli opuscoli di vari autori raccolti et in due parti diviso da Giacinto Bartolomeo Fabri*. Bologna: Per Girolamo Corciolani.

Fantuzzi, G. (1781–94). *Notizie degli scrittori bolognesi* (Vols. 1–9). Bologna: San Tommaso d'Aquino.

Faraday, M. (1839a). Experimental researches in electricity—Fifteenth series. Notice on the character and direction of the electric force of the Gymnotus. *Philosophical Transactions of the Royal Society of London, 129* (Pt. 1), 1–12.

Faraday, M. (1839b). *Experimental researches in electricity*. London: R. & J. E. Taylor.

Fermin, P. (1769). *Description générale, historique, géographique et physique de la colonie de Surinam, contenant ce qu'il y a de plus curieux & de plus remarquable* (Vols. 1–2). Amsterdam: E. van Harrevelt.

Ferrari, G. (1987). Public anatomy lessons and the carnival. The anatomy theatre of Bologna. *Past and Present, 117*, 50–106.

Field, A. J. (1994). French optical telegraphy, 1793–1855. Hardware, software, administration. *Technology and Culture, 35*, 325–347.

Findlen, P. (1993). Science as a career in Enlightenment Italy. The strategies of Laura Bassi. *Isis, 84*, 441–469.

Finger, S., & Piccolino, M. (2011). *The shocking history of electric fishes: From ancient epochs to the birth of modern neurophysiology*. New York: Oxford University Press.

Finger, S., Piccolino, M., & Stahnisch, F. W. (2013a). Alexander von Humboldt: Galvanism, Animal Electricity, and Self-Experimentation. Part 1: Formative Years, Naturphilosophie, and Galvanism. *Journal of the History of the Neurosciences*, DOI 10.1080/0964704X.2012.732727

Finger, S., Piccolino, M., & Stahnisch, F. W. (2013b). Alexander von Humboldt: Galvanism, Animal Electricity, and Self-Experimentation. Part 2: The Electric Eel, Animal Electricity, and Later Years. *Journal of the History of the Neurosciences*, DOI 10.1080/0964704X.2012.732728

Finger, S., & Wade, N. J. (2002). The neuroscience of Helmholtz and the theories of Johannes Müller. Part I: Nerve cell structure, vitalism, and the nerve impulse. *Journal of the History of the Neurosciences*, *11*, 136–155.

Focaccia, M., & Simili, R. (2007). Luigi Galvani, physician, surgeon, physicist: From animal electricity to electro-physiology. In H. Whitaker, C. U. M. Smith, & S. Finger (Eds.), *Brain, mind and medicine. Essays in eighteenth century neuroscience* (pp. 145–159). New York: Springer.

Fontana, F. (1757). Dissertation épistolaire [...] adressée au r.p. Urbain Tosetti. In A. von Haller (Ed.), *Mémoires sur la nature sensible et irritable des parties du corps animal* (Vols. 1–4, pp. 157–243). Lausanne, Switzerland: M-M. Bousquet.

Fontana, F. (1781). *Traité sur le venin de la vipère, sur les poisons américains, sur le laurier-cerise et sur quelques autres poisons végétaux* (Vols. 1–2). Florence: s.n.

Fontana, F. (1792). Articolo di Lettera all'Abate Mangili. *Giornale Fisico-Medico*, *4*, 116–118.

Fordyce, G. (1788). The Croonian lecture on muscular motion. *Philosophical Transactions of the Royal Society of London*, *78*, 23–36.

Foster, M. (1901). *Lectures on the history of physiology during the 16th, 17th and 18th centuries*. Cambridge, England: Cambridge University Press.

Fowler, R. (1793). *Experiments and observations relative to the influence lately discovered by Mr. Galvani and commonly called animal electricity*. Edinburgh: T. Duncan.

Fox, R. (1974). The rise and fall of Laplacian physics. *Historical Studies in the Physical Sciences*, *4*, 89–136.

Frängsmyr, T., Heilbron, J. L., & Rider, R. E. (Eds.). (1990). *The quantifying spirit in the 18th century*. Berkeley, CA: University of California Press.

Franklin, B. (1941). *Benjamin Franklin's experiments. A new edition of Franklin's experiments and observations on electricity*. (I. B. Cohen, Ed.). Cambridge, MA: Harvard University Press.

Franklin, B. (1751–1754). *Experiments and observations on electricity, made at Philadelphia in America*. London: E. Cave.

Fregonese, L. (1999). *Volta. Teorie ed esperimenti di un filosofo naturale*. Milano: Le Scienze.

Fulton, J. F. (1926). *Muscular contraction and the reflex control of movement*. Baltimore, MD: Williams & Wilkins.

Fulton, J. F. (1966). *Selected readings in history of physiology* (2nd ed.). Springfield, IL: Thomas.

Galilei, G. (1623). *Il Saggiatore: Nel quale con bilancia esquisita e giusta si ponderano le cose contenute nella Libra astronomica e filosofica di Lotario Sarsi*. Rome: Giacomo Mascardi.

Galilei, G. (1632). *Dialogo di Galileo Galilei sopra i due massimi sistemi del mondo*. Florence: Gio. Batista Landini.

Galvani, L. (1762). *De ossibus. Theses physico-medico-chirurgicae*. Bononiae: S. Thomas Aquinatis.

Galvani, L. (1767). De renibus atque ureteribus volatilium. *De Bononiensi Scientiarum et Artium Instituto atque Academia Commentarii*, *5*, 500–508.

Galvani, L. (1783). De volatilium aure. *De Bononiensi Scientiarum et Artium Instituto atque Academia Commentarii*, *6*, 420–424.

Galvani, L. (1787). *L'elettricità naturale*. In L. Galvani, *1967*, *Opere scelte* (pp. 190–212). Turin: UTET.

Galvani, L. (1791). *De viribus electricitatis in motu musculari commentarius*. *De Bononiensi Scientiarum et Artium Instituto atque Academia Commentarii*, 7, 363–418.

Galvani, L. (1792). *De viribus electricitatis in motu musculari commentarius, cum Ioannis Aldini dissertatione et notis*. Mutina: Apud Societatem Typographicam.

Galvani, L. (1794a). *Dell'uso e dell'attività dell'arco conduttore nelle contrazioni dei muscoli*. Bologna: San Tommaso d'Aquino.

Galvani, L. (1794b). *Supplemento al Trattato dell'uso e dell'attività dell'arco conduttore nelle contrazioni de' muscoli*. Bologna: San Tommaso d'Aquino.

Galvani, L. (1797). *Memorie sulla elettricità animale di Luigi Galvani P. Professore di Notomia nella Università di Bologna al celebre Abate Lazzaro Spallanzani Pubblico professore nella Università di Pavia. Aggiunte alcune elettriche esperienze di Gio. Aldini P. prof. di Fisica*. Bologna: Sassi.

Galvani, L. (1841). *Opere edite ed inedite del professore Luigi Galvani. Raccolte e pubblicate per cura dell'Accademia delle Scienze dell'Istituto di Bologna* (S. Gherardi, Ed.). Bologna: Dall'Olmo.

Galvani, L. (1888). *Orazione di Luigi Galvani letta nel 25 novembre 1782 per la laurea del nipote Giovanni Aldini*. Bologna: Monti.

Galvani, L. (1937a). *Memorie ed esperimenti inediti*. Bologna: Cappelli.

Galvani, L. (1937b). *Il Taccuino di Luigi Galvani*. Bologna: Zanichelli.

Galvani, L. (1953a). *Commentary on the effects of electricity on muscular motion* (I. B. Cohen, Ed., M. Glover Foley, Trans.). Norwalk, CT: Burndy Library.

Galvani, L. (1953b). *Commentary on the effect of electricity on muscular motion: A translation of Luigi Galvani's De viribus electricitatis in motu musculari commentarius*. (R. Montraville Green, Trans.). Cambridge, MA: Elizabeth Licht.

Galvani, L. (1965). *Lezioni inedite di ostetricia* (L. Giardina, Ed.). Bologna: CLUEB.

Galvani, L. (1966). *De ossibus, lectiones quattuor* (M. Pantaleoni & G. Calboli, Eds.). Bologna: Compositori.

Galvani, L. (1967). *Opere scelte* (G. Barbensi, Ed.). Turin: UTET.

Galvani, L. (2009). *Un laboratorio sperimentale di ostetricia*. (M. Focaccia, Ed.). Bologna: Pendragon.

Gandolfi, G. (1817). *Elogio a Germano Azzoguidi*. Padua: Tipografia del Seminario.

Ganot, A. (1859). *Cours de physique purement expérimentale à l'usage des gens du monde*. Paris: Chez l'auteur.

Gardini, F. G. (1774). *L'applicazione delle nuove scoperte del fluido elettrico agli usi della ragionevole medicina, per A*. Genoa: Scionico.

Gardini, F. G. (1780). *De effectis electricitatis in homine dissertatio*. Genoa: Haeredes Adae Scionici.

Garelli, A. (Ed.). (1885). *Lettere inedite alla celebre Laura Bassi, scritte da illustri Italiani e stranieri, con biografia*. Bologna: Cenerelli.

Geddes, L. A., & Hoff, H. E. (1971). The discovery of bioelectricity and current electricity. *IEEE Spectrum*, *8*(12), 38–46.

Geison, G. (1978). *Michael Foster and the Cambridge school of physiology*. Princeton, NJ: Princeton University Press.

Gherardi, S. (1841). *Rapporto sui manoscritti del celebre professore Luigi Galvani.* In S. Gherardi, (Ed.), *Opere edite ed inedite del professore Luigi Galvani. Raccolte e pubblicate per cura dell'Accademia delle Scienze dell'Istituto di Bologna* (pp. 3–106). Bologna: Dall'Olmo.

Ghiretti, F. (1996). *I pesci elettrici.* Bologna: Calderini.

Gigli Berzolari, A. (1993). *Alessandro Volta e la cultura scientifica e tecnologica tra Settecento e Ottocento.* Milan: Cisalpino.

Gilbert, W. (1600). *De magnete, magneticisque corporibus, et de magno magnete tellure.* London: P. Short.

Gillis, J. (1973). Kekulé von Stradonitz, (Friedrich) August. In C. C. Gillispie (Ed.), *Dictionary of scientific biography* (Vol. 7., pp. 279–283). New York: Scribner.

Gillispie, C. C. (Ed.). (1970–80). *Dictionary of scientific biography* (Vols. 1–18). New York: Scribner.

Gillispie, C. C. (1980). *Science and polity in France at the end of the old regime.* Princeton, NJ: Princeton University Press.

Girardi, M. (1786). Saggio di Osservazioni Anatomiche intorno agli Organi Elettrici della Torpedine. *Memorie di Matematica e di Fisica della Società Italiana delle Scienze, 3,* 553–570.

Gliozzi, M. (1937). *L'elettrologia fino al Volta* (Vols. 1–2). Naples: Loffredo.

Glisson, F. (1677). *Tractatus de ventriculo et intestinis. Cui praemittitur alius, De partibus continentibus in genere & in specie, de iis abdominis.* London: Apud Henricum Brome.

Goethe, J. W. (1810). *Zur Farbenlehre* (Vols. 1–2). Tübingen: J. G. Cotta.

Goldman, D. E. (1943). Potential, impedance, and rectification in membranes. *Journal of General Physiology, 27,* 37–60.

Golinski, J. (1992). *Science as public culture. Chemistry and Enlightenment in Britain, 1760–1820.* Cambridge, England: Cambridge University Press.

Gooding, D., Pinch T., & Schaffer, S. (Eds.). (1989). *The uses of experiment. Studies in the natural sciences.* Cambridge, England: Cambridge University Press.

Gotch, F. (1902). The submaximal electrical response of nerve to a single stimulus. *Journal of Physiology, 28,* 395–416.

Govoni, P. (2002). *Un pubblico per la scienza. La divulgazione scientifica nell'Italia in formazione.* Rome: Carocci.

Grapengiesser, C. J. C. (1802). *Essai sur le Galvanisme. Extrait du huitième volume de la Bibliothèque Germanique Médico-Chirurgicale, par Brewer et Delaroche.* Paris: Croullebois et Fuchs.

Grmek, M. (1968). First steps in Claude Bernard's discovery of the glycogenic function of the liver. *Journal of the History of Biology, 1,* 141–154.

Grmek, M. (1982). La théorie et la pratique de l'expérimentation biologique au temps de Spallanzani. In G. Montalenti & P. Rossi (Eds.), *Lazzaro Spallanzani e la biologia del Settecento* (pp. 321–352). Florence: Olschki.

Grmek, M. (1991). *Claude Bernard et la méthode expérimentale.* Paris: Payot.

Grundfest, H. (1965). Julius Bernstein, Ludimar Hermann and the discovery of the overshoot of the spike. *Archivio Italiano di Fisiologia, 103,* 483–490.

Guillemin, C-M. (1860). Mémoire sur la propagation des courants dans les fils télégraphiques. *Annales de Chimie et de Physique, 60,* 385–448.

Guisan, F. L. (1819). *De gymnoto electrico.* Tubingae: Typis Reisianis.

Hackmann, W. D. (1978). *Electricity from glass. The history of the frictional electrical machine, 1600–1850.* Alphen aan den Rijn: Sijthoff & Noordhoff.

Hales, S. (1733). *Statical essays, containing haemastaticks, or, an account of some hydraulick and hydrostatical experiments* (Vols. 1-2). London: W. Innys & R. Manby.

Haller, A. von. (1753). De partibus corporis humani sensibilibus et irritabilibus. *Commmentarii Societatis regiae Scientiarum Gottingensis, 2,* 114-158.

Haller, A. von. (1755). *Dissertation sur les parties irritables et sensibles des animaux [...] Traduit par M. Tissot.* Lausanne: M-M. Bousquet.

Haller, A. von. (Ed.). (1756-60). *Mémoires sur la nature sensible et irritable des parties du corps animal* (Vols. 1-4). Lausanne: M-M. Bousquet.

Haller, A. von. (1757-1766). *Elementa physiologiae corporis humani* (Vols. 1-8). Lausanne: Sumptibus M-M. Bousquet et Sociorum.

Haller, A. von. (1762). *Elementa physiologiae corporis humani. Tomus IV: Cerebrum. Nervi. Musculi.* Lausanne: Sumptibus Francisci Grasset et Sociorum.

Hamill, O. P., Marty, A., Neher, E., Sakmann, B., & Sigworth, F. J. (1981). Improved patch-clamp techniques for high-resolution recording from cells and cell-free membrane patches. *Pflügers Archive, 391,* 85-100.

Hartline, H. K. (1934). Intensity and duration in the excitation of single photoreceptor units. *Journal of Cellular and Comparative Physiology, 5,* 229-239.

Heathcote, N. H. de V. H. (1955). Review of the "Commentary on the effects of electricity on muscular motion by Luigi Galvani." *Isis, 46,* 305-309.

Heilbron, J. L. (1978) Volta's path to the battery. In G. Dubpernell & J. H. Westbrook (Eds.), *Proceedings of the Symposium on Selected Topics in the History of Electrochemistry* (pp. 39-65). Princeton, NJ: The Electrochemical Society.

Heilbron, J. L. (1979). *Electricity in the 17th and 18th centuries. A study of early modern physics.* Berkeley, CA: University of California Press (New Edition 1999, Mineola, NY: Dover).

Heilbron, J. L. (1991). The contributions of Bologna to galvanism. *Historical Studies in the Physical Sciences, 22,* 57-85.

Heilbron, J. L. (1993). *Weighing imponderables and other quantitative science around 1800.* Berkeley, CA: University of California Press.

Heilbron, J. L. (2000). Analogy in Volta's exact natural philosophy. In F. Bevilacqua & L. Fregonese (Eds.), *Nuova Voltiana. Studies on Volta and his times* (pp. 1-23). Milan: Hoepli.

Helmholtz, H. von. (1850a). Vorläufiger Bericht über die Fortpflanzungsgeschwindigkeit der Nervenreizung. *Archiv für Anatomie, Physiologie und Wissenschaftliche Medicin,* 71-73.

Helmholtz, H. von. (1850b). Messungen über den zeitlichen Verlauf der Zuckung animalischer Muskeln und die Fortpflanzungsgeschwindigkeit der Reizung in den Nerven. *Archiv für Anatomie, Physiologie und Wissenschaftliche Medicin,* 276-364.

Helmholtz, H. von. (1850c). Note sur la vitesse de propagation de l'agent nerveux dans les nerfs rachidiens. *Compte Rendu de l'Académie des Sciences, 30,* 204-206.

Helmholtz, H. von. (1851). Deuxième note sur la vitesse de propagation de l'agent nerveux. *Compte Rendu de l'Académie des Sciences, 32,* 262-265.

Helmholtz, H. von. (1852). Messungen über Fortpflanzungsgeschwindigkeit der Reizung in den Nerven. Zweite Reihe. *Archiv für Anatomie, Physiologie und Wissenschaftliche Medicin,* 199-216.

Helmholtz, H. von. (1874). On the later views of the conduction of the electricity and magnetism. In *Annual Report of the Smithsonian Institution for 1873* (pp. 247-253). Washington, DC: Smithsonian.

Herlitzka, A. (1937a). Elettrofisiologia. In F. Bottazzi (Ed.), *Trattato di fisiologia, Vol. 1. Fisiologia generale e fisiologia dei tessuti* (pp. 531–630). Milan: Vallardi.

Herlitzka, A. (1937b). Fisiologia generale dei nervi. In F. Bottazzi (Ed.), *Trattato di fisiologia, Vol. 1. Fisiologia generale e fisiologia dei tessuti* (pp. 428–529). Milan: Vallardi.

Hermann, L. (Eds.). (1879). *Handbuch der physiologie, Vol. 2. Physiologie des Nervensystems*. Leipzig: Vogel.

Hermann, L. (1899). Zur Theorie der Erregungsleitung und der elektrischen Erregung. *Archiv für die Gesamte Physiologie des Menschen und der Thiere, 75*, 574–590.

Hill, A. V. (1921). The temperature coefficient of the velocity of a nervous impulse. *Journal of Physiology, 54*, 332–334.

Hill, A. V. (1932). *Chemical wave transmission in nerve*. Cambridge, England: Cambridge University Press.

Hille, B. (2001). *Ionic channels of the excitable membranes*. Sunderland, MA: Sinauer.

Hodgkin, A. L. (1937). Evidence for electric transmission in nerve. Part II. *Journal of Physiology, 90*, 211–232.

Hodgkin, A. L. (1951). The ionic basis of electrical activity in nerve and muscle. *Biological Reviews, 26*, 339–409.

Hodgkin, A. L. (1964). *The conduction of the nervous impulse*. Liverpool, England: Liverpool University Press.

Hodgkin, A. L. (1972). The ionic basis of nervous conduction. The Nobel lecture for 1963. In *Nobel lectures. Physiology and medicine, 1963–1970* (pp. 32–48). Amsterdam: Elsevier.

Hodgkin, A. L. (1976). Chance and design in electrophysiology. An informal account of certain experiments on nerve carried out between 1934 and 1952. *Journal of Physiology, 263*, 1–21.

Hodgkin, A. L. (1977). Lord Adrian, 1889–1977. *Nature, 269*, 543–544.

Hodgkin, A. L. (1979). Edgar Douglas Adrian, Baron Adrian of Cambridge, 1889–1977. *Biographical Memoirs of Fellows of the Royal Society, 25*, 1–74.

Hodgkin, A. L. (1992). *Chance & design: Reminiscences of science in peace and war*. Cambridge, England: Cambridge University Press.

Hodgkin, A. L., & Huxley, A. F. (1939). Action potentials recorded from inside a nerve fibre. *Nature, 144*, 710–711.

Hodgkin, A. L., & Huxley, A. F. (1952a). Currents carried by sodium and potassium ions through the membrane of the giant axon of *Loligo*. *Journal of Physiology, 116*, 449–472.

Hodgkin, A. L., & Huxley, A. F. (1952b). The components of membrane conductance in the giant axon of *Loligo*. *Journal of Physiology, 116*, 473–496.

Hodgkin, A. L., & Huxley, A. F. (1952c). The dual effect of membrane potential on sodium conductance in the giant axon of *Loligo*. *Journal of Physiology, 116*, 497–506.

Hodgkin, A. L., & Huxley, A. F. (1952d). A quantitative description of membrane current and its application to conduction and excitation in nerve. *Journal of Physiology, 117*, 500–544.

Hodgkin, A. L., Huxley, A. F., & Katz, B. (1952). Measurement of current-voltage relations in the membrane of the giant axon of *Loligo*. *Journal of Physiology, 116*, 424–448.

Hodgkin, A. L., & Katz, B. (1949). The effect of sodium ions on the electrical activity of the giant axon of the squid. *Journal of Physiology, 108*, 37–77.

Hoff, H. E. (1936). Galvani and the pre-galvanian electrophysiologists. *Annals of Science, 1*, 157–172.

Hoff, H. E. (1942). The history of the refractory period. A neglected contribution of Felice Fontana. *Yale Journal of Biology and Medicine, 14,* 635–672.

Holmes, F. L. (1974). *Claude Bernard and animal chemistry. The emergence of a scientist.* Cambridge, MA: Harvard University Press.

Holmes, F. L. (1981). The fine structure of scientific creativity. *History of Science, 19,* 60–71.

Holmes, F. L. (1985). *Lavoisier and the chemistry of life. An exploration of scientific creativity.* Madison: University of Wisconsin Press.

Holmes, F. L. (1990). Laboratory notebooks. Can the daily record illuminate the broader picture? *Proceedings of the American Philosophical Society, 134,* 349–366.

Holmes, F. L. (1993). The old martyr of science. The frog in experimental physiology. *Journal of the History of Biology, 26,* 311–328.

Holmes, F. L. (1998). *Antoine Lavoisier. The next crucial year, or, the sources of his quantitative method in chemistry.* Princeton, NJ: Princeton University Press.

Holmes, F. L. (1999). *Galvani on respiration and inflammation.* In M. Bresadola & G. Pancaldi (Eds.), *Luigi Galvani International Workshop. Proceedings* (pp. 99–113). Bologna: CIS.

Holmes, F. L. (2004). *Investigative pathways. Patterns and stages in the careers of experimental scientists.* New Haven, CT, and London: Yale University Press.

Home, R. W. (1992). *Electricity and experimental physics in eighteenth-century Europe.* Aldershot, England: Variorum.

Humboldt, A. von. (1797). *Versuche über die gereizte Muskel- und Nervenfaser. Nebst Vermuthungen über den chemischen Process des Lebens in der Thier- und Pflanzenwelt* (Vols. 1–2). Posen: Decker und Compagnie.

Hunter, J. (1773). Anatomical observations on the torpedo. *Philosophical Transactions of the Royal Society of London, 63,* 481–489.

Hunter, J. (1775). An account of the Gymnotus electricus. *Philosophical Transactions of the Royal Society of London, 65,* 395–407.

Huxley, A. F. (1980). *Reflections on muscle.* Liverpool, England: Liverpool University Press.

Huxley, A. F. (1992). Kenneth Stewart Cole. July 10, 1900–April 18, 1984. *Biographical Memoirs of Fellows of the Royal Society of London, 38,* 96–110.

Huxley, A. F. (2000). Sir Alan Lloyd Hodgkin, O.M., K.B.E., February 5, 1914-December 20, 1998. *Biographical Memoirs of Fellows of the Royal Society of London, 46,* 219–241.

Ingenhousz, J. (1775). Extract of a letter from Dr. John Ingenhousz to Sir John Pringle, Bart. p.r.s., containing some experiments on the torpedo, made at Leghorn, January 1st 1773 (after having been informed of those made by Mr. Walsh). Dated Salzburg, March 27th 1773. *Philosophical Transactions of the Royal Society of London, 65,* 1–4.

Ingenhousz, J. (1782). *Vermischte Schriften phisisch-medizinischen Inhalts.* Vienna: J. P. Krauss.

Innes Williams, B. (2000). *The matter of motion and Galvani's frogs.* Bletchingdon, England: Rana.

Jang, Y., Lee, A., Chen, J., Ruta, V., Cadene, M., Chait, B. T., & MacKinnon, R. (2003). X-ray structure of a voltage-dependent K$^+$ channel. *Nature, 423,* 33–41.

Kato, G. (1924). *The theory of decrementless conduction.* Tokyo: Nankodo.

Kato, G. (1970). The road a scientist followed. *Annual Review of Neuroscience, 32,* 1–20.

Keilin, D. (1966). *The history of cell respiration and cytochrome.* Cambridge, England: Cambridge University Press.

Kekulé, A. (1867). *Chemie der Benzolderivate oder der aromatischen Substanzen*. Erlangen: Enke.

Kellaway, P. (1946). The part played by electric fish in the early history of bioelectricity and electrotherapy. *Bulletin of the History of Medicine, 20*, 112–137.

Kelvin, W. T. (1855). On the theory of electric telegraph. *Proceedings of the Royal Society of London, 7*, 382–389.

Keynes, R. D. (1956). The generation of electricity by fishes. *Endeavour, 15*, 215–222.

Keynes, R. D., & Martins-Ferreira, H. (1953). Membrane potentials in the electroplates of the electric eel. *Journal of Physiology, 119*, 315–351.

Kimura, J. 1989 *Electrodiagnosis in diseases of nerve and muscles. Principles and practice*. Oxford, England: Oxford University Press.

Kipnis, N. (1987). Luigi Galvani and the debate on animal electricity, 1791–1800. *Annals of Science, 44*, 107–142.

Knoefel, P. K. (1984). *Felice Fontana. Life and works*. Trento: Società di Studi Trentini di Scienze Storiche.

Koehler, P., Finger, S., & Piccolino, M. (2009). The "eels" of South America: Mid-18th-century Dutch contributions to the theory of animal electricity. *Journal of the History of Biology, 42*, 715–763.

Koenigsberger, L. (1906). *Hermann von Helmholtz*. Braunschweig: Vieweg.

Krüger, J. G. (1744). *Der Weltweisheit und Artzneygelahrheit Doctors und Professors auf der Friedrichs Universität Zuschrifft an seine Zuhörer, worinnen er Ihnen seine Gedancken von der Electricität mittheilet und Ihnen zugleich seine künftige Lectionen bekant macht*. Halle: Verlegts Carl Hermann Hemmerde.

Kuhn, T. S. (1962). *The structure of scientific revolutions*. Chicago, IL: University of Chicago Press.

Laghi, T. (1757a). Cl. Viro D. Cesareo Pozzi [epistola]. In G. B. Fabri (Ed.), *Sulla insensitività ed irritabilità halleriana. Opuscoli di vari autori raccolti da Giacinto Bartolomeo Fabri* (Vol. 2, pp. 110–116). Bologna: Per Girolamo Corciolani.

Laghi, T. (1757b). *De sensitivitate, atque irritabilitate halleriana. Sermo alter*. In G. B. Fabri (Ed.), *Sulla insensitività ed irritabilità halleriana. Opuscoli di vari autori raccolti da Giacinto Bartolomeo Fabri* (Vol. 2, pp. 326–344). Bologna: Per Girolamo Corciolani.

La Roche, D. de. (1778). *Analyse des fonctions du système nerveux, pour servir d'introduction à un examen pratique des maux de nerfs* (Vols. 1–2). Geneva: Du Villard fils & Nouffer.

Le Cat, C. N. (1765). *Traité de l'existence, de la nature et des propriétés du fluide des nerfs et principalement de son action dans le mouvement musculaire*. Berlin: s. n.

Le Dru, N. P. (1776). Osservazioni e sperienze del Sig. Comus su l'elettricità medica. *Scelta di Opuscoli Interessanti Tradotti da Varie Lingue, 21*, 70–85.

Lenoir, T. (1982). *The strategy of life. Theology and mechanics in nineteenth century German biology*. Dordrecht: Reidel.

Lenoir, T. (1986). Models and instruments in the development of electrophysiology, 1845–1912. *Historical Studies in the Physical and Biological Sciences, 17*, 1–54.

Le Roy, J-B. (1776). Lettre adressée à l'auteur de ce Recueil par M. Le Roy. *Observations sur la Physique, 8*, 331–335.

Le Roy, J-B. (1777). Estratto di lettera del Signor Le Roy al Sig. abate Rozier sulla scintilla elettrica. *Scelta di Opuscoli Interessanti Tradotti da Varie Lingue, 25*, 106–108.

Lesch, J. E. (1984). *Science and medicine in France. The emergence of experimental physiology, 1790–1855*. Cambridge, MA: Harvard University Press.

Leslie, J. (1835). Dissertation fourth: Exhibiting a general view of the progress of mathematical and physical science, chiefly during the eighteenth century. In *Dissertations on the history of methaphysical and ethical, and mathematical and physical science* (pp. 575–677). Edinburgh: Adams and Charles Black.

Levi, P. (1984, May 3). Casa Galvani. *La Stampa, 104*, p. 3.

Levi, P. (1985). *L'altrui mestiere*. Torino: Einaudi.

Levinson, S. (2012). La beffa di Periergopoulos. *Sapere, 78*(5), 70–75.

Linari, S. (1836). Vera scintilla elettrica, Suppl. *L'Indicatore Sanese, 4*(50).

Linari, S. (1838). Sperimenti sopra le proprietà elettriche della torpedine, Suppl. 2. *L'Indicatore Sanese, 4*(50).

Lindeboom, G. A. (1968). *Herman Boerhaave. The man and his work*. London: Methuen.

Lorenzini, S. (1678). *Osservazioni intorno alle torpedini*. Florence: Onofri.

Lucas, K. (1905). On the gradation of the activity in a skeletal muscle fibre. *Journal of Physiology, 33*, 125–137.

Lucas, K. (1909). The all or none contraction of the amphibian skeletal muscle fibre. *Journal of Physiology, 38*, 113–133.

Lucas, K. (1917). *The conduction of nervous impulse* (E. D. Adrian, Ed.). London: Longmans Green.

Ludwig, C. (1861). Über die Krafte der Nervenprimitivenrohr. *Wiener Medizinische Wochenschrift, 729*, 129.

Mach, E. (1865). Über die Wirkung der räumlichen Vertheilung des Lichtreizes auf die Netzhaut. *Sitzungsberichte der mathematisch-naturwissenschaftlichen Classe der Kaiserlichen Akademie der Wissenschaften, 52*(2), 303–322.

Majorana, Q. (1937). Commemorazione di Luigi Galvani del Professore Quirino Majorana. In *Celebrazione del secondo centenario della nascita di Luigi Galvani* (pp. 51–64). Bologna: Luigi Parma.

Malagola, C. (1879). *Luigi Galvani nell'Università, nell'Istituto e nell'Accademia delle Scienze di Bologna*. Bologna: Romagnoli.

Malpighi, M. (1661). *De pulmonibus observationes anatomicae*. Bononiae: Typis Jo. Baptistae Ferronii.

Malpighi, M. (1665a). *De externo tactus organo anatomica observatio […] ad […] Iacobum Ruffum*. Neapoli: Apud Aegidium Longum.

Malpighi, M. (1665b). *Tetras anatomicarum epistolarum de lingua, et cerebro*. Bononiae: Benati.

Malpighi, M. (1666). De renibus. In *De viscerum structura, cui accedit dissertatio eiusdem De polypo cordis* (pp. 71–100). Bononiae: Montii. [1669 ed., Amsterdam: Apud Petrum Le Grand, pp. 66–95].

Malpighi, M. (1686). *Opera omnia*. London: Scott & Wells.

Malpighi, M. (1697). Risposta del Dottor Marcello Malpighi alla Lettera Intitolata.… In *Opera posthuma* (pp. 99–197). London: A. & J. Churchill.

Malpighi, M. (1967). *Opere scelte* (L. Belloni, Ed.). Turin: UTET.

Mamiani, M. (1983). Il galvanismo di Galvani. In L. Galvani, *Memorie sulla elettricità animale* (pp. 23–29). Rome: Theoria.

Mamiani, M. (1998). *Storia della scienza moderna*. Rome: Laterza.

Manzoni, T. (2001). *Il cervello secondo Galeno*. Ancona: Il Lavoro Editoriale.

Maquet, P., Laureys, S., Peigneux, P., Fuchs, S., Petiau, C., Phillips, C., ... Cleeremans, A. (2000). Experience-dependent changes in cerebral activation during human REM sleep. *Nature Neuroscience, 3*, 831–836.

Marchand, J. F., & Hoff, H. E. (1955). Felice Fontana. The laws of irritability. *Journal of the History of Medicine and Allied Sciences, 10*, 197–206.

Marey, E-J. (1876). Des excitations électriques du cœur. *Travaux du Laboratoire de M. Marey, 2*, 63–86.

Marmont, G. (1949). Studies on the axon membrane, I. A new method. *Journal of Cellular and Comparative Physiology, 34*, 351–382.

Marsili, L. F. (1725). *Histoire physique de la mer*. Amsterdam: Au Dépens de la Compagnie.

Marx, C. (1969). Le neurone. In C. Kaiser (Ed.), *Physiologie* (Vol. 2, pp. 7–285). Paris: Flammarion.

Mascheroni, L. (1793). *L'Invito, versi sciolti di Dafni Orobiano a Lesbia Cidonia*. Milan: G. Galeazzi.

Matteucci, C. (1837). Recherches physiques, chimiques et physiologiques sur la torpille. *Annales de Chimie et de Physique, 66*, 396–437.

Matteucci, C. (1838). Sur le courant électrique ou propre de la grenouille. *Bibliothèque Universelle de Genève, 15*, 157–168.

Matteucci, C. (1842). Expériences rapportées dans un paquet cacheté déposé par M. Dumas, au non de M. Matteucci, et dont l'auteur, présent à la séance, désire aujourd'hui l'ouverture. *Compte Rendu de l'Académie des Sciences, 15*, 797–798.

Matteucci, C. (1843). De l'existence et des lois du courant électrique musculaire, dans les animaux vivants ou recemment tués. *Archives de l'Electricité, 3*, 5–28.

Matteucci, C. (1844). *Traité des phénomènes électro-physiologiques des animaux suivi d'études anatomiques sur le système nerveux et sur l'organe électrique de la Torpille par Paul Savi*. Paris: Fortin, Masson et C.ie.

Matteucci, C. (1846). Recherches électrophysiologiques. *Annales de Chimie et de Physique, Ser. 3, 18*, 4–32.

Matteucci, C. (1847). *Lezioni dei fenomeni fisico-chimici dei corpi viventi*. Firenze: Ricordi.

Matteucci, C. (1856) *Lezioni di elettro-fisiologia. Corso dato nell'Università di Pisa nell'anno 1856*. Turin: Paravia.

Mauduyt, P. J. C. (1778). Lettera del Signor Mauduit sulle precauzioni necessarie nelle malattie, che si curano coll'elettricità. *Opuscoli Scelti Sulle Scienze e Sulle Arti, 1*, 267–270.

Mauro, A. (1969). The role of voltaic pile in the Galvani-Volta controversy concerning animal vs. metallic electricity. *Journal of the History of Medicine and Allied Sciences, 24*, 140–150.

Maxwell, J. C. (1879). *The electrical researches of Henry Cavendish*. Cambridge, England: Cambridge University Press.

Mazzarello, P. (2000). What dreams may come? *Nature, 408*, 523.

Mazzarello, P. (2004). *Costantinopoli 1786: la congiura e la beffa: l'intrigo Spallanzani*. Turin: Bollati Boringhieri.

Mazzei, R. (1984). Esperimenti sulla elettricità medica a Bologna nel Settecento. *Il Carrobbio, 10*, 200–207.

Mazzetti, S. (1840). *Repertorio di tutti i professori antichi, e moderni [...] di Bologna*, Bologna: San Tommaso d'Aquino.

Mazzolini, R. G. (1985) Il contributo di Leopoldo Nobili all'elettrofisiologia. In In G. Tarozzi (Ed.), *Leopoldo Nobili e la cultura scientifica del suo tempo* (pp. 183–199). Bologna: Istituto peri Beni Artistici, Culturali, Naturali della Regione Emilia-Romagna.

Mazzotti, M. (2007). *The world of Maria Gaetana Agnesi, mathematician of God*. Baltimore, MD: The John Hopkins University Press.

McNaughton, P. A. (1990). Light response of vertebrate photoreceptors. *Physiological Reviews, 70*, 847–884.

Medici, M. (1845). *Elogio di Luigi Galvani*. Bologna: Tipografia Governativa della Volpe e del Sassi.

Medici, M. (1857). *Compendio storico della scuola anatomica di Bologna dal rinascimento delle scienze e delle lettere a tutto il XVIII secolo*. Bologna: Tipografia Governativa della Volpe e del Sassi.

Medici, M. (1861). Elogio di Tommaso Laghi. *Memorie dell'Accademia delle Scienze dell'Istituto di Bologna, 11*, 355–383.

Mesini, C. (1958). *Luigi Galvani*. Bologna: San Francesco.

Mesini, C. (1971). *Nuove ricerche galvaniane*. Bologna: Tamari.

Messbarger, R. (2010). *The lady anatomist: The life and work of Anna Morandi Manzolini*. Chicago, IL: The University of Chicago Press.

Meulders, M. (2001). *Helmholtz. Des lumières aux neurosciences*. Paris: Jacob. [English edition, *Helmholtz: From enlightenment to neuroscience* (L. Garey, Trans.) Cambridge, MA: MIT Press, 2010].

Millar, D., Millar, I., Millar, J., & Millar, M. (1996). *The Cambridge dictionary of scientists*. Cambridge, England: Cambridge University Press.

Moller, P. (1995). *Electric fishes. History and behaviour*. London: Chapman & Hall.

Moller, P., & Fritzsch, B. (1993). From electrodetection to electroreception. The problem of understanding a nonhuman sense. *Journal of Comparative Physiology, 173*, 734–737.

Mollon, J. D., & Perkins, A. J. (1996). Errors of judgement at Greenwich in 1796. *Nature, 380*, 101–102.

Montalenti, G., & Rossi, P. (Eds.). (1982). *Lazzaro Spallanzani e la biologia del Settecento*. Florence: Olschki.

Monti, M. T. (1990). *Congettura ed esperienza nella fisiologia di Haller*, Florence: Olschki.

Moravia, S. (1970). *La scienza dell'uomo nel Settecento*. Rome: Laterza.

Morgagni, G. B. (1761). *De sedibus et causis morborum per anatomen indagatis*. Venetiis: Remondini.

Morus, I. R. (1998a). *Frankenstein's children. Electricity, exhibition, and experiment in early-nineteenth-century London*. Princeton, NJ: Princeton University Press.

Morus, I. R. (1998b). Galvanic cultures. Electricity and life in the early nineteenth century. *Endeavour, 22*, 7–11.

Morus, I. R. (2009). Radicals, romantics and electrical showmen: placing galvanism at the end of the English enlightenment. *Notes and Records of the Royal Society, 63*, 263–275.

Moruzzi, G. (1964). L'opera elettrofisiologica di Carlo Matteucci. *Physis, 4*, 101–140.

Muir, H. (1994). *Larousse dictionary of scientists*. Edinburgh: Larousse.

Müller, J. (1844). *Handbuch der Physiologie des Menschen für Vorlesungen* (4th ed.). Coblenz: Hölscher.

Musitelli, S. (1999). L'elettricità animale in età pregalvaniana. *Atti dell'Accademia delle Scienze dell'Istituto di Bologna, 286*, 21–43.

Musitelli, S. (2002). *L'elettricità animale dalle origini alla polemica Galvani-Volta*. Pavia: La Goliardica Pavese.

Musschenbroek, P. van. (1746). Lettre à M. de Réaumur du 20 janvier 1746. *Procès-verbaux de l'Académie royale des Sciences de Paris, 66*, 6.

Nachmansohn, D., & Neumann, E. 1975 *Chemical and molecular basis of nerve activity*, New York: Academic Press.

Needham, D. M. (1971). *Machina carnis*. Cambridge, England: Cambridge University Press.

Neher, E., & Sakmann, B. (1976). Single-channel currents recorded from membranes of denervated frog muscle fibres. *Nature, 260*, 799–802.

Nernst, W. (1888). Zur Kinetik der in Lösung befindlichen Körper. Theorie der Diffusion. *Zeitschrift für Chemie und Physik, 2*(9), 613–637.

Nernst, W. (1908). Zur Theorie des elektrischen Reizes. *Archiv für die Gesamte Physiologie des Menschen und der Thiere, 122*, 275–314.

Neuburger, M. (1897). *Die historische Entwicklung der experimentellen Gehirn- und Rückenmarksphysiologie vor Flourens*. Stuttgart: Enke. [English edition, *The historical development of experimental brain and spinal cord physiology before Flourens* (E. Clarke, Ed.). Baltimore, MD: The Johns Hopkins University Press, 1981].

Neville Bonner, T. (1995). *Becoming a physician. Medical education in Great Britain, France, Germany, and the United States, 1750–1945*. New York: Oxford University Press.

Newton, I. (1726). *Philosophiæ naturalis principia mathematica*. (3rd ed.). London: Apud G. & J. Innys.

Nicholson, W. (1797). Observations on the electrophore, tending to explain the means by which the torpedo and other fish communicate the electric shock. *Journal of Natural Philosophy, Chemistry and the Arts, 1*, 355–359.

Nicholson, W. (1801). Account of the new electrical or Galvanic apparatus of Sig. Alex. Volta, and experiments performed with the same. *Journal of Natural Philosophy, Chemistry and the Arts, 4*, 179–187.

Nilius, B. (2003). Milestones in physiology: Pflügers archive and the advent of modern electrophysiology: From the first action potential to patch clamp. *Pflügers Archive, European Journal of Physiology, 447*, 267–271.

Nobili, L. (1828). Comparaison entre les deux galvanomètres les plus sensibles, la grenouille et le moltiplicateur à deux aiguilles, suivie de quelques resultats nouveaux. *Annales de Chimie et de Physique, 38*, 225–245.

Nollet, J-A. (1743–48). *Leçons de physique expérimentale* (Vols. 1–6). Paris: Guérin.

Nollet, J-A. (1746). *Essai sur l'électricité des corps*. Paris: Guérin.

O'Malley, C. D. (1956). Review of "Commentary on the effect of electricity on muscular motion." A translation of "De viribus electricitatis in motu musculari commentaries." *Isis, 47*, 454–455.

Overton, E. (1902). Beiträge zur allgemeinen Muskel- und Nervenphysiologie. *Archiv für die Gesamte Physiologie des Menschen und der Thiere, 92*, 346–386.

Pacini, F. (1853). Sur la structure intime de l'organe électrique de la torpille, du gymnote et d'autres poissons. *Archives des Sciences Physiques et Naturelles, 24*, 313–336.

Pagel, W. (1967). *William Harvey's biological ideas. Selected aspects and historical background*. Basel: Karger.

Pancaldi, G. (1989). Luigi Galvani. In W. Tega (Ed.), *Storia illustrata di Bologna* (Vol. 8, pp. 281–300). Milan: NEA.

Pancaldi, G. (1990). Electricity and life. Volta's path to the battery. *Historical Studies in the Physical and Biological Sciences, 21*, 123–160.

Pancaldi, G. (Ed.). (1993). *Le università e le scienze*. Bologna: CIS.

Pancaldi, G. (2003). *Volta. Science and culture in the age of Enlightenment*. Princeton, NJ: Princeton University Press.

Pantaleoni, M., & Bernabeo, R. (1963). Pier Paolo Molinelli e l'istituzione della cattedra di medicina operatoria in Bologna. In *Atti della Biennale della Marca e dello Studio di Fermo* (pp. 369–385). Fermo: Stabilimento Tipografico Sociale.

Pera, M. (1986). *La rana ambigua. La controversia sull'elettricità animale tra Galvani e Volta*. Turin: Einaudi. [English edition, *The ambiguous frog: The Galvani-Volta controversy on animal electricity* (J. Mandelbaum, Trans). Princeton, NJ: Princeton University Press, 1992].

Petrini, G. V. (Ed.). (1755). *Sull'insensività e irritabilità di alcune parti animali*. Rome: G. Zempel.

Pfaff, C. H. (1854). *Lebenserinnerungen*. Kiel: Schwers.

Pflüger, E. (1859). *Untersuchungen über die Physiologie des Electrotonus*. Berlin: Hirshwald.

Piccolino, M. (1997). Luigi Galvani and animal electricity. Two centuries after the foundation of electrophysiology. *Trends in Neurosciences, 20*, 443–448.

Piccolino, M. (1998). Animal electricity in the birth of electrophysiology. The legacy of Luigi Galvani. *Brain Research Bulletin, 46*, 381–407.

Piccolino, M. (1999). Marcello Malpighi and the difficult birth of modern life sciences. *Endeavour, 23*, 175–179.

Piccolino, M. (2000). The bicentennial of the Voltaic battery (1800–2000). The artificial electric organ. *Trends in Neurosciences, 23*, 47–51.

Piccolino, M. (2001). Lazzaro Spallanzani e le ricerche sui pesci elettrici nel secolo dei lumi, in Spallanzani 1984–2013: *Edizione nazionale delle opere di Lazzaro Spallanzani*, 33 vols. (P. Di Pietro et al., Eds.). Modena: Mucchi; Parte 4, *Opere edite direttamente dall'autore, 5 17821791*, Tomo. 3 *Supplemento e Carteggi*, pp. 359–456.

Piccolino, M. (2003a). *The taming of the ray. Electric fish research in the Enlightenment, from John Walsh to Alessandro Volta*. Florence: Olschki.

Piccolino, M. (2003b). Nerves, alcohol and drugs. The Adrian-Kato controversy on nervous conduction: deep insights from a "wrong" experiment? *Brain Research Reviews, 43*, 257–265.

Piccolino, M. (2003c). A "lost time" between science and literature: The "temps perdu" from Hermann von Helmholtz to Marcel Proust. *Audiological Medicine, 1*, 261–270.

Piccolino, M. (2008). Visual images in Luigi Galvani's path to animal electricity. *Journal of the History of the Neurosciences, 17*, 335–348.

Piccolino, M. (2011). Carlo Matteucci (1811–1868): tra il Risorgimento dell'Italia e la Rinascita dell'elettrofisiologia. *Accademia Nazionale di Scienze Lettere e Arti di Modena Serie 8, 14*, 261–318.

Piccolino, M., & Bresadola, M. (2002). Drawing a spark from darkness. John Walsh and electric fish. *Trends in Neurosciences, 25*, 51–57.

Piccolino, M., Finger, S. & Barbara, J. G. (2011). Discovering the African freshwater "torpedo": Legendary Ethiopia, religious controversies, and a catfish capable of reanimating dead fish. *Journal of the History of the Neurosciences, 20*, 210–235.

Piccolino, M., & Navangione, A. (1998). Un système sensoriel à haute performance. La rétine des vertébrés. In D. Paupardin-Tritsch, D. Chesnois-Marchais, & A. Feltz (Eds.), *Physiologie du neurone* (pp. 605–653). Paris: Doin.

Piccolino, M., & Wade, N. J. (2012). The frog's dancing master: Science, séances and the transmission of myths. *Journal of the History of the Neurosciences*, doi: 10.1080/0964704X.2012.671020.

Pivati, G. (1747). *Della elettricità medica. Lettera [...] al celebre Signore Francesco Maria Zanotti*. Lucca: s.n.

Poirier, J-P. (2008). *L'abbé Bertholon*. Paris: Hermann.

Polanyi, M. (1958). *Personal knowledge: Towards a post-critical philosophy*. Chicago, IL: The University of Chicago Press.

Poli, M. (Ed.). (1998). *Luigi Galvani (1737–1798)*. Bologna: Fondazione del Monte di Bologna e Ravenna.

Polvani, G. (1942). *Alessandro Volta*. Pisa: Domus Galilaeana.

Pomata, G. (1994). *La promessa di guarigione: Malati e curatori in Antico Regime: Bologna XVI–XVIII secolo*. Rome: Laterza. [English edition, *Contracting a cure: Patients, healers, and the law in early modern Bologna*. Baltimore, MD: The Johns Hopkins University Press, 1998].

Porter, R. (Ed.). (1995). *Medicine in the Enlightenment*. Amsterdam: Rodopi.

Porter, R. (2000). *The creation of the modern world. The untold story of the British Enlightenment*. New York: Norton.

Porter, R., & Ogilvie, M. (Ed.). (2000). *The Hutchinson dictionary of scientific biography* (Vols. 1–2). Oxford, England: Helicon.

Priestley, J. (1767). *The history and present state of electricity, with original experiments* (Vols. 1–2, 3rd ed.). London: J. Dodsley, J. Johnson, B. Davenport, and T. Cadell.

Priestley, J. (1774–77). *Experiments and observations on different kinds of air* (Vols. 1–3). London: J. Johnson.

Prosperi, A. (Ed.). (2008). *Storia di Bologna. III. Bologna nell'età moderna (secoli XVI–XVIII)* (Vols. 1–2). Bologna: Bononia Unversity Press.

Pupilli, G. C., & Fadiga, E. (1963). The origins of electrophysiology. *Journal of World History, 7*, 547–589.

Ranvier, M. L. (1878). *Leçons sur l'histologie du système nerveux recuellies par M. Ed. Weber* (Vols. 1–2). Paris: Savy.

Réaumur, R. A. F. (1714). Des effets que produit le poisson appellé en français torpille, ou trembleur, sur ceux qui le touchent; et de la cause dont ils dépendent. In *Histoire de l'Académie Royale des Sciences pour l'Année 1714* (pp. 344–360). Paris: Imprimerie Royale.

Redi, F. (1671). *Esperienze intorno a diverse cose naturali e particolarmente intorno a quelle che ci son portate dall'Indie, scritte al Reverendissimo Padre Atanasio Chircher della Compagnia di Giesù*. Florence: All'insegna della Nave.

Ritter, J. W. (1798). *Beweis, dass ein beständiger Galvanismus den Lebenproces in dem Thierreich begleite*. Weimar: Industrie-Comptoir.

Ritter, J. W. (1801–05). *Beyträge zur näheren Kentnis des Galvanismus und der Resultate seiner Untersuchungen* (Vols. 1–2). Jena: F. Frommann.

Ritterbush, P. C. (Ed.). (1964). *Ouvertures to biology. The speculations of eighteenth-century naturalists*. New Haven, CT: Yale University Press.

Roger, J. (1963). *Les sciences de la vie dans la pensée française du xviii siècle. La génération des animaux de Descartes à l'Encyclopédie*. Paris: Colin.

Rosa, M. (Ed.). (1981). *Cattolicesimo e lumi nel Settecento italiano*. Rome: Herder.

Rossi, P. (1998). *La nascita della scienza moderna in Europa*. Rome: Laterza.

Rostand, J. (1963). *Lazzaro Spallanzani e le origini della biologia sperimentale*. Turin: Einaudi.

Rothschuh, K. E. (1953). *Geschichte der Physiologie*. Berlin: Springer. [English edition, *History of physiology* (G. B. Risse, Ed.). Huntington, England: Krieger, 1973].

Rothschuh, K. E. (1960). Von der Idee bis zum Nachweis der tierischen Elektrizität. *Sudhoffs Archiv für Geschichte der Medizin und der Naturwissenschaften, 44*, 25–44.

Roversi, G. (Ed.). (1987). *L'Archiginnasio. Il palazzo, l'università, la biblioteca*. Bologna: Grafis.

Rowbottom, M., & Susskind, C. (1984). *Electricity and medicine. History of their interaction*. San Francisco, CA: San Francisco University Press.

Sakmann, B., & Neher, E. (1995). *Single channel recording*. New York: Plenum.

Sanlorenzo, O. (1988). *L'insegnamento di ostetricia nell'Università di Bologna*. Bologna: Lorenzini.

Santucci, A. (Ed.). (1979). *Interpretazioni dell'Illuminismo*. Bologna: Il Mulino.

Sarkar, T., Mailloux, R., Oliner, A. A., Salazar-Palma, M., & Sengupta, D. L. (2006). *History of wireless*. Hoboken, NJ: Wiley-Interscience.

Sars, M. de. (1948). *Le Noir, lientenant de police (1732–1807)*. Paris: Hachette.

Sbaraglia, G. G. (1689). *De recentiorum medicorum studio dissertatio epistolaris ad amicum*. S.L.: s.n.

Schatzmann, A. J. (1953). Herzglykoside als Hemmstoffe für den aktive kalium and natriumtransport durch die erythrocytenmembran. *Helvetica Physiologica et Pharmacologica Acta, 11*, 101–122.

Schuetze, S. (1983). The discovery of the action potential. *Trends in Neurosciences, 6*, 164–168.

Segrè, E. (1996). *Personaggi e scoperte della fisica contemporanea: Da Galileo ai quark* (Vols. 1–2). Milan: Mondadori. [English edition, *From X-rays to quarks: Modern physicists and their discoveries* (Vols. 1–2). Mineola, NY: Dover, 2007].

Seligardi, R. (1997). Luigi Galvani e la chimica del Settecento. In F. Calascibetta (Ed.), *Atti del VII Convegno nazionale di storia e fondamenti della chimica* (pp. 147–162). Rome: Accademia Nazionale delle Scienze detta dei XL.

Seligardi, R. (2002). *Lavoisier in Italia. La comunità scientifica italiana e la rivoluzione chimica*. Florence: Olschki.

Sguario, E., & Wabst, C. X. (attributed to). (1746). *Dell'elettricismo: o sia delle forze elettriche de' corpi, svelate dalla fisica sperimentale*. Venice: Presso Gio. Battista Recurti. [Other ed, *A spese di Giuseppe Ponzelli nella stamperia di Giovanni di Simone*, Napoli, 1747].

Shaefer, H. (1936). Untersuchungen über den Muskelaktionstrom. *Archiv für die Gesamte Physiologie des Menschen und der Thiere, 237*, 329–355.

Shapin, S. (2008). *The scientific life: A moral history of a late modern vocation*. Chicago, IL: University of Chicago Press.

Shapin, S., & Schaffer, S. (1985). *Leviathan and the air-pump. Hobbes, Boyle, and the experimental life*. Princeton, NJ: Princeton University Press.

Sherrington, C. (1949). *Man and his nature: Broadcast talks in religion and philosophy*. Cambridge, England: CUP Archive.

Simeoni, L., & Sorbelli, A. (1940). *Storia della Università di Bologna* (Vols. 1–2). Bologna: Zanichelli.

Simili, R. (1995). Il caso di Luigi Galvani e delle sue rane. Astuzie, tranelli, incantesimi. In A. Guagnini & G. Pancaldi (Eds.), *Cento anni di radio. Le radici dell'invenzione* (pp. 19–64). Turin: Seat.

Simili, R. (2001). Luigi Galvani: Animal electricity and medical therapy. In M. Beretta & K. Grandin (Eds.), *A galvanized network. Italian-Swedish scientific relations from Galvani to Nobel* (pp. 13–46). Stockholm: The Royal Swedish Academy of Science.

Simili, R. (2005). Two special doctors: Erasmus Darwin and Luigi Galvani. In C. U. M. Smith & R. Arnott (Eds.), *The genius of Erasmus Darwin* (pp. 145–158). Aldershot, England: Ashgate.

Sirol, M. (1939). *Galvani et le galvanisme. L'électricité animale*. Paris: Vigot.

Shaw, P. E. (1904). The sparking distance between electrically charged surfaces—Preliminary note. *Proceedings of the Royal Society of London, 73*, 337–342.

Skou, J. C. (1957). The influence of some cations on an adenosine triphosphatase from peripheral nerves. *Biochimica et Biophysica Acta, 23*, 394–401.

Snyder, C. D. (1908). A comparative study of the temperature coefficients of the velocities of various physiological actions. *American Journal of Physiology, 22*, 309–334.

Soave, F. (1792). Transunto della Dissertazione del Sig. Dott. Luigi Galvani P. Prof. nell'Università di Bologna sulle forze dell'elettricità ne' moti muscolari. *Opuscoli Scelti Sulle Scienze e Sulle Arti, 15*, 113–141.

Spallanzani, L. (1768). *Dell'azione del cuore ne' vasi sanguigni*. Modena: G. Montanari.

Spallanzani, L. (1780). *Dissertazioni di Fisica animale, e vegetabile*. Modena: Società Tipografica.

Spallanzani, L. (1783). Lettera dell'abate Spallanzani [...] al Sig. Marchese Lucchesini, Ciamberlano di S.M il Re di Prussia. *Opuscoli Scelti Sulle Scienze e Sulle Arti, 7*, 73–104.

Spallanzani, L. (1784). Lettera prima relativa a diverse produzioni marine al Sig. Carlo Bonnet scritta il giorno 15 gennaio 1784. *Memorie di Matematica e di Fisica della Società Italiana delle Scienze, 2*, 603–661. [also *Opuscoli Scelti Sulle Scienze e Sulle Arti, 7*, 340–360].

Spallanzani, L. (1803). *Mémoires sur la respiration* (J. Senebier, Ed.). Geneva: Paschoud.

Spallanzani, L. (1929). *Viaggi ed escursioni scientifiche di Lazzaro Spallanzani* (G. Pighini, Ed.). Reggio Emilia: Officine Grafiche Reggiane.

Spallanzani, L. (1932–36). *Le opere di Lazzaro Spallanzani* (F. Bottazzi & M. L. Patrizi, Eds., Vols. 1–6). Milan: Hoepli.

Spallanzani, L. (1985). *Edizione nazionale delle opere di Lazzaro Spallanzani, Carteggi V, Carteggi con Fossombroni... Lucchesini* (P. Di Pietro et al., Eds.). Modena: Mucchi.

Spillane, J. D. (1981). *The doctrine of the nerves. Chapters in the history of neurology*. Oxford, England: Oxford University Press.

Stanhope, C. Lord Mahon. (1779). *Principles of electricity*. London: P. Elmsy.

Steinke, H. (2005). *Irritating experiments. Haller's concept and the European controversy on irritability and sensibility, 1750–90*. Amsterdam: Rodopi.

Stensen, N. (1662). *Observationes anatomicae*. Leiden: Apud Jacobum Chouët.

Stensen, N. (1664). *De musculis et glandulis observationum specimen, cum epistolis duabus anatomicis*. Amsterdam: Apud Petrum Le Grand.

Stensen, N. (1673). *Ova viviparum spectantes observationes*. Acta Medica et Philosophica Hafniensia, 2, 219-232.

Stoye, J. (1994). *Marsigli's Europe, 1680-1730: The life and times of Luigi Ferdinando Marsigli, soldier and virtuoso*. New Haven, CT: Yale University Press.

Sue, P. (1802-05). *Histoire du Galvanisme et analyse des differens ouvrages publiées sur cette découverte depuis son origine jusqu'à ce jour* (Vols. 1-4). Paris: Bernard.

Susskind, C. (1964). Observations of electromagnetic wave radiations before Hertz. Isis, 55, 32-42.

Sutton, G. V. (1995). *Science for a polite society. Gender, culture, and the demonstration of Enlightenment*. Boulder, CO: Westview Press.

Swammerdam, J. (1737-38). *Bybel der natuure*. (Vols. 1-2). Leyden: I. Severinus, B. Vander Aa, Pieter Vander Aa.

Sydenham, T. (1700). *Opera omnia medica*. Patavii: Apud Joannem Manfrè.

Tabarroni, G. (1971). Luigi Galvani. In *Grande antologia filosofica* (pp. 96-104, 129-154). Milan: Marzorati.

Tega, W. (Ed.). (1986-87). *Anatomie accademiche* (Vols. 1-2). I Commentari dell'Accademia delle Scienze di Bologna; L'Enciclopedia scientifica dell'Accademia delle Scienze di Bologna. Bologna: Il Mulino.

Tega, W. (Ed.). (1987-91). *Storia illustrata di Bologna* (Vols. 1-8). San Marino and Milan: AIEP/ NEA.

Teichmann, J. (2001). *Volta and quantitative conceptualisation of electricity from electric capacity to the preconceptions of Omhs' law*. In F. Bevilacqua & L. Fregonese (Eds.), Nuova Voltiana. Studies on Volta and his times (pp. 53-80). Milan: Hoepli.

Temkin, O. (1964). The classical roots of Glisson's doctrine of irritation. Bulletin of the History of Medicine, 38, 297-328.

Tissot, S. A. A. D. (1755). Discours preliminaire. In *Dissertation sur les parties irritables et sensibles des animaux [...] Traduit par M. Tissot* (pp. iii-xli). Lausanne, Switzerland: M-M. Bousquet.

Tissot, S. A. A. D. (1778-80). *Traité des nerfs et de leurs maladies* (Vols. 1-4). Paris: chez P. F. Didot le Jeune.

Trumpler, M. (1992). *Questioning nature. Experimental investigations of animal electricity in Germany, 1791-1810*. Unpublished Ph.D. dissertation, Yale University, New Haven, CT.

Università degli Studi di Bologna. (1979). *I materiali dell'Istituto delle Scienze*, Bologna: Accademia delle Scienze.

Valli, E. (1793). *Experiments on animal electricity with their applications to physiology and some pathological and medical observations*. London: J. Johnson.

Valli, E. (1794). *Lettera XI sull'elettricità animale*. Mantua: G. Braglia.

Vecchi, M. (1986). *Luigi Galvani nella storiografia scientifica*. Unpublished university dissertation, Università di Bologna.

Vendola, F. R. (2000). *Giambattista Beccaria nella storia della fisica piemontese del Settecento*. Turin: Crisis.

Venturoli, G. (1841). Elogio del celebre Professore Luigi Galvani. In S. Gherardi, (Ed.), *Opere edite ed inedite del professore Luigi Galvani. Raccolte e pubblicate per cura dell'Accademia delle Scienze dell'Istituto di Bologna* (pp. 107-119). Bologna: Dall'Olmo.

Veratti, G. (1748). *Osservazioni fisico-mediche intorno alla elettricità*. Bologna: Lelio dalla Volpe.

Veratti, G. (1752). *Osservazione fatta in Bologna l'anno MDCCLII dei fenomeni elettrici nuovamente scoperti in America, e confermati a Parigi*. Bologna: Lelio dalla Volpe.

Veratti, G. (1755). De electricitate medica. *De Bononiensi Scientiarum et Artium Instituto atque Academia Commentarii, 3*, 454–478.

Volta, A. (1800). On the electricity excited by the mere contact of conducting substances of different species. Letter to Sir Joseph Banks, March 20, 1800. *Philosophical Transactions of the Royal Society of London, 90*, 403–431 [text in French; an English translation with the same title was published in the same year in *Philosophical Magazine, 7*, 289–311.]

Volta, A. (1918–29). *Le opere di Alessandro Volta. Edizione nazionale sotto gli auspici della Reale Accademia dei Lincei e del Reale Istituto Lombardo di Scienze e Lettere* (Vols. 1–7). Milan; Hoepli.

Volta, A. (1949–55). *Epistolario di Alessandro Volta. Edizione nazionale* (Vols. 1–5). Bologna: Zanichelli.

Walsh, J. (1772). *Experiments made in La Rochelle and Isle de Ré. June and July 1772*. Mansucript #609. London: Royal Society of London.

Walsh, J. (1773). On the electric property of torpedo: In a letter to Ben. Franklin. *Philosophical Transactions of the Royal Society of London, 63*, 461–479.

Whitaker, H., Smith, C. U. M., Finger. S. (Eds.). (2007). *Brain, mind, and medicine. Essays in eighteenth century neuroscience*. New York: Springer.

Whittham, R. (1958). Potassium movement and *ATP* in human red cells. *Journal of Physiology, 140*, 479–497.

Wilkinson, C. H. (1804). *Elements of galvanism, in theory and practice* (Vols. 1–2). London: Murray.

Wollstonecraft, M. (1818). *Frankenstein, or, The modern Prometheus*. London: Printed for Lackington, Hughes, Harding, Mayor & Jones.

Wu, C. H. (1984). Electric fish and the discovery of animal electricity. *American Scientist, 72*, 598–606.

Young, J. Z. (1936). Structure of nerve fibres and synapses in some invertebrates. *Cold Spring Harbor Symposia on Quantitative Biology, 4*, 1–6.

Zanotti, F. M. (1757). De quibusdam animalium partibus, an sensu sint compotes, et unde irritabilitas. *De Bononiensi Scientiarum et Artium Instituto atque Academia Commentarii, 4*, 48–57.

Zett, L., & Nilius, B. (1983). *Bernstein-Symposium. Anlässlich des 100 jahrigen Bestehens des Physiologischen Instituts der Martin-Luther-Universität Halle-Wittenberg*. Halle-Wittenberg: Martin Luther Universität.

INDEX

Notice: In this Index, the page numbers followed by n denote occurrence of the term in footnotes; those followed by f appear in figure legends.

Académie des Sciences of Paris, 4, 33, 49, 63, 144
Academy of Milan, 142
Academy of Sciences of Bologna, 2, 43, 70, 109–110, 113, 211
Academy of Sciences of Göttingen (Germany), 45
Acetylcholine, as nerve signal "messenger," 288
Action current, 16, 275, 280
Action mechanism of will, 185
Action potential, 9, 293, 309–310
Adenosine triphosphate (ATP), 301
Adrian, Edgar Douglas (1889–1977), 283–289, 291, 312
Aepinus, Franz Ulrich (1724–1802), 134
African catfish. *See* Electric fish
Airs. *See* Gases ("airs") studies
Aldini, Giovanni (1762–1834)
    De viribus second edition edited by, 144
    De viribus sent to Volta by, 142
    as Galvani collaborator, 74
    revitalization demonstrations of, 17–18, 118
    Two Dissertations on Animal Electricity, 179
    Volta's letters to, 159, 211–214
Algarotti, Francesco (1712–1764), 46
Alibert, Jean-Louis (1768–1837), 5
"All-or-none" law

electrical excitation following, 282–284
    mechanism of, 97, 105–106
    membrane permeability and, 296
    nerve impulses obeying, 310
Altieri-Biagi, Maria Luisa (1930 -), 327
Ambiguous Frog, The (Pera), 9–10, 81, 174, 216, 265
Ampère, André (1775–1836), 7
Anatomical dissection, 29
Anatomy, pathological, 31
Animal economy
    chemical research connected to, 110
    electricity role in, 12, 66–67, 257–259
    as structure and function connection, 40
Animal electricity hypothesis. *See also* Electrophysiology; Galvani, animal electricity theory of; Galvani-Volta controversy, first stage; Galvani-Volta controversy, second stage
    in electrophysiology development, 8–10
    Galvani's contradictory results on, 100–102
    ionic pumps in, 14
    Nernst equilibrium to confirm, 306
    in nervous conduction, 3, 20–21
    as "secret of life," 2
    validation problems, 12

361

Animal electricity in nerve conduction, 289–298. *See also* Nerve conduction and muscular motion
  local nerve blockade studies, 289–291
  membrane electrochemical gradient, 296–297
  "patch clamp" technique, 297–298
  subthreshold potentials and full-blown response problem, 291–294
  "voltage clamp" technique, 294–295
Animal electricity measurement, 270–280
  Bessel's correction of astronomical errors, 278
  du Bois-Reymond nerve and muscle excitation, 275–276, 279–280
  Helmholtz muscle contraction recording, 276–278
  Matteucci "induced twitch" experiment, 271–274
  by Nobili, 271
  physical investigations of electricity, 270–271
Animal electrometer, prepared frog as, 153, 201
Animal experimentation, 29
"Animal Leyden jar" model
  atmospheric electricity *versus*, 123–127
  disequilibrium in organism shown by, 127–128
  electric circuits analogous to, 126
  electric fish analogy, 129–131
  Galvani description of, 83–84
  Galvani explaining all based on, 322–323
  metal arc effects, 106, 128–129
  muscle as, 166–167, 199–201
  perfected, 169
  Volta's criticism of, 157–158
"Animal machine" concept
  animal electricity distinguished in, 189–190
  electric characteristics governing, 145
  membranes as, 320
  patch clamp technique for study of, 13–14
  structure and function understanding from, 173–175
Animal spirits hypothesis, 43, 139–140. *See also* Nerve conduction and muscular motion
Annali di chimica e storia naturale, 212, 226, 248, 260, 262
Anti-vitalist tradition, 266
Arago, Dominique-François-Jean (1786–1853), 1, 4, 92, 97
Argelander, Friedrich Wilhelm (1799–1875), 278
Arterial pressure measurement, 109
Artificial and natural electricity, 56–62
  atmospheric electricity compared to, 145

  electric fish and eels, 57–58
  Leyden jar to compare, 63
  medical electricity use of, 58–60
  nerves as conductors, 60–62
  weather-related, 56–57
Artificial intelligence, 9
"Artificial torpedo," 65–66, 169, 252–253, 259
Astatic electromagnetic galvanometer, 22, 271, 323
Atmospheric electricity, 116–127
  "animal Leyden jar" model *versus*, 123–127
  artificial electricity compared to, 145
  climate related to, 79
  human health effects of, 79–80, 220
  lightning effects, 116–119
  metal arc effects, 119–123
  sensitivity to, 147
ATP (adenosine triphosphate), 301
"Autoregenerative" cycle, in membranes, 296, 312–313
Avicenna (880–943), 29
Azzoguidi, Germano (1740–1814), 48, 101–102

Bacon, Francis (1561–1626), 32
Baglivi, Giorgio (1668–1707), 137
Bancroft, Edward (1744–1821), 64
Banks, Joseph (1743–1820), 214, 226, 242, 250
Barbensi, Gustavo (1875–1974), 16
Baronio, Giuseppe (1758–1811), 246
Bassi, Laura (1711–1778), 38, 39f, 46, 55, 59, 72, 74
Bath application of electricity, 220
Battery
  in "chain" experiments of Volta, 242–243
  as "column apparatus," 260
  concentration, 306
  electric fish basis of, 249–253, 259–261
  path to, 321
  to stimulate sensory systems, 237–238
  torpedo electric organ as inspiration for, 65
  "trough" version of, 220–221
  Volta's announcement of invention of, 214
  Volta's discovery of, 3–4, 7
"Beast-machine" (Descartes), 77
Beccari, Jacopo Bartolomeo (1682–1766)
  animal economy research, 110
  Beccaria relationship with, 55
  food chemical analysis interest, 46
  as Galvani's teacher, 29, 33

Institute of Sciences of Bologna influenced by, 173
on nervous fluid concept, 44
"rational" view of medicine and, 31
Beccaria, Giambattista (1716–1781)
on artificial and natural electricity, 56–58
"electrician" laboratory described by, 73
electricity theory of, 55–56
Becquerel, Antoine Caesar (1788–1878), 115n
Benedictines, at Institute of Sciences of Bologna, 38
Benedict XIV (pope; see also Lambertini, Prospero), 27, 28f, 33–35, 38
Bennett, Abraham (1750–1799), 79
Bennet-type electrometer, 243–244
Bergman, Torbern (1735–1784, 110
Bernard, Claude (1813–1868), 31, 210
Bernardi, Walter, (1948 -),102, 141, 144
Bernstein, Julius (1839–1917)
"action current" recording by, 16
differential rheotome technique of, 279, 280f, 309
electric potentials theory applied by, 281
membrane permeability increase finding of, 281–282, 289
membrane theory of, 304
Bertholon, Pierre (1741–1800)
conjectural character of writings of, 18
electro-animal economy doctrine, 145–147
electrological hypothesis of human health, 116, 220
Bessel, Friedrich Wilhelm (1784–1846), 278
Bianconi, Carlo (1758–1812), 142
Biolelectric potentials, 304–306
Black, Joseph (1728–1799), 10
Bloch, Marc (1886 -1944), 333
Bloodletting practice, 35
Boerhaave, Herman (1668–1738), 32f, 43–44, 47, 137
Bonaparte Prix from Institut de France, to Volta, 221–222
Bohr, Niels Henrik David (1885–1962), 172, 215
Bonnet, Charles (1720–1793), 114, 265
Borelli, Giovanni Alfonso (1608–1679), 210
Booklets on Hallerian Insensitivity and Irritability (Fabri), 45
"Book of the Universe" (Galileo), 19
Boscovich, Roger Joseph (1711–1787), 36
Bowditch, Henry Pickering (1840–1911), 106, 282f
Boyle, Robert (1626–1691), 34, 147–148

Brain involvement, 8, 204–206
Breschet, Gilbert (1784–1845), 115n
Brugnatelli, Luigi Valentino (1761–1818), 226–227, 256, 260
Buffon, Georges-Louis Leclerc, de (1707–1788), 36.
Busch, August Ludwig (1804–1855), 278
Byron, Lord, 18

Cable equation (Lord Kelvin), 112
Caldani, Leopoldo Marc'Antonio (1725–1813)
on electricity to produce muscular contractions, 60, 101–102
frog preparation developed by, 75–76
Galvani variations on experiments of, 78
in Hallerian debate, 46–48
Institutiones physiologicae, 102
Laghi's research criticized by, 144
Cambridge University Press (UK), 6
Cambridge University (UK), 270, 289
Canguilhem, Georges (1904–1995), 210
Canterzani, Sebastiano (1734–1819), 142, 147
Capacity concept, 66, 153
Capillary electrometer (Lippmann), 284
Carlisle, Anthony (1768–1840), 270
Carminati, Bassiano (1750–1830), 155
Cartesian philosophy, 43, 173
Catholic Enlightenment, 26–28
Cavallo, Tiberio (or Tiberius; 1749–1809)
Complete Treatise on Electricity, 55–56, 73, 82
on electric atmospheres, 92
on electric shock danger, 94
Galvani's research reported by, 144
Volta's letters to, 162, 223
Volta's vision experiments described to, 232
voluntary movement studies, 227
weak electricity studies, 79
Cavendish, Henry (1731–1810)
artificial torpedo of, 65–66, 169, 252–253, 259
on bodily fluids as poor conductors, 313
on electric fish shock phenomenon, 124
hydrogen discovered by, 111
modern chemistry and, 10
Cell membranes. See Membranes
"Chain" experiments. See also Volta, Alessandro, electrophysiological work of
of Galvani, 74, 205–207
of Volta, 241–249
battery impact on, 242–243
electrical resistance concept from, 245–248

Index

"Chain" experiments (*Cont.*)
    metallic and electrical machine electricity differences, 248–249
    at progressively lower tensions, 243–245
    variations in, 241–242
Chance, role of, in electrophysiology discovery, 5–7, 91, 121, 139, 151
Claudianus Claudius (*ca.* 370- *ca.* 404), 267
Cohen, Ierome Bernard (1914–2003), 92, 327
Cole, Kenneth, (1900–1984) 291, 294
Complete Treatise on Electricity (Cavallo), 55–56, 82
Concentration battery, 306
Condensatore (Volta), 3, 80, 153, 193, 247
Condensing electroscope (Volta), 213–214
Conduction of Nervous Impulse, The (Adrian), 283
"Conductors of the first class," 183
"Conductors of the second class," 183
Configliachi, Pietro (1779–1844), 208, 245, 256, 258f, 261–262
Contractions at a distance, 88–99
    description of, 88–91
    nerve involvement, 95–99
    observation of, 71
    "return stroke" as alternative explanation of, 91–94
    Volta on, 157–158
Contrary electricity, 129, 133–134, 138
Convulsive motions, 175
Copley medal, to Volta, 214
Corpuscular matter theory, 175
Cours de Physique (Ganot), 1
Croone, William (1633–1684), 43
Current. *See also* Electric circuits
    frog muscle contractions from, 199
    magnetic processes associated with, 270–271
    Volta's description of, 160
Curtis, Howard (1906–1972), 291
Cussini, Ercole Antonio (*d.* 1758), 29

Darwin, Erasmus (1731–1802), 25
Davies, David (1741–1819), 64
Davy, Humphrey (1778–1829), 270
Davy, John (1790–1868), 115n
De Bononiensi Scientiarum et Artium Instituto atque Academia Commentarii (Institute of Sciences of Bologna), 33, 36, 40, 43,143
de Brosses, Charles (1709–1777), 26–27
Delfico, Orazio (1769–1842), 185, 227
Dell'elettricismo (Sguario and Wabst), 52f

Deluc, Jean-André (1727–1817), 153
De magnete (Gilbert), 11
Descartes, René (1596–1650), 77
De viribus electricitatis in motu musculari (Galvani). *See also* Galvani-Volta controversy, second stage
    debate based on, 141
    intellectual turmoil caused by, 2, 18
    lightning experiments in, 118
    publication of, 143
    translations of, 324
    "truth and usefulness" themes in, 41
Dictionary of Scientists (Cambridge University Press), 6
Dictionary of Scientists (Larousse), 6
Differential rheotome technique, 279, 309
Disequilibrium, animal electricity from
    in "animal Leyden jar" model, 124, 127–128
    in cellular membranes, 137
    electricity in state of, 270
    Galvani's conclusions on, 3, 8–9
    Galvani's hypothetical seat of, 155
    Hallerian objection to, 127
    heterogeneous conductors touching for, 85
    microscopic level of, 135–137
    muscle as electricity accumulator in, 133
    stimulus requirements for, 320
Dissimilar armatures (metals), to produce contractions, 157, 159, 185
Dissimilar metals, metallic electricity of, 232
Double electricity, 134–135, 296
Dove, Heinrich Wilhelm (1803–1879), 22
du Bois-Reymond, Emil (1818–1896)
    animal electricity measurement, 16, 20–23
    on Galvani, 325, 327
    Hermann controversy with, 280
    nerve and muscle excitation studies, 273–275, 279
    on Volta, 217
Dufay, Charles (1698–1759), 53–54
Duplicator device (Nicholson), 193, 247

"Electrical chain." *See* "Chain" experiments
Electrical equipment and instruments. *See also* Battery; Leyden jar
    artificial torpedo, 65–66, 169, 252–253, 259
    astatic galvanometer, 22, 271, 323
    Bennet-type electrometer, 243–244
    capillary electrometer (Lippmann), 284

condensatore (Volta), 3, 80, 153, 193, 247
condensing electroscope (Volta), 213–214
differential rheotome, 279, 309
duplicator device (Nicholson), 193, 247
electric machine, 125
electrophorus, 125, 153
electroscopes, 12
electrostatic machine, 84
"Franklin's square," 58, 78, 84, 103–104, 125, 128
in Galvani laboratory, 72–73
galvanometer, 7
micro-electroscope (Volta), 193
multiplier device (Nicholson), 193, 247
patch clamp technique, 13
"straw micro-electrometer," 80
two-flask apparatus, 99–100
voltage clamp, 294–295
Electrical resistance, 245–248, 282
Electric atmospheres, 4, 55, 91–92. *See also* Atmospheric electricity
Electric circuits
"animal Leyden jar" analogy to, 126–127
animal machine concept, 174–175
contractions on closure or interruption of, 189
in Galvani's response to Volta's criticism, 162–165
medical applications, 175–176
metal arc effects, 166–167
between metals, 160
muscle-nerve direct contact experiment, 167–171, 173–174
"occult arc" experiment, 171–172
in parallel and series, 261–262
in *Trattato dell'arco conduttore* (Galvani), 165–166
Electric eel of the Guyana. *See* Electric fish
Electric fields, 66, 91
Electric fish research. *See also* "Artificial torpedo"; Volta, Alessandro, electrophysiological work of
"animal Leyden jar" model analogous to, 129–131
battery comparison to, 13–14
contrary electricity accumulation in, 129
18th century studies of, 57–58
by Galvani, 201–211
animal electricity distinct from other electrical forms, 207–211

brain involvement, 204–206
electricity accumulation, 318
frog muscles on torpedo experiments, 206–207
nerve fluid and electric fish organ relation, 201–204
nerve role in shock generation, 85
Hallerian irritability negated by, 144
shock of
measurement of, 65
nature of, 62–65
spark from, 66–67
by Volta, 249–268
"animal economy encompassing electricity" concept, 257–259
battery based on, 249–253, 259–261
electricity involvement in, 254–256, 318
inanimate substances affected by, 263–265
nerve relationship to electricity in, 256–257
Nicholson's view, 253–254
parallel and series electric circuits, 261–262
physics and physiology differences, 267–268
polarity, 262–263
residual electricity in separated organs, 265–267
by Walsh, 13–14
Electricity. *See also* Medical electricity; Weak electricity
artificial and natural. *See also* Electric fish research
electric fish and eels, 57–58
medical electricity, 58–60
weather-related, 56–57
18th century studies, 48–56
animal spirits' electrical nature, 48–50
Beccaria's theory of, 55–56
fascination with, 50
Franklin's theory of, 54–55
Leyden jar invention, 53–54
Electric machine, 125, 248–249
"Electric molecules," theory of, 22–23, 279
"Electric storm," nerve signals as, 8
Electrodynamics, 49, 223
Electrolocation mechanism, 318
Electromagnetic galvanometer, 271
Electromotive power of metals. *See* Metals, electromotive power of
Electrophorus, 125, 153

Electrophysiology, 1–25, 61. *See also* Volta, Alessandro, electrophysiological work of
 animal electricity hypothesis, 8–10, 12, 14, 20
 "capital" experiment in, 191
 chance in discovery of, 5–7
 du Bois-Reymond's work in, 21–23
 electric fish studies, 13–14
 Galvani, stereotypes about, 23–24
 Galvani's *De viribus* on, 18
 Galvani's discovery of, 1–3, 5
 Galvani-Volta controversy in, 14–17, 24–25
 Hodgkin and Huxley's research on, 7–8
 interdisciplinary scientific approach and, 10–11
 vital spirit, electricity as, 17–18
 Volta's battery, 12–13
 Volta's experiments on, 3–4
Electrophysiology of last two centuries, 269–298
 animal electricity in nerve conduction, 289–298
  local nerve blockade studies, 289–291
  membrane electrochemical gradient, 296–297
  "patch clamp" technique, 297–298
  subthreshold potentials and full-blown response problem, 291–294
  "voltage clamp" technique, 294–295
 animal electricity measurement, 270–280
  Bessel error correction, 278
  du Bois-Reymond nerve and muscle excitation, 275–276, 279–280
  Helmholtz muscle contraction recording, 276–278
  Matteucci "induced twitch" experiment, 271–274
  by Nobili, 271
  physical investigations of electricity, 270–271
 nervous conduction, 280–289
  "all-or-none" electrical excitation, 282–284
  depolarization state to increase membrane permeability, 281–282
  "local-circuit" theory, 281
  negativity of external fiber surface for, 280
  "pulse-code-modulation," 284–288
  temperature effects, 288–289
 overview, 269–270
Electroreception mechanism, 318
Electroscopes, 12

Electrostatic induction, 55, 91
Electrostatic machine, 84
Elements of Physiology of the Human Body (von Haller), 61
Eloge historique de Galvani (Alibert), 5
Essay on the Nervous Force and Its Relations With Electricity of 1782 (Galvani), 99–108
 animal electricity corollaries, 101–102
 experimental methods, 103–104
 observations, 104–105
 stimulus-response lack of proportionality, 105–107
 two-flask apparatus, 99–100
"Ether," 93
"Experimental determinism," 87
Experimental physiology, 31
Experiments and Observations on Electricity (Franklin), 54

Fabri, Giacinto Bartolomeo (*ca.* 1726–1783/94), 45
Fadiga, Ettore (1927–1978), 16
Faraday, Michael (1791–1867), 270–271
"First experiment" of Galvani (January 1781). *See* Galvani, early muscular motion research of
First Memoir on Animal Electricity (Volta), 42, 151–152, 156, 254, 313
"Flicker-fusion" in luminous sensation, 233
Flourens, Marie Jean Pierre (1794–1867), 21f
Foley, Margaret Glover (*fl.* 1930–1994), 324
Fontana, Felice (1730–1805)
 on "all-or-none" character of electrical excitation, 282
 on animal spirits electrical nature, 48–49
 on combustion and respiration, 113
 "explosion" analogy of, 287
 frog preparation developed by, 75–76
 Galvani variations on experiments of, 78
 in Hallerian debate, 46–48
 Hallerian experiments replicated by, 60
 heart studies of, 225
 Laghi's research criticized by, 144
 on nerve fiber composition, 168
 Volta's claims contradicted by, 160
 Walsh experiments' impact on, 67–70
Foschi, Barbara Caterina (1705?–1774), 28
Fowke, Arthur (*fl.* 1756–1775), 64
Frank, Johann Peter (1745–1821), 228, 254
Frankenstein, myth of (Shelley), 17–18, 118

Index

Frankenstein, or the Modern Prometheus (Wollstonecraft Shelley), 18
Franklin, Benjamin (1706-1790)
   electric fish findings and, 64
   electricity theory of, 53-55
   lightning studies of, 56-57, 117
   metallic electricity championed by, 221
   one-fluid theory of electricity, 60, 118
   Volta on lightning studies of, 143, 152
"Franklin's square," 58, 78, 84, 103-104, 125, 128
Frog, as research animal, 75-78
Fulton, John Farquhar (1899-1960), 16

Galeazzi, Domenico Gusmano (1686-1775), 23, 31-34, 36-38, 72
Galeazzi, Lucia (1743-1790), 5, 23, 38-40, 73-74
Galen (129-201), 29, 31
Galileo (Galilei, Galileo), 12, 18-19, 30, 34, 41, 62-63, 77, 299, 329, 332-333
Galli, Giovanni Antonio (1708-1782), 33, 35-36, 72, 108
Galvani, Camillo (1753-1828), 74, 119
Galvani, Domenico (1690-1777), 28
Galvani, Luigi (1737-1798). *See also De viribus electricitatis in motu musculari* (Galvani)
   "animal Leyden jar" model of, 322
   atmospheric electricity experiments of, 57
   early anatomy-physiology investigations, 40-41
   education, 26-35
      Catholic Enlightenment, 27-28
      in hospitals, 34-35
      Institute of Sciences of Bologna, 32-34
      University of Bologna, 28-32
   electrophysiology research of, 1-3, 5, 61
   Hallerian irritability influence on, 311
   ignorance of physics, 4
   Nerst equilibrium and, 306
   as obstetrics professor, 108
   professional career, 36-40
   "Public Anatomy" lectures of, 61-62
   saturation, observations on, 312
   stereotypes about, 23-24
   Volta's physiological work influenced by, 217-219
Galvani, early muscular motion research of, 69-107
   contractions at a distance, 88-99
      description of, 88-91
      nerve involvement, 95-99

      "return stroke" as alternative explanation of, 91-94
   early electrophysiological experiments, 71-88
      "animal Leyden jar," 83-84
      collaborators, 74-75
      "first experiment" (January 1781), 84-87
      frog as research animal, 75-78
      home laboratory, 72-74
      laws derived from, 81-82
      research methods, 81
      variability in, 87-88
      on weak electricity, 79-80
   Essay on the Nervous Force and Its Relations With Electricity of 1782, 99-107
      animal electricity corollaries, 101-102
      experimental methods, 103-104
      observations, 104-105
      stimulus-response lack of proportionality, 105-107
      two-flask apparatus, 99-100
   overview, 69-71
Galvani, animal electricity theory of, 108-140
   "airs" studies, 109-115
      inflammable animal air, 111-114
      physiological chemistry in, 110-111
      Priestley's discoveries, 109-110
      unitary view of natural phenomena from, 114-115
   "animal Leyden jar" model, 127-131
      disequilibrium in organism shown by, 127-128
      electric fish analogy, 129-131
      metal arc effects, 128-129
   atmospheric electricity effects on muscular motion, 116-127
      "animal Leyden jar" model *versus*, 123-127
      lightning effects, 116-119
      metal arc effects, 119-123
   final elaboration of, 132-140
      contrary electricity in nerves and muscles, 133-134
      electrical nature of animal spirits established, 139-140
      Hallerian objections overcome, 137-139
      metal arc effects, 132-133
      microscopic level as disequilibrium site, 135-137
      tourmaline analogy, 134-135
   overview, 108-109

Galvanism, 20–21, 170, 184, 212–213, 221, 228, 248, 319, 333
Galvani-Volta controversy, first stage, 141–176
  animal preparation differences and, 77
  Galvani's circuit of animal electricity, 162–176
    animal machine concept, 174–175
    medical applications, 175–176
    metal arc effects, 166–167
    muscle-nerve direct contact experiment, 167–171, 173–174
    "occult arc" experiment, 171–172
    response to Volta's criticism, 162–165
    Trattato dell'arco conduttore, 163–176
  overview, 14–17, 24–25
  scientific culture of late 18th century, 142–152
    Bertholon's work *versus*, 145–147
    Galvani-Volta relationship, 142–144
    literary style, 151–152
    reaction to *De viribus*, 144–145
    replication of experiments, 147–148
    "virtual witnessing," 148–151
  sensation *versus* muscular motion distinction in, 75
  Volta's animal electricity research, 152–162
    on contractions at a distance, 157–158
    Galvani's theory discarded by, 159–161
    quantification of, 154–155
    seat of animal electricity, 155–157
    special theory of contact electricity, 161–162
    taste sensations, 158–159
    weak electricity research, 152–154
Galvani-Volta controversy, second stage, 177–214
  conclusion to, 211–214
    condensing electroscope (Volta), 213–214
    Galvani's death ending, 214
    Galvanism term coined, 212–213
    Volta's reaction to *Memoirs on Animal Electricity*, 211–212
  Galvani's electric fish research, 201–211
    animal electricity distinct from other electrical forms, 207–211
    brain involvement, 204–206
    frog muscles on torpedo experiments, 206–207
    nerve fluid and electric fish organ relation, 201–204
  Galvani's reply to Volta's criticisms, 186–201
    experimental setup to remove mechanical stimulation, 189–192

experimentation as, 195–198
Galvani's theory, 198–199
Memoirs on Animal Electricity (Galvani), 187–208
muscle as "animal Leyden jar," 199–201
reconciliation difficulties, 194–195
Volta's metallic electricity overshadowing, 192–194
general theory of contact electricity (Volta), 179–186
  judgments of other scholars, 182–183
  in letters to Anton Maria Vassalli, 179–181
  limitations of, 181–182
  nerve-muscle contact experiment impact, 184–186
  phenomena accounted for by, 183–184
overview, 177–179
Galvanization, 7
Galvanometer, 7, 22
Ganot, Adolphe (1804–1887), 1–6, 8, 216f
Gardini, Francesco Giuseppe (1740–1816), 18, 116, 145–146, 151, 210, 220
Gardos, George (1927–), 302n
Gases ("airs") studies
  Galvani research on, 81
  inflammable animal air, 111–114
  physiological chemistry in, 110–111
  Priestley's discoveries' impact, 109–110
  unitary view of natural phenomena from, 114–115
General theory of contact electricity (Volta)
  judgments of other scholars, 182–183, 213
  in letters to Anton Maria Vassalli, 179–181
  limitations of, 181–182
  nerve-muscle contact experiment impact, 184–186
  phenomena accounted for by, 183–184
Gherardi, Silvestro (1802–1879), 20
Giant nerve fiber in squid, 291
Gilbert, William (1544–1603), 11
Giornale fisico-medico (journal), 155, 157, 179
Girardi, Michele (1731–1797), 205
Glisson, Francis (1597–1677), 43
Goethe, Johann Wolfgang (1749–1832), 299
Gold-leaved Bennet-type electrometer, 244
Goldman-Hodgkin-Katz equation, 308

Gotch, Francis (1853–1913), 106, 283
Göttingen University (Switzerland), 43
Green, Robert Montraville (1880 - 1955), 324n
Gren, Carl Friedrich (1760–1798), 192–193, 212, 256.
Gren's Neues Journal der Physik, 212
Grmek, Mirko (1924–2000), 211

Hales, Stephen (1677–1761), 11, 57, 109–110
Haller, Albrecht, *von* (1708–1777)
  animal spirits and electric fluid as separate, 61
  experiments of, 41
  "fiber" concept, 137
  Franklin knowledge of, 53–54
  literary connotations of, 25
  living anatomy method of, 174
  neuromuscular physiology experiments of, 43
Hallerian irritability, 43–48.
  debate on, 46–47
  experiments on, 44
  Galvani and
    criticized by supporters of, 144
    influences on, 311
    observations *versus*, 104
    overcoming objections of, 137–139
    research on, 79
  Laghi rebuttal to, 47–48
  replication of experiments on, 60
  sensation and, 43–44
  stimulus-response lack of proportionality, 97–98
  unitary view of organisms *versus*, 44–46
Hartline, Haldan Keffer (1903–1983), 286n, 346
Harvey, William (1578–1657), 30, 40, 44, 258f
Hawthorne, Nathaniel (1804–1864), 17
Hearing sensation experiments, of Volta, 234–235, 238–239
Heart studies, 205–206, 225–226
Helmholtz, August Ferdinand (1792 -1858), 277
Helmholtz, Hermann, *von* (1821–1894)
  medical, physics, and physiology contributions of, 11
  muscle contractions recorded, 276–279
  nerve conduction speed measurement, 287
  temperature effects on conduction speed, 288
Henly, William (*d.* 1779), 79

Hermann, Ludimar (1838–1914), 280–281, 289–290
HH (Hodgkin-Huxley) model of nerve-fiber membrane electrical behavior, 296–297
Hippocrates (*ca.* 460 BC - *ca.* 370 BC), 29–30
History and Present State of Electricity (Priestley), 50
History of science, 333
Hodgkin, Alan (1914–1998)
  "all-or-none" law, 106
  frog preparation used by, 75
  Goldman-Hodgkin-Katz equation, 308
  local nerve blockade effects, 289–291
  membrane sodium permeability and nerve impulses, 292–293
  nerve conduction experiments of, 7–8, 14
  nerve signals as electrical phenomenon shown by, 270
  transmembrane potential measurement, 291
  "voltage clamp" used by, 294
Hodgkin cycle, 296, 307
Hodgkin-Huxley (HH) model of nerve-fiber membrane electrical behavior, 296–297
Holmes, Frederic Lawrence (1932–2003), 70
Home laboratory of Galvani, 72–74
Hospitals, Galvani's education in, 34–35
Humboldt, Alexander, *von* (1769–1859), 7, 25, 92, 170, 266
Hunter, John (1728–1793), 64–65, 69, 203
Hutchinson Dictionary of Scientific Biography (Porter), 7
Huxley, Andrew Fielding (1917–2012)
  HH model of nerve-fiber membrane electrical behavior, 296–297
  nerve conduction experiments of, 7–8, 14
  transmembrane potential measurement, 291
  "voltage clamp" used by, 294
Hydrodynamic model of electricity (Beccaria), 55
Hydrogen (inflammable air), 111–114, 263

Imaginary Invalid, The (Moliere), 31
"Induced twitch" experiment
  Galvani experimental setup and, 201–202
  Matteucci's progress with, 271–274
  nerve excitation by, 207
Inflammable air (hydrogen), 111–114, 263
Ingenhousz, Jan (1730–1799), 11
Institut de France, 23, 221–222

Institute of Sciences of Bologna
  Academy of Sciences as part of, 36
  Beccaria's membership in, 55
  Galvani's manuscripts at, 20, 22, 32–33
  Galvani's membership in, 23, 38, 55, 108
  interdisciplinary approach of, 10
  Marsili's founding of, 32–33
  Volta's visit to, 142–143
Institutiones medicae (Azzoguidi), 102
Institutiones physiologicae (Caldani), 102
Interdisciplinary scientific approach, 10–11, 36
"Internal electric nervous fluid" hypothesis (Galvani), 101
Introduction à l'étude de la médecine expérimentale (Bernard), 79
Ionic channels, 168, 297, 303–304. *See also* Membranes
Ionic diffusion, theory of (Nernst), 281

Journal de chimie et de physique, 259
Journal of Physiology, 289

Kato, Gen-Ichi (1890–1979), 285n
Katz, Bernard (1911–2003)
  Goldman-Hodgkin-Katz equation, 308
  membrane sodium permeability and nerve impulses, 292–293
  "voltage clamp" used by, 294
Keilin, David (1887–1963), 6
Keio University of Tokyo, 285n
Kekulé, August (1829–1896), 258
Kelvin, Lord (William Thomson; 1824–1907), 112, 281
Keynes, Richard (1919–2010), 318n
Kinnebrook, David (1772–1802), 278
Kuhn, Thomas (1922–1996), 11
"Kymographion," 276

Laboratory, in Galvani home, 72–74
Laboratory medicine, 31
Laghi, Tommaso (1709–1764),
  animal spirits' electrical nature, 57
  Caldani disagreement with, 60
  Galvani's research and, 144
  in Hallerian debate, 46–48
La malade imaginaire (Moliere), 31
Lambertini, Prospero (*see also* Benedict XIV, pope), 27, 28f, 33–35, 38
Landriani, Marsilio (1751–1816), 113, 220

Laplace, Pierre-Simon, *de* (1749–1827), 80, 114
La Roche, Daniel, *de* (1743–1812), 69
Lavoisier, Antoine-Laurent (1743–1794)
  on combustion and respiration, 10, 110–111, 113–114
  scientific community debates, 141
  on water evaporation, 80
Leak conductance for potassium, 308
Lelli, Ercole (1702–1766), 39
Le Noir de Nanteuil, Anne-Pauline (1757/62-post 1791), 257
Leopardi, Giacomo (1798–1837), 25
Leopoldo, Pietro (Grand Duke of Tuscany; 1747–1792), 48
Leyden jar. *See also* "Animal Leyden jar" model
  as artificial torpedo, 66
  electrical capacity with metal armatures, 130
  Galvani experiments with, 104
  invention of, 53–54
  precautions in handling, 94
  respiration phenomenon analogy to, 113
  in Science Museum of London, 73
Licht, Sidney (1908–1979), 327n
Lichtenberg, Georg Christoph (1742–1799), 153
L'identità del fluido elettrico col così detto fluido galvanico (Volta), 208
Life sciences work, of Volta. *See also* Volta, Alessandro, electrophysiological work of
  Bonaparte Prix from Institut de France, 221–222
  Galvani's influence on, 217–219
  heart studies, 225–226
  intermittence of contractions, 224–225
  medicine, 219–221
  polarity of current, 222–224
  voluntary movements, 226–229
Lightning studies, 56–57, 117
Linari, Santi (1777–1858), 178
Lippmann, Gabriel (1845–1921), 284
Living machine, torpedo as, 265
"Local-circuit" theory, 281, 289–290
Local nerve blockade studies, 289–291
Lorenzini, Stefano (*fl.* 1652–1700), 63
"Lost time," in perception, 278
Lucas, Keith (1876–1916), 283–288, 291, 312
Lucchesini, Girolamo (1751–1825), 114, 124
Lyon Académie des Sciences, Belle-Lettres et Arts, 145

Mach, Ernst (1838–1916), 230
Machine vision, 9
MacKinnon, Roderick (1956 -), 297
"Magic square." *See* "Franklin's square"
Magnetic processes, electrical currents associated with, 270–271
Mahon, Lord (*see* Charles Stanhope)
Majorana, Quirino (1871–1957), 92n
Malpighi, Marcello (1628–1694)
  anatomy views of, 40
  animal spirits and, 43
  frog preparation used by, 75
  as Institute of Bologna "authority," 34, 173
  kidney filtration discoveries of, 20
  mechanical view of vital phenomena, 77
  "rational" view of medicine of, 29–31, 36, 59
Manzolini, Giovanni (1700 -1755), 39
Marmont, George (1914–1983), 294
Marsili, Luigi Ferdinando (1658–1730), 31–32
Martins-Ferreira, His (1920–2009), 318n
Marum, Martinus, van (1750–1837), 225, 227, 231, 240–241
Maskelyne, Nevil (1732–1811), 278
Matteucci, Carlo (1811–1868)
  animal electricity measurement by, 16, 20, 22
  "induced twitch" experiment, 202, 207, 271–274
  spark from torpedo shock, 178
Medical electricity
  Bertholon work on, 145–147
  convulsive motion treatment, 175–176
  examples of, 58–60
  shock therapies, 94
  Veratti's interest in, 46
  Volta's interest in, 219–221
Medicine and natural philosophy in 18th century, 26–41. *See also* "Rational" view of medicine
  atmospheric electricity health effects, 79
  Galvani's early anatomy-physiology investigations, 40–41
  Galvani's education, 26–35
    Catholic Enlightenment, 27–28
    in hospitals, 34–35
    Institute of Sciences of Bologna, 32–34
    University of Bologna, 28–32
  Galvani's professional career, 36–40
Membranes, 300–310
  action potential, 309–310
  amplifying mechanism in, 97
  bioelectric potentials, 304–306
  characteristics, 300–301
  depolarization state to increase permeability of, 281–282
  electric potential across, 15
  electrochemical gradient of, 296–297
  HH model of nerve-fiber membrane electrical behavior, 296–297
  historical perspective, 306–307
  ionic channels, 297, 303–304
  as machine, 320
  membrane accommodation, 310–312
  molecular switch in, 8
  resistance in, 291
  sodium permeability, 292–293, 295, 307–309
  sodium-potassium pump, 301–303
Membrane theory, 183, 282, 289, 304
Memoir on the Identity of Electric and Galvanic Fluids (Volta), 248–249
Memoirs on Animal Electricity (Galvani), 143, 147, 150, 187–208, 211–212, 256
Metal arc effects
  in "animal Leyden jar" model, 128–129
  atmospheric electricity and, 119–123
  electric circuits shown with, 166–167
  in Galvani's animal electricity theory, 132–133
Metal laminas, experiments with, 130
Metallic electricity
  of dissimilar metals, 157, 159, 186, 213, 214, 232
  electrical machine electricity differences, 248–249
  experiments with exclusive use of, 192–193
  Franklin and, 221
  nerve excitation from, 14
Metals, electromotive power of
  as electrical disequilibrium source, 175
  electrodynamics emergence and, 160–161
  in general theory of contact electricity, 183
  Volta's introduction of, 152
  Volta's objections to animal electricity based on, 14, 170
  weak, 213
Micro-electroscope (Volta), 193
Microscope, as discovery tool, 29
Minimum electrical stimulus. *See* Weak electricity
Mocchetti, Francesco (1766–1839), 185–186, 223, 227, 244, 246

Molière (Jean-Baptiste Poquelin; 1622–1673), 27, 31
Molinelli, Pier Paolo (1702–1764), 46
Morandi, Anna (1714–1774), 39
Morgagni, Giovanni Battista (1682–1771), 31, 205
Moruzzi, Giuseppe (1910–1986), 273
Müller, Johannes (1801–1858), 239, 276
Multiplier device (Nicholson), 193–194, 213, 247
Muratori, Ludovico Antonio (1672–1750), 27
Muscle-nerve direct contact experiment (Galvani "third experiment"), 167–171, 173–174
Muscular motion. *See* Galvani, early muscular motion research of; Nerve conduction and muscular motion
Museum of Physics and Natural History (Florence, Italy), 48
Musschenbroek, Jan, *van*, 34
Myelinated segments, of nerves, 315
"Myographion," 276

Nachmansohn, David (1899–1983), 288
Nanotechnology, 9
Napoleonic authorities, Galvani and, 23
"Natural electromotor," 263
Nature group of journals, 333
Naturphilosophie, in 19th century, 17
Negative Schwankung (variation), 275, 280
Neher, Erwin (1944 -), 13–14, 297–298
Nernst, law of, 304–306, 308
Nernst, Walther (1864–1941), 281
Nernst's theory of electrochemical potential, 183
Nerve blockade studies, 289–291
Nerve conduction and muscle excitation mechanism, 310–318
  autoregenerative character of, 312–313
  membrane accommodation, 310–312
  muscle fiber generation of electrical impulses, 316–318
  physical considerations, 313–316
Nerve conduction and muscular motion, 42–68, 280–289. *See also* Animal electricity in nerve conduction; Galvani, early muscular motion research of; Neuromuscular excitability
  "all-or-none" electrical excitation, 282–284
  animal electricity and, 3
  artificial and natural electricity, 56–62
  electric fish and eels, 57–58
  medical electricity, 58–60
  nerves as conductors, 60–62
  weather-related, 56–57
  depolarization state to increase membrane permeability, 281–282
  electrical process complexity of, 15
  electric fish, 62–68
    measurement of shock, 65
    nature of shock, 62–65
    spark from shock of, 66–67
  electricity studies in 18th century, 48–56
    animal spirits' electrical nature, 48–50
    Beccaria's theory of, 55–56
    fascination with, 50
    Franklin's theory of, 53–55
    Leyden jar invention, 53–54
  Hallerian irritability
    debate on, 46–47
    experiments on, 44
    Laghi rebuttal to, 47–48
    sensation and, 43–44
    unitary view of organisms *versus*, 44–46
  Hodgkin and Huxley experiments, 7–8
  "local-circuit" theory, 281
  negativity of external fiber surface for, 280
  "pulse-code-modulation," 284–288
  temperature effects, 288–289
Neumann, Franz Ernst (1798–1895), 278
Neural networks, 9
Neuroelectric hypothesis, 109
Neuromuscular excitability, 299–318
  cell membrane and ions, 300–310
    action potential, 309–310
    bioelectric potentials, 304–306
    historical perspective, 306–307
    ionic channels, 303–304
    membrane characteristics, 300–301
    membrane sodium permeability triggering, 307–309
    sodium-potassium pump, 301–303
  nerve conduction and muscle excitation mechanism, 310–318
    autoregenerative character of, 312–313
    membrane accommodation, 310–312
    muscle fiber generation of electrical impulses, 316–318
    physical considerations, 313–316
Neurosciences, modern, 8
New Atlantis (Bacon), 32

Index

Newton, Isaac (1642–1727), 14, 18, 34, 41, 46, 54–58, 62, 67, 102, 163, 329
Newtonianesimo per le dame (Algarotti), 46
"Newton of electricity," Volta as, 153
Nicholson, William (1753–1815)
    capacitors as torpedo electrical organ, 253
    dissociation of water, 270
    on electric fish discharge mechanism, 129
    "multiplier" or "duplicator" device, 193–194, 213, 247
    Philosophical Magazine, 250
Nobel Prize in Physiology and Medicine to Neher and Sakmann, 13
Nobel Prize in Physiology and Medicine, to Hodgkin, 294
Nobili, Leopoldo (1784–1835).
    animal electricity measurement, 16, 271
    astatic galvanometer of, 22, 323
    on polarity of fish shock, 115n
Nodes of Ranvier, in nerves, 315–317
Nollet, Jean-Antoine (1700–1770), 49, 152

"Occult arc" experiment, 171–172, 180
Oersted, Hans Christian (1777–1851), 271
Of Artificial and Natural Electricity (Beccaria), 55
Of the Use and Activity of the Conducting Arc in the Contractions of the Muscles (attributed to Galvani), See *Trattato dell'arco conduttore*
On Experiments and Observations To Be Carried Out on Torpedoes (Volta), 262
On the Kidneys and Ureters of Birds (Galvani), 40
On the Way in Which the Most Feeble Electricity, Both Natural and Artificial, Can Be Detected (Volta), 116
Opuscoli scelti sulle scienze e sulle arti (journal), 110, 114, 125, 151, 160, 265
"Oratorio" of the Philippine Fathers, 27
"Overshoot," 279
Oxidation, 263

Pain sensation experiments, of Volta, 223, 235–237, 262
Pancaldi, Giuliano (1946 -), 193, 321
"Parallel," Leyden jars in, 66
"Patch clamp" technique, 13, 297–298
Pathological anatomy, 31
Pera, Marcello (1943 -), 9–10, 265
Perception, 230, 278

Personaggi e scoperte nella fisica contemporanea (Segrè), 15
Pfaff, Christian Heinrich (1773–1852), 170
Pflüger, Eduard (1829–1910), 222
Philippine Fathers, 27
Philosophical Magazine, 250
Philosophical Transactions (Royal Society of London), 144, 162, 232
Physico-Medical Observations on Electricity (Veratti), 59
Physiological chemistry, in "airs" studies, 110–111
Pivati, Gianfrancesco (1689–1764), 58
Pneumatic physics, 81, 109
Polar excitation, law of, 222
Polarity, in electricity, 222–224, 262–263
Polidori, John (1795–1821), 18
Polvani, Giovanni (1892–1970), 10, 217, 257, 259, 265
Porter, Roy (1946–2002), 7
Potassium, leak conductance for, 308
Priestley, Joseph (1733–1804), 10, 49–53, 56, 109–110
Principia (Newton), 67
Proust, Marcel (1871–1922), 277
Psychophysics of perception, 230
"Public Anatomy" lectures, 37–39, 61–62, 71
"Public happiness," theory of, 27
"Pulse-code-modulation," 284–288
Pupilli, Giulio Cesare (1893–1973) 16, 324n, 327n

Ranvier, Louis-Antoine (1835–1922), 315
"Rational" view of medicine, 29–31, 36, 59, 77
Réaumur, René-Antoine Ferchault, de (1683–1757), 63
Redi, Francesco (1626–1694), 63
"Repetitive experimentation," of Haller, 44
Republic of Letters of 18th century, 36, 66, 153
Residual electricity, 265–267
"Resinous" electricity, 49
Respiration phenomenon, 110–111, 113
Retinal illumination, temporal variations in, 233
"Return stroke" explanation of contractions at a distance, 4, 91–94, 96
Rialp, Raymond (*or* Raimundo; 1750-*post* 1791), 74, 126, 147, 323
Richman, George Wilhelm (1711–1753), 94
Ritter, Johann Wilhelm (1776–1810), 17, 170
Robotics, 9

Index

Royal Institution of London, 23, 270
Royal Society of London
  Beccari as member of, 29
  Copley medal to Volta, 214
  electricity discussed by, 49
  Institute of Sciences of Bologna compared to, 33
  Volta as member of, 152–153
  Volta's battery presented to, 12
  Volta's research published by, 114, 144, 162, 232, 250
  Walsh studies presented to, 64
Russian Academy of Science, 94

Sakmann, Bert (1942 -), 13–14, 297–298
San Giobbe hospital, 34
San Lazzaro hospital, 34
Santa Maria della Morte hospital, 34
Santa Maria della Vita hospital, 34
Sant'Orsola hospital, 34–35
Saussure, Horace-Bénédict, *de* (1740–1799), 79
Scarpa, Antonio (1752–1832), 234
Scheele, Carl Wilhelm (1742–1786), 110
Science Museum of London, 72–73
Scientific approach, interdisciplinary, 10–11
Scientific instruments, in Galvani laboratory, 72–73
"Second experiment" of Galvani (1791). *See* Galvani, animal electricity theory of
Second Memoir on Animal Electricity (Volta), 4, 91, 157–159, 180
Segrè, Emilio (1905–1989), 15
Sensation, in Hallerian irritability, 43–44
Sensation experiments, of Volta, 229–241. *See also* Volta, Alessandro, electrophysiological work of
  battery to stimulate sensory systems, 237–238
  hearing, 234–235, 238–239
  specific nerve energies doctrine, 239–241
  taste, 230–232
  touch, 235–237
  vision, 232–233
Sensors, high-performance, 9
Sguario, Eusebio (*fl.* 1717–1764), 52f
Shelley, Mary Wollstonecraft (1797–1851), 18, 118
Shelley, Percy Bysshe (1792–1822), 18
Sherrington, Charles Scott (1792–1822), 9
Shocking eel of South America. *See* Electric fish research

"Small machine" (*macchinetta*) (Galvani), 169
Social management of illness, 34
Sodium permeability of membranes, 292–293, 307–309
Sodium-potassium pump, 301–303
Spallanzani, Lazzaro (1729–1799)
  experimental methods of, 163, 211
  Galvani's *Memoirs on Animal Electricity* sent to, 147
  Galvani's *Trattato dell'arco conduttore* sent to, 177–178
  reproduction research of, 75
  residual electricity studies of, 265–266
  torpedo studies of, 114–115
  Volta's replication of experiments of, 216
  Walsh studies reported by, 124
Spatial diffusion, 66, 261
Special theory of contact electricity (Volta), 3, 160–162
Specific nerve energies doctrine, 239–241
Spincterometer, 260
Squid, giant nerve fiber in, 16, 291. *See also* Electric fish research
Stahl, Georg Ernst (1660–1734), 9, 216n
Stanhope, Charles (Lord Mahon; 1753–1816), 4, 91–93, 96
Steensen, Niels (1638–1687), 43
Stimulus-response, lack of proportionality in, 105–107
"Straw micro-electrometer," 80
Supplemento al Trattato dell'arco conduttore (Galvani), 172–176, 182
Swammerdam, Jan (1637–1680), 75
Sydenham, Thomas (1624–1689), 31

Taste sensations
  Volta's experiments on, 75, 158–159, 230–232
Temporal variations in retinal illumination, 233
Tension concept, 153
Testimony strategy, in reports, 147
Tetanus, 273
Theory of Colours (Goethe), 299
"Third experiment," of Galvani. *See* Galvani-Volta controversy, first stage; Galvani-Volta controversy, second stage
Third Memoir on Animal Electricity (Volta), 160–161
Thomson, William (Lord Kelvin), 112, 281
"Threshold" concept, 105

Tissot, Samuel (1728–1797), 61
Tongue experiments, of Volta. *See* Sensation experiments, of Volta
Torpedo ray. *See* Electric fish research
"Torrent" of electricity, 160, 168
Touch sensation experiments, of Volta, 235–237
Tourmaline analogy, 117, 128, 134–135, 137, 144
"Transflow" of electricity, 160
Transmembrane potential measurement, 291
Trattato dell'arco conduttore (Galvani), 147, 151, 163–166, 169, 184
Treatise on Nerves and Their Diseases (Tissot), 61
Trinity College, Cambridge University (UK), 284f, 289
"Trough" version of battery, 220–221
Trumpler, Maria (1960–), 141
Truth and usefulness. *See* Medicine and natural philosophy in 18th century
Truthfulness of experiment, testimony for, 147
Turin Academy of Sciences, 179
Two Dissertations on Animal Electricity (Aldini), 179
Two-flask apparatus (Galvani), 99–100

Unitary view of natural phenomena, 44–46, 114–115
University "La Sapienza" (Rome), 27
University of Bologna, 28–32, 36
University of Pavia, 2, 152, 177, 228, 254
Untersuchungen über thierische Elektricität (du Bois-Reymond), 20, 23

Vacuum, electrical effects in, 100
Valli, Eusebio (1755–1816), 144, 182, 255.
Vallisneri, Antonio (1661–1730), 40
Valsalva, Antonio Maria (1666–1723), 40
Vassalli, Anton Maria (1761–1825) 79, 179–182, 187–189, 191, 194, 221
Veratti, Giuseppe (1707–1793)
    Beccaria relationship with, 55
    electrical therapy tested by, 59–60
    Galvani's experiments similar to, 78–79, 81, 110
    as Galvani's mentor, 173
    gases studies of, 110
    Institute of Sciences of Bologna influenced by, 173
    laboratory of, 46, 72
    lightning experiments of, 116
    respiration experiments of, 110
    Volta meeting with, 142
"Virtual witnessing," of experiments, 148–151, 326
Vision sensation experiments, of Volta, 232–233
Visual testimony, to verify experiments, 147
Vital spirit or fluid, electricity as, 17–18
"Vitreous" electricity, 49
Volta, Alessandro (1745–1827). *See also* Galvani-Volta controversy, first stage; Galvani-Volta controversy, second stage
    animal electricity research, 152–162
    on contractions at a distance, 157–158
    Galvani's theory discarded by, 159–161
    quantification of, 154–155
    seat of animal electricity, 155–157
    special theory of contact electricity, 161–162
    taste sensations, 158–159
    weak electricity research, 152–154
    animal soul belief of, 78
    battery discovery of, 7, 12–13, 321
    Beccaria's influence on, 55
    Bologna visit in 1780, 142
    on combustion and respiration, 113–114
    electric current concept, 49
    electricity in neuromuscular physiology, 324
    fast electrical stimuli for reactions, 311
    First Memoir on Animal Electricity, 42, 151–156, 254, 313
    Galvani response to criticism from, 162–165
    Galvani's influence on, 2–4
    on metallic electricity of dissimilar metals, 312
    metals as electricity source, 85
    muscular motion explanation of, 42
    On the Way in which the Most Feeble Electricity, both Natural and Artificial, can be Detected (Volta), 116
    Second Memoir on Animal Electricity, 4, 157–159
    source and detector of electricity distinction, 323
    Third Memoir on Animal Electricity, 160–161
    torpedo electric organ and, 65
    weak electricity studies of, 79–80, 82–83
Volta, Alessandro, electrophysiological work of, 215–268
    "chain" experiments, 241–249
    battery impact on, 242–243

Volta, Alessandro, electrophysiological work (*Cont.*)
    electrical resistance concept from, 245–248
    metallic and electrical machine electricity differences, 248–249
    at progressively lower tensions, 243–245
    variations in, 241–242
  electric fish experiments, 249–268
    "animal economy encompassing electricity" concept, 257–259
    battery invention based on, 249–253, 259–261
    effects on inanimate substances, 263–265
    electricity involvement in, 254–256
    nerve relationship to electricity in, 256–257
    Nicholson's view, 253–254
    parallel and series electric circuits, 261–262
    physics and physiology differences, 267–268
    polarity, 262–263
    residual electricity in separated organs, 265–267
  life sciences and, 217–229
    Bonaparte Prix from Institut de France, 221–222
    Galvani's influence on, 217–219
    heart studies, 225–226
    interrmittance of contractions, 224–225
    medicine, 219–221
    polarity of current, 222–224
    voluntary movements, 226–229
  overview, 215–217
  on sensations, 229–241
    battery to stimulate sensory systems, 237–238
    hearing, 234–235, 238–239
    specific nerve energies doctrine, 239–241
    taste, 230–232
    touch, 235–237
    vision, 232–233
"Voltage clamp" technique, 294–295
Voltaic battery, 306
Voltaire (François-Marie Arouet; 1694–1778), 36
"Voltian alternatives," 168, 240
Voluntary movements, Volta's studies of, 226–229

Wabst, Christian Xavier (*or* Xaver) (*fl.* 1710–1760), 52f
Walbeck, Henrik Johan (1793–1822), 278
Walsh, John (1726–1795)
  electric fish shock phenomenon, 124, 126
  electric fish studies of, 13–14, 64–66
  Galvani and Volta linked to, 267
  Galvani view of, 115
  La Roche's reports on research of, 69
  nerve fluid and torpedo shock relation, 203
  Spallanzani's reports on research of, 114
Water, dissociation of, 270
Water, electrification of, 53
Weak electricity
  distinguishing electricity to measure, 321–322
  frog's response to, 218
  Galvani's experiments on, 104, 265
  heart contraction lag, 206–207
  research on, 79–80
  subthreshold stimuli combined, 288
  Volta research on, 152–154
  Volta's rejection of animal electricity based on, 161
Wellcome, Henry Solomon (1853–1936), 72
Will, action mechanism of, 185
Willis, Thomas (1621–1675), 43
Wollstonecraft Shelley, Mary, *see* Shelley, Mary Wollstonecraft.

Young, John Zachary (1907–1997), 291
Young, Thomas (1773–1829), 11

Zambeccari, Jacopo (1723-*post* 1791), 149
Zanotti, Francesco Maria (1692–1777), 43, 48, 58–59
Zotterman, Yngve (1898–1982), 284